Lectures in Applied Mathematics

Proceedings of the Summer Seminar, Boulder, Colorado, 1960

VOLUME 1 LECTURES IN STATISTICAL MECHANICS
G. E. Uhlenbeck and G. W. Ford with E. W. Montroll

VOLUME 2 MATHEMATICAL PROBLEMS OF RELATIVISTIC PHYSICS
I. E. Segal with G. W. Mackey

VOLUME 3 PERTURBATION OF SPECTRA IN HILBERT SPACE
K. O. Friedrichs

VOLUME 4 QUANTUM MECHANICS
R. Jost

Proceedings of the Summer Seminar, Ithaca, New York, 1963

VOLUME 5 SPACE MATHEMATICS, PART 1
J. Barkley Rosser, Editor

VOLUME 6 SPACE MATHEMATICS, PART 2
J. Barkley Rosser, Editor

VOLUME 7 SPACE MATHEMATICS, PART 3
J. Barkley Rosser, Editor

Proceedings of the Summer Seminar, Ithaca, New York, 1965

VOLUME 8 RELATIVITY THEORY AND ASTROPHYSICS
1. RELATIVITY AND COSMOLOGY
Jürgen Ehlers, Editor

VOLUME 9 RELATIVITY THEORY AND ASTROPHYSICS
2. GALACTIC STRUCTURE
Jürgen Ehlers, Editor

VOLUME 10 RELATIVITY THEORY AND ASTROPHYSICS
3. STELLAR STRUCTURE
Jürgen Ehlers, Editor

Proceedings of the Summer Seminar, Stanford, California, 1967

VOLUME 11 MATHEMATICS OF THE DECISION SCIENCES, PART 1
George B. Dantzig and Arthur F. Veinott, Jr., Editors

VOLUME 12 MATHEMATICS OF THE DECISION SCIENCES, PART 2
George B. Dantzig and Arthur F. Veinott, Jr., Editors

Proceedings of the Summer Seminar, Troy, New York, 1970

VOLUME 13 MATHEMATICAL PROBLEMS IN THE GEOPHYSICAL SCIENCES
1. GEOPHYSICAL FLUID DYNAMICS
William H. Reid, Editor

VOLUME 14 MATHEMATICAL PROBLEMS IN THE GEOPHYSICAL SCIENCES
2. INVERSE PROBLEMS, DYNAMO THEORY, AND TIDES
William H. Reid, Editor

Mathematical Problems in the Geophysical Sciences

1. Geophysical Fluid Dynamics

William H. Reid, Editor

The University of Chicago

LECTURES IN APPLIED MATHEMATICS, VOLUME 13

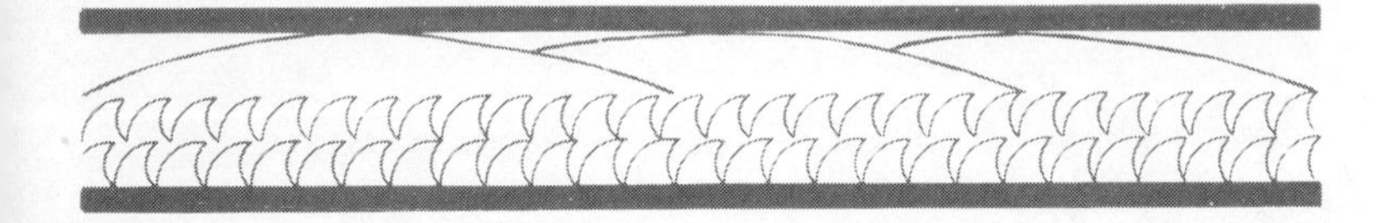

AMERICAN MATHEMATICAL SOCIETY, PROVIDENCE, RHODE ISLAND, 1971

Prepared by the American Mathematical Society under Contract N00014–69–C–0381 with the Office of Naval Research, Grant SSF(70)–16 of the New York State Science and Technology Foundation, and Grant GZ–1509 of the National Science Foundation

International Standard Book Number 0–8218–1113–4
Library of Congress Catalog Number 62–21481
AMS 1970 Primary Subject Classification 86–02

Printed in the United States of America

Contents

1. Geophysical Fluid Dynamics

2. Inverse Problems, Dynamo Theory, and Tides

Preface

The Sixth Summer Seminar on Applied Mathematics, sponsored jointly by the American Mathematical Society and the Society for Industrial and Applied Mathematics, was held at the Rensselaer Polytechnic Institute from July 6 to 31, 1970.

The seminar program was intended to provide an opportunity for graduate students and recent recipients of the Ph.D. degree to become familiar with current developments in a number of areas of the geophysical sciences in which applied mathematics plays a central role. In the geophysical sciences at the present time, there is an important interplay between the observational and experimental work on the one hand and the mathematical developments on the other, and this aspect of the subject is clearly evident in many of the papers which appear in these proceedings.

The program for the seminar was organized by a committee which included Hirsh G. Cohen (I.B.M., T. J. Watson Research Center), Richard C. DiPrima (Rensselaer Polytechnic Institute), Dave Fultz (The University of Chicago), C. C. Lin (Massachusetts Institute of Technology), and William H. Reid.

Thanks are due to the Office of Naval Research, the National Science Foundation, and the New York State Science and Technology Foundation for their financial support of the seminar. I am also grateful to Dr. Gordon L. Walker, executive director of the American Mathematical Society, and Mrs. Lillian R. Casey for their generous help in the planning and administration of the seminar; to Professor Lester A. Rubenfeld for his efficient handling of the local arrangements; and to Mrs. Mary A. Coccoli for her devoted work in the day-to-day operation of the conference office.

I also wish to thank the speakers for preparing their lecture notes in advance so that they could be distributed at the time of the seminar and thus permit the rapid publication of these proceedings.

WILLIAM H. REID, EDITOR
Departments of Mathematics and
the Geophysical Sciences
The University of Chicago

Geophysical Fluid Dynamics

J. Pedlosky

LECTURE I

1. **Geophysical fluid dynamics.** The atmosphere and the oceans move ceaselessly in complex patterns and on awesome scales. Geophysical Fluid Dynamics is the study of the dynamical processes involved in those motions, and the purpose of these lectures will be to touch on and present some of the basic concepts of the subject. Of necessity our treatment will be incomplete and somewhat sketchy, but the essential flavor of the subject should by the end be apparent.

The present state of the science is such that an understanding of planetary motions requires the study of rather specific problems rather than the development of very general theories. More often than not this requires the study of highly idealized model problems. An idealized model problem, while not being faithful in detail to the geophysical phenomenon which suggests the study of the model, is meant to focus attention on the underlying dynamical mechanism.

Now motions which are "geophysical," i.e. naturally occurring on the earth, range in scale from small ripples on the surface of a brook to the majestic current systems of the oceans (e.g. the Gulf Stream) and the atmosphere (e.g. the Jet Stream). It is clearly impossible for us to discuss the dynamical character of this enormous spectrum of motions, and in these lectures I will discuss only dynamical concepts which are appropriate for the description of large scale motions and let the grandeur of the phenomena be sufficient compensation for the loss of generality. Other lecturers in this series will, I understand, discuss smaller scale (although no less important) geophysical motions.

It is naturally somewhat difficult to define unambiguously what we mean by "large scale," but for our purposes large scale motions are those which are significantly influenced, if not dominated, by the effects of the

---

AMS 1970 *subject classifications*. Primary 76–02, 76C05, 76C10, 76C15, 76C20, 76D10, 76D30, 76E20, 76U05, 76V05.

earth's rotation. There are several important parametric measures of the importance of rotation, among the most important of which is the *Rossby number* defined as follows. Let L be a typical horizontal length scale of the motion (e.g. the distance between major low pressure centers on a weather map), and let U be similarly a characteristic horizontal velocity scale measured relative to the earth's surface; then ε, the Rossby number, is defined by

$$\varepsilon = U/\Omega L$$

where Ω is the planetary frequency of rotation ($\Omega = 7.3 \times 10^{-5}\,\mathrm{sec}^{-1}$). A necessary but not sufficient (as we shall see) condition for rotation to be dominant (which defines what we mean by large scale) is

$$\varepsilon \ll 1. \tag{1.1}$$

Typical values of ε for atmospheric motions with scales of $O(1000\,\mathrm{km})$ range between 0.1 to 0.5 while the much more sluggish oceanic círculation yields, for the same length scales, values of ε which are $O(10^{-3})$.

Indeed the speeds of both the atmosphere and the oceans relative to the earth, when viewed on a large scale, are small, and it is consequently convenient to describe the motions kinematically in a coordinate frame rotating with the planet. Since such a frame is accelerating and not inertial, certain D'Alembert forces arise. They are the familiar centrifugal force (which merely redefines the effective gravitational potential) and the important and subtle Coriolis force. Indeed the importance of the latter characterizes large scale geophysical motions, and whenever the Rossby number is small the Coriolis force is a dominant participant in the balance of forces acting on a fluid element. The dynamics of large scale motions have two other features of importance. First, the motions occur in a thin spherical shell, that is the vertical scale of the motion is much less than the horizontal scale. The general oceanic circulation with a horizontal scale of thousands of kilometers has a vertical scale of only about one kilometer! Thus the fluid element trajectories are very shallow and the flow is confined to a shell of very small aspect ratio. Second, the atmosphere and the oceans both possess significant density variations. This density stratification is, when averaged over a large scale, almost always stable in the sense that heavy fluid underlies lighter fluid. An important consequence of this stable stratification is that motion parallel to the gravitational vector is inhibited, which reinforces the tendency toward flat fluid trajectories. It will become apparent later that because of the rotation an intimate relation exists between the density and velocity fields which may lead to important and rather unexpected modes of motion. However, while stratification is frequently significant there are modes of motion for which the effects of stratification are of secondary importance. We will often capitalize on this fact by directing our attention to the motion of a homogeneous rotating fluid. Frequently, as in the oceanic circulation, or in discussing the properties of certain atmospheric wave phenomena, a homogeneous model yields useful information concerning the dynamics of the vertical mean of the motion.

2. **Equations of motion.** As we noted above, large scale atmospheric or oceanic motions are really only small deviations from the solid body rotation imposed upon the fluid due to the planetary rotation. In a co-ordinate frame rotating with the planet (angular velocity $\mathbf{\Omega}$) the Navier-Stokes equations are (for either air or water)

$$\rho \frac{D\boldsymbol{q}}{Dt} = -\nabla p + \rho \nabla \Phi - \rho 2\mathbf{\Omega} \times \boldsymbol{q} + \mu \nabla^2 \boldsymbol{q} + \frac{\mu}{3} \nabla(\nabla \cdot \boldsymbol{q}) \tag{2.1}$$

where

$\boldsymbol{q}$ = velocity vector as observed in the earth frame,
p = fluid pressure,
ρ = fluid density,
Φ = effective gravitational potential,
μ = coefficient of viscosity,

while the operator D/Dt, the substantial derivative, is

$$D/Dt = \partial/\partial t + (\boldsymbol{q} \cdot \nabla). \tag{2.2}$$

We shall, without question, assume the validity of the Navier–Stokes equations. For large scale motions, the coefficient of viscosity is often assumed to also represent or parameterize the momentum transport of small scale random eddies. This parameterization (sic) of the interaction of the small and large scale motions is accomplished by increasing the size of μ from its molecular value, supposedly because of the greater efficiency of smoothing of momentum on a large scale by the transport of great blocks of fluid across the mean momentum gradient. This rather disreputable and desperate measure cannot be justified a priori. In some cases it is clearly wrong to imagine that small scale turbulence acts on the large scale flow as a beefed-up group of molecules. In those cases when it is qualitatively and empirically correct it is still difficult to assign with confidence numerical values to the turbulent viscosity. However, it is also absurd to rely totally on molecular viscosity as the agent for the dissipation of momentum. More sensible results are obtained from the use of the larger turbulent viscosities, but any such results, while viewed with sympathy, must also be viewed with suspicion. Any dynamical theory which depends critically on a specific value of the turbulent viscosity must be considered as having a shaky foundation indeed.

The dynamical system is closed with the consideration of the equation of conservation of mass:

$$(D\rho/Dt) + \rho\nabla \cdot \boldsymbol{q} = 0, \tag{2.3}$$

the first law of thermodynamics,

$$De/Dt = Q + (k\nabla^2 T/\rho) - [p(D\rho^{-1}/Dt)] + \mathscr{D}, \tag{2.4}$$

where

e = the internal energy per unit mass,
Q = the rate of heat addition per unit mass due to internal heat sources,
k = coefficient of thermal diffusion,

while

$$\mathscr{D} = \frac{\mu}{2\rho}\left\{\frac{\partial q_i}{\partial x_j} + \frac{\partial q_j}{\partial x_i} - \frac{2}{3}\delta_{ij}\frac{\partial q_k}{\partial x_k}\right\}^2$$

is the production of thermal energy by the viscous dissipation of mechanical energy. With the two constitutive relations,

$$e = e(p, T), \qquad p = p(\rho, T), \tag{2.5}$$

there are just enough equations (seven) to match unknowns. It is frequently useful, however, to introduce a new variable, the specific entropy s, by the equilibrium state relation

$$T(Ds/Dt) = (De/Dt) + p(D\rho^{-1}/Dt). \tag{2.6}$$

For the purposes of these lectures, the simplest sensible thermodynamic state relations will be used in order that we may focus our attention on the interesting mechanical as opposed to thermodynamical geophysical phenomena.

The gas law

$$p = \rho RT \tag{2.7a}$$

and a linear density-temperature relation

$$\rho = \rho_0(1 - \alpha(T - T_0)) \tag{2.7b}$$

are frequently applied to the atmosphere and the oceans respectively, while $e = CT$ is a simplified law for the relation between temperature and internal energy, where C is the appropriate specific heat at constant volume.

3. **Vorticity.** A derived quantity of central importance in Geophysical Fluid Dynamics is the vorticity $\boldsymbol{\omega}$

$$\boldsymbol{\omega} = \nabla \times \boldsymbol{q}. \tag{3.1}$$

If we take the curl of equation (2.1), after using the vector identities,

$$\begin{aligned} \boldsymbol{q} \cdot \nabla \boldsymbol{q} &= (\boldsymbol{\omega} \times \boldsymbol{q}) + \nabla \boldsymbol{q} \cdot \boldsymbol{q}/2, \\ \nabla \times (\boldsymbol{\omega} \times \boldsymbol{q}) &= \boldsymbol{q} \cdot \nabla \boldsymbol{\omega} - \boldsymbol{\omega} \cdot \nabla \boldsymbol{q} + \boldsymbol{\omega} \nabla \cdot \boldsymbol{q}, \end{aligned} \tag{3.2}$$

we obtain the equation expressing the rate of change of relative vorticity (vorticity observed in the earth frame) following an individual fluid element, viz:

$$D\boldsymbol{\omega}/Dt = (\boldsymbol{\omega} + 2\boldsymbol{\Omega}) \cdot \nabla \boldsymbol{q} - (\boldsymbol{\omega} + 2\boldsymbol{\Omega}) \nabla \cdot \boldsymbol{q} + [(\nabla \rho \times \nabla p)/\rho^2] + \nu \nabla^2 \boldsymbol{\omega}. \tag{3.3}$$

We have assumed that the kinematic viscosity $\nu = \mu/\rho$ may be approximated as a constant.

The last two terms on the right-hand side of (3.3) express the well-known mechanisms of vorticity change due to density inhomogeneities and vorticity diffusion. The first two terms require some comment. They express the change of vorticity at a point due to the contraction and twisting of the vortex tubes, and the main element of interest is that even if the relative vorticity of the fluid $\boldsymbol{\omega}$ is originally zero, contraction of the fluid in planes perpendicular to $\boldsymbol{\Omega}$ and the tilting of fluid lines which were once parallel to $\boldsymbol{\Omega}$ will produce relative vorticity. For this reason large scale geophysical flows are almost always vorticity-ful and the interaction of the relative vorticity and the so-called planetary vorticity $2\boldsymbol{\Omega}$ is a central feature of geophysical fluid dynamics.

4. **Ertel's theorem of potential vorticity.** A beautiful and unusually useful theorem can be derived from the vorticity equation (3.3) as follows. Use of the equation of mass conservation (2.3) allows us to rewrite (3.3)

$$\frac{D\boldsymbol{\omega}/\rho}{Dt} = \frac{(\boldsymbol{\omega} + 2\boldsymbol{\Omega})}{\rho} \cdot \nabla \boldsymbol{q} + \frac{\nabla \rho \times \nabla p}{\rho^3} + \frac{\nu}{\rho} \nabla^2 \boldsymbol{\omega}. \tag{4.1}$$

Suppose there is some fluid property λ which is conserved by the fluid (it might be the entropy, for example, in a fluid with negligible dissipation), i.e.

$$D\lambda/Dt = 0. \tag{4.2}$$

Then a simple calculation shows that

$$(\boldsymbol{\omega} + 2\boldsymbol{\Omega}) \cdot (D/Dt) \nabla \lambda = -\nabla \lambda \cdot \{(\boldsymbol{\omega} + 2\boldsymbol{\Omega}) \cdot \nabla\} \boldsymbol{q}. \tag{4.3}$$

Dotting (4.1) with $\nabla\lambda$ thus leads to

$$\frac{D}{Dt}\frac{(\boldsymbol{\omega} + 2\boldsymbol{\Omega})}{\rho}\cdot\nabla\lambda = \nabla\lambda\cdot\frac{(\nabla\rho \times \nabla p)}{\rho^3} + \frac{\nu}{\rho}\nabla\lambda\cdot\nabla^2\boldsymbol{\omega}. \tag{4.4}$$

If (i) the fluid is inviscid (or nearly so) and *either*

(ii) (a) the fluid is barotropic ($\nabla\rho \times \nabla p = 0$)

or

(ii) (b) λ is a function of p and ρ only ($\nabla\lambda\cdot(\nabla\rho \times \nabla p) = 0$), then the scalar quantity

$$\Pi = \frac{(\boldsymbol{\omega} + 2\boldsymbol{\Omega})}{\rho}\cdot\nabla\lambda, \tag{4.5}$$

which is called the potential vorticity, is conserved by each fluid element. The proper application of this fact provides the dynamical backbone of much of the theory of geophysical fluid dynamics.

Lecture II

5. **The geostrophic approximation.** We are now in a position to estimate more accurately the importance of the various terms in the momentum balance for large scale motions. If U and L are characteristic values for the horizontal velocity and horizontal length scales, then

$$D\mathbf{q}/Dt = O(U^2/L), \tag{5.1}$$

assuming also that the appropriate time scale for the motion is L/U.

Thus the ratio of the relative acceleration to the Coriolis force is

$$(U^2/L)/2\Omega U = U/2\Omega L = \varepsilon, \tag{5.2}$$

i.e., the Rossby number. Similarly the ratio of the viscous force to the Coriolis force can be estimated as

$$(\nu U/D^2)/2\Omega U = \nu/2\Omega D^2 = E/2 \tag{5.3}$$

where D is the characteristic vertical scale of the motion. The parameter E is called the Ekman number. An important feature of large scale motions is that both the Rossby and Ekman numbers are small. Naturally the Ekman number is as difficult to estimate as the coefficient of viscosity if it is the turbulent viscosity which is used to define E. Typical values of E range from 10^{-3} in the earth's troposphere ($D = 10$ km), using a value of turbulent viscosity of 10^5 cm^2/sec, to a ludicrous 10^{-9} if the value of molecular kinematic viscosity 10^{-1} cm^2/sec is used. In any event, with both ε and E small the primary balance of forces is between the pressure gradient, the gravitational force, and the Coriolis force, i.e. to an excellent

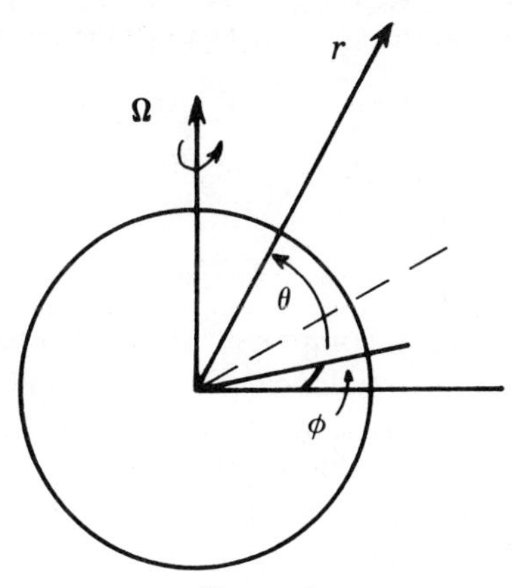

FIGURE 1.

approximation

$$2\boldsymbol{\Omega} \times \boldsymbol{q} = -(\nabla p/\rho) + \nabla\Phi. \tag{5.4}$$

Let us examine this force balance more closely by rewriting (5.4) in component form in a spherical co-ordinate system with its origin at the earth's center, with (r, θ, φ) as the spherical co-ordinates (θ is the latitude, φ the longitude), the components of (5.4) are (u is the eastward velocity, v the northward velocity and w the vertical velocity)

$$(2\Omega \sin\theta)u = -(1/\rho r)(\partial p/\partial\theta), \tag{5.5a}$$

$$-(2\Omega \sin\theta)v + (2\Omega\cos\theta)w = -(1/\rho r\cos\theta)(\partial p/\partial\varphi), \tag{5.5b}$$

$$-(2\Omega\cos\theta)u = -(1/\rho)(\partial p/\partial r) + \partial\Phi/\partial r, \tag{5.5c}$$

where we have assumed that the effective gravitational force is purely radial. Now, as we observed earlier the trajectory of a fluid parcel, constrained to flow in the thin spherical shell of either the ocean or atmosphere, is very, very flat. Consequently

$$w/v = O(D/L) \ll 1$$

and, therefore, except in the immediate vicinity of the equator, the component of the Coriolis force in (5.5b) which is proportional to w is completely negligible.[1] Similarly, with

$$\begin{aligned} \partial\Phi/\partial r &= -g, \\ 2\Omega U\cos\theta/g &= O(2\Omega U/g). \end{aligned} \tag{5.6}$$

[1] Near the equator, the terms proportional to $2\Omega\sin\theta$ are small. It does not, however, imply that the Coriolis force due to the vertical motion is significant. It is still tiny, and other dynamical terms which were ignored to obtain (5.4) must be retained near the equator.

For the atmospheric motions, with $U = 10$ meters/sec,

$$2\Omega U/g = 1.4 \times 10^{-4} \times 10^3/10^3 = 1.4 \times 10^{-4},$$

and thus to an excellent approximation the vertical force balance is the hydrostatic one

$$\partial p/\partial r = -\rho g. \tag{5.7}$$

Hence, only the local normal component $2\Omega \sin \theta$ of the planet's rotation is of importance in the force balance as a consequence of the flatness of the fluid trajectories. For the same reason, r, which never significantly departs from the value it has at the earth's surface may be replaced by R, the earth's radius, so that finally we have the *geostrophic approximation* to the equations of motion

$$fu = -(1/\rho R)(\partial p/\partial \theta), \tag{5.8a}$$

$$fv = (1/\rho R \cos \theta)(\partial p/\partial \varphi), \tag{5.8b}$$

$$0 = -(\partial p/\partial r) - \rho g, \tag{5.8c}$$

where the notation

$$f = 2\Omega \sin \theta, \tag{5.9}$$

the local normal component of the planetary vorticity, has been introduced. f is called the *Coriolis parameter.*

Further simplifications are possible if the north-south extent of the motion is sufficiently small (say of the order of 1,000 km) so that a locally Cartesian co-ordinate system can be used, with x measured eastward, y northward, and z upward. Then (5.8) becomes the traditional form of the geostrophic approximation

$$fu = -(1/\rho)(\partial p/\partial y), \tag{5.9a}$$

$$fv = (1/\rho)(\partial p/\partial x), \tag{5.9b}$$

$$0 = -(\partial p/\partial z) - \rho g. \tag{5.9c}$$

6. **Properties of the geostrophic approximation and the Taylor–Proudman theorem.** Flows which satisfy (5.8) or (5.9) to lowest order are called geostrophically balanced. Since most large scale motions are in geostrophic balance we take this opportunity to list some of the properties of *geostrophy* (the state of geostrophic balance) which are derivable from (5.9).

(i) The horizontal, geostrophic velocities are perpendicular to the horizontal pressure gradient, i.e. from (5.9a, b)

$$u(\partial p/\partial x) + v(\partial p/\partial y) = 0. \tag{6.1}$$

Thus if the motion is in geostrophic balance, the horizontal velocity flows along and not across lines of constant pressure. For this reason, isobars on a weather map are excellent approximations to the streamlines of the flow.

(ii) Within the context of the scale approximation which led from (5.8) to (5.9), f in (5.9) may be considered constant so that p/f is a stream function for the horizontal momentum

$$\rho \boldsymbol{q}_H = \hat{k} \times \nabla(p/f). \tag{6.2}$$

If the horizontal variation of the density is sufficiently small, the function $p/\rho f$ will be a stream function for the horizontal velocity itself.

From (5.9a) and (5.9c)

$$f\frac{\partial u}{\partial z} = \frac{g}{\rho}\frac{\partial \rho}{\partial y} + \frac{1}{\rho^2}\frac{\partial \rho}{\partial z}\frac{\partial p}{\partial y}, \tag{6.3}$$

but

$$\frac{\partial p}{\partial y} = -\frac{\partial p}{\partial z}\left(\frac{\partial z}{\partial y}\right)_p = \rho g\left(\frac{\partial z}{\partial y}\right)_p; \tag{6.4}$$

hence

$$f\frac{\partial u}{\partial z} = \frac{g}{\rho}\left[\frac{\partial \rho}{\partial y} + \frac{\partial \rho}{\partial z}\left(\frac{\partial z}{\partial y}\right)_p\right] \tag{6.5}$$

or

$$f\frac{\partial u}{\partial z} = \frac{g}{\rho}\left(\frac{\partial \rho}{\partial y}\right)_p. \tag{6.6}$$

Similarly

$$f\frac{\partial v}{\partial z} = -\frac{g}{\rho}\left(\frac{\partial \rho}{\partial x}\right)_p. \tag{6.7}$$

Thus if the motion is geostrophically balanced, the variation of the velocity with height is proportional to the gradient of the density within surfaces of constant pressure. If ρ is constant in surfaces of constant p, i.e. if $\rho = \rho(p)$ so that $\nabla\rho \times \nabla p = 0$, the velocity is independent of height.

We have then the Taylor–Proudman theorem

Velocities are independent of the co-ordinate parallel to the rotation vector if the fluid is in geostrophic balance and if the fluid is barotropic.

In the present case although the rotation vector is not, strictly speaking, in the z direction, the approximations derived from the slimness of the fluid layer allow us to identify the local normal component of the earth's rotation with the rotation vector specified by the Taylor–Proudman theorem.

(iii) Consistent with (6.2) the horizontal divergence of the velocity field vanishes if the horizontal variations of density are small, i.e., from (6.2) and (6.3)

$$\nabla \cdot \boldsymbol{q}_H = \frac{f}{g}\hat{k} \cdot \left[\boldsymbol{q} \times \frac{\partial \boldsymbol{q}}{\partial z} \right]. \tag{6.8}$$

Each term in the divergence on the left-hand side of (6.8) is $O(U/L)$ while the right-hand side is $O[(f/g)(U^2/D)]$ whose ratio is

$$\varepsilon f^2 L^2/gD.$$

For most atmospheric and oceanic large scale phenomena, $f^2L^2/gD = O(1)$ or less, and hence to be consistent with the geostrophic approximation (which requires $\varepsilon \ll 1$) we must conclude that in the geostrophic approximation

$$\nabla \cdot \boldsymbol{q}_H = 0. \tag{6.9}$$

(iv) For similar reasons the vertical component of the relative vorticity

$$\zeta = \hat{k} \cdot \nabla \times \boldsymbol{q}_H \tag{6.10}$$

is given in terms of the pressure, viz:

$$\zeta = \nabla_H^2 p/\rho f. \tag{6.11}$$

7. **Geostrophic degeneracy.** The geostrophic approximation is extraordinarily appealing because of its simplicity. Once the pressure field is determined (perhaps by direct observation) the velocity fields (at least to O (ε or E)) are determined as well as the density field. (In oceanographic circumstances the density field is usually measured and the pressure field inferred from it.) As a diagnostic tool for analyzing the motion field at any instant, it is simplicity itself. Its very simplicity, however, presents fundamental difficulties, for the balance presented by the geostrophic approximation is quasi-static and there is no way, at the level of approximation suitable for (5.4), to describe and determine the evolution of the

pressure field with time and hence of the geostrophic approximation to the velocities. The geostrophic approximation alone simply does not contain enough information to specify the dynamics. Any pressure field can yield a consistent geostrophic motion so long as its Rossby number is small. The geostrophic approximation is therefore degenerate in the sense that alone it is incapable of choosing from among the infinite array of a priori possible motions the correct one. Since the geostrophic approximation is not enough to close the dynamical framework, higher order dynamical effects are required to achieve dynamical closure. Whether these higher order effects involve small nonlinear accelerations, dissipation or some combination depends on the relative size of the Rossby and Ekman numbers and the detailed physical nature of the problem.

In the remainder of these lectures we shall examine a few such typical problem types suggested by specific geophysical phenomena to see how the geostrophic approximation can be utilized in a dynamically predictive framework.

In a crude sense the difficulty comes about because of the very near balance of the Coriolis force and the pressure gradient, each of which dominates the other forces in the momentum equation. This difficulty can be removed by examining the balance of vorticity, for the pressure gradient is ineffective as a contributor to the vorticity as is the Coriolis force since its curl is proportional to the divergence of the geostrophic velocity which is $O(\varepsilon)$. The hint this heuristic reasoning supplies leads us to consider the vorticity balance, and in nondissipative problems, the balance which is inferred from the conservation of potential vorticity.

Lecture III

8. **The removal of the geostrophic degeneracy by viscous forces.** We remarked in the previous lecture that the difficulty associated with the degeneracy of the geostrophic balance could be removed only by the consideration of higher order dynamical effects, either viscous or inertial. It is quite clear that near solid surfaces where the no-slip boundary condition must hold regardless of the size of ε or E, viscous effects,[2] elsewhere disregarded because of the smallness of the Ekman number, must become important to frictionally slow the inviscid geostrophic flow to the speed of the boundary. This does not imply that because viscosity becomes important in regions near solid surfaces that the removal of geostrophic

[2] It is also clear that even the boundary condition on the normal flow to the boundary may not be satisfied by the geostrophically balanced flow. In such circumstances the inertial terms are possible candidates for relief. In any event, viscosity must ultimately be considered to satisfy the no-slip condition.

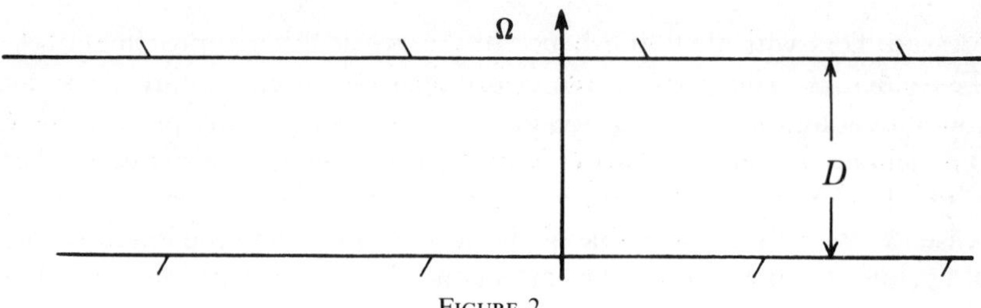

FIGURE 2.

degeneracy will be effected throughout the fluid. We must see if it is so and under what parametric conditions the consideration of the viscous forces (as opposed to the inertial forces) in fact determines the geostrophically balanced bulk of the fluid. We will investigate this problem by consideration of a simple example.

Consider the motion of a layer of homogeneous, incompressible fluid confined between two infinite horizontal surfaces separated by a distance D. We suppose that when viewed in the rotating frame each plate is observed to have a relative velocity, the upper $\boldsymbol{q}_T(x, y)$ and the lower $\boldsymbol{q}_L(x, y)$, in its own plane and that further both $\nabla \cdot \boldsymbol{q}_T$ and $\nabla \cdot \boldsymbol{q}_L = 0$. The simplest such situation, easily reproduced in the laboratory, has the upper plate rotating somewhat faster than the lower plate. Let U be a velocity characteristic of $\boldsymbol{q}_L$ and $\boldsymbol{q}_T$ (perhaps their maximum). Let L be a typical horizontal scale associated with the given velocity fields. We then introduce nondimensional variables, denoted by primes, as follows:

$$(u, v) = U(u', v'), \qquad (x, y) = L(x', y'),$$

$$w = U\frac{D}{L}w', \qquad z = Dz',$$

$$p = p_0 + 2\rho U\Omega L p' - \rho g z,$$

and the Navier–Stokes equations then become, for steady flow,

$$\text{(8.1a)} \quad \varepsilon(uu_x + vu_y + wu_z) - v = -p_x + (E/2)(u_{zz} + \delta^2[u_{xx} + u_{yy}]),$$

$$\text{(8.1b)} \quad \varepsilon(uv_x + vv_y + wv_z) + u = -p_y + (E/2)(v_{zz} + \delta^2[v_{xx} + v_{yy}]),$$

$$\text{(8.1c)} \quad \delta^2\varepsilon(uw_x + vw_y + ww_z) = -p_z + (\delta^2 E/2)(w_{zz} + \delta^2[w_{xx} + w_{yy}]),$$

$$\text{(8.2)} \quad u_x + v_y + w_z = 0,$$

$$\text{(8.3)} \quad \begin{aligned} (u, v, w) &= (u_T, v_T, 0), \qquad z = 1, \\ (u, v, w) &= (u_L, v_L, 0), \qquad z = 0. \end{aligned}$$

Primes have been dropped from the nondimensional quantities. When we wish to refer to dimensional quantities we shall use an asterisk to distinguish them from the dimensionless variables. The nondimensional parameters

$$\delta = D/L, \qquad \varepsilon = U/2\Omega L, \qquad E = \nu/\Omega D^2,$$

have been naturally introduced into the dimensionless mathematical problem. Our interest is centered in this model problem to the case where both ε and E are small, for as we have seen that is precisely the parameter situation for geophysically interesting large scale motions.

For small ε and E a perturbation expansion is obviously in order, and in the region outside the viscous boundary layers which we expect to exist near $z = 0$ and $z = 1$, the appropriate asymptotic expansion must have the form, for example for the velocity in the x direction,

$$u = u_I(x, y, z, \varepsilon, E) = u_I^{(0)}(x, y, z) + R(\varepsilon, E)u_I^{(1)}(x, y, z), \tag{8.4}$$

where $R(\varepsilon, E)$ goes to zero in the limit $\varepsilon \to 0$, $E \to 0$. Precisely what the gauge function R is, i.e. how the expansion proceeds in ε and E, will depend on the relative sizes of ε and E and can be determined only as we go deeper into the problem. The subscript I reminds us that the representation is valid outside the regions affected by viscosity, the so-called interior region. Substitution of (8.4) into (8.1) yields the $O(1)$ balances

$$v_I^{(0)} = \partial p_I^{(0)}/\partial x, \tag{8.5a}$$

$$u_I^{(0)} = -\partial p_I^{(0)}/\partial y, \tag{8.5b}$$

$$0 = \partial p_I^{(0)}/\partial z, \tag{8.5c}$$

$$(\partial u_I^{(0)}/\partial x) + (\partial v_I^{(0)}/\partial y) + (\partial w_I^{(0)}/\partial z) = 0. \tag{8.5d}$$

Here we find ourselves in a, by now, familiar situation. The $O(1)$ velocity field is in geostrophic balance. Because the fluid is homogeneous, (8.5c) implies that $u_I^{(0)}$ and $v_I^{(0)}$ are independent of z (the Taylor–Proudman theorem). Furthermore, since

$$(\partial v_I^{(0)}/\partial y) + (\partial u_I^{(0)}/\partial x) \equiv 0, \tag{8.6}$$

(8.5d) yields

$$\partial w_I^{(0)}/\partial z = 0. \tag{8.7}$$

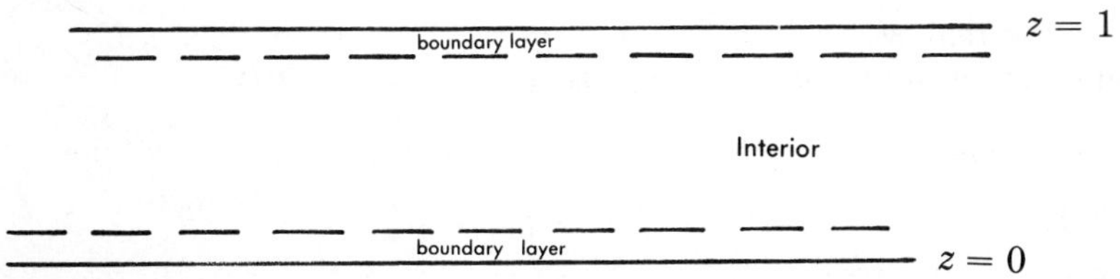

$z = 1$

boundary layer

Interior

boundary layer

$z = 0$

FIGURE 3.

Indeed this is all we can determine at this stage and we are again faced with the indeterminateness of the geostrophic motion.

9. **The Ekman layer.** Since $u_I^{(0)}$ and $v_I^{(0)}$ are independent of z they clearly cannot satisfy (8.3) except in the trivial case $\boldsymbol{q}_T = \boldsymbol{q}_L$. In general the interior velocities are matched to the velocities of the solid surfaces within a boundary layer whose scale is sufficiently small so that the viscous forces, hitherto neglected, become as important as the Coriolis force because of the large shears developed in the boundary layer. In order for the viscous force to balance the Coriolis force near $z = 0$, say, we require that, for example,

$$Eu_{zz} = O(v).$$

Since both u and v are $O(1)$ this requires a vertical scale of variation

$$\delta = E^{1/2} = (1/D)(\nu/\Omega)^{1/2}, \tag{9.1}$$

or a dimensional boundary layer thickness

$$\delta_* = (\nu/\Omega)^{1/2} - D\delta. \tag{9.2}$$

The resulting boundary layer is called the Ekman layer whose thickness $(\nu/\Omega)^{1/2}$ is *independent* of the relative motion of the fluid.

Near $z = 0$, within the Ekman layer we represent the velocity fields

$$u = u_B(x, y, \eta, \varepsilon, E) = u_B^{(0)}(x, y, \eta) + \cdots, \tag{9.3a}$$

$$v = v_B(x, y, \eta, \varepsilon, E) = v_B^{(0)}(x, y, \eta) + \cdots, \tag{9.3b}$$

$$w = w_B(x, y, \eta, \varepsilon, E) = 0 + E^{1/2}w_B^{(1)}(x, y, \eta) + \cdots, \tag{9.3c}$$

where

$$\eta = z/E^{1/2}$$

is the natural co-ordinate scaling in the boundary layer. The fact that the expansion (9.3c) starts with an $O(E^{1/2})$ term for w_B is in consequence of the continuity equation and the condition that w_B vanishes on $z = 0$. If $w_B^{(0)}$ is assumed different from zero, (8.2) would quickly yield

$$\partial w_B^{(0)}/\partial\eta = 0, \tag{9.4}$$

implying that $w_B^{(0)}$, which vanishes on $\eta = 0$, must be identically zero. Substitution of (9.3) into (8.1) yields, as long as $\varepsilon < O(1)$,

$$-v_B^{(0)} = -p_{Bx}^{(0)} + \tfrac{1}{2}u_{B\eta\eta}^{(0)}, \tag{9.5a}$$

$$u_B^{(0)} = -p_{By}^{(0)} + \tfrac{1}{2}v_{B\eta\eta}^{(0)}, \tag{9.5b}$$

$$0 = -p_{B\eta}^{(0)}, \tag{9.5c}$$

$$u_{Bx}^{(0)} + v_{By}^{(0)} = -w_{B\eta}^{(1)}. \tag{9.5d}$$

Since $p_B^{(0)}$ is independent of η, (9.5a) and (9.5b) are two simple, coupled ordinary differential equations and the solutions which remain finite for large η ($z = O(1)$) are

$$u_B^{(0)} = -(\partial p_B^{(0)}/\partial y) + e^{-\eta}(A\cos\eta + B\sin\eta), \tag{9.6a}$$

$$v_B^{(0)} = (\partial p_B^{(0)}/\partial x) + e^{-\eta}(-A\sin\eta + B\cos\eta). \tag{9.6b}$$

On $\eta = 0$,

$$u_B^{(0)} = u_L(x, y), \tag{9.7a}$$

$$v_B^{(0)} = v_L(x, y); \tag{9.7b}$$

hence:

$$u_B^{(0)} = -(\partial p_B^{(0)}/\partial y) + e^{-\eta}([u_L + (\partial p_B^{(0)}/\partial y)]\cos\eta + [v_L - (\partial p_B^{(0)}/\partial x)]\sin\eta), \tag{9.8a}$$

$$v_B^{(0)} = (\partial p_B^{(0)}/\partial x) + e^{-\eta}(-[u_L + (\partial p_B^{(0)}/\partial y)]\sin\eta + [v_L - (\partial p_B^{(0)}/\partial x)]\cos\eta), \tag{9.8b}$$

and using (9.5d),

$$w_B^{(1)} = -[(\partial v_L/\partial x - \partial u_L/\partial y) - \nabla_H^2 p_B^{(0)}][(1 - e^{-\eta}(\sin\eta + \cos\eta))/2]. \tag{9.9}$$

The boundary layer solutions must merge smoothly into the interior solutions for large η, e.g.

$$\lim_{\eta\to\infty} p_B^{(0)}(x, y, \eta) = \lim_{z\to 0} p_I^{(0)}(x, y, z). \tag{9.10}$$

Since both $p_B^{(0)}$ and $p_I^{(0)}$ are independent of η and z respectively,

$$p_B^{(0)}(x, y) = p_I^{(0)}(x, y), \tag{9.11}$$

hence using (8.5a, b)

$$u_B^{(0)} = u_I^{(0)} + e^{-z/E^{1/2}}\{(u_L - u_I^{(0)})\cos z/E^{1/2} + (v_L - v_I^{(0)})\sin z/E^{1/2}\}, \tag{9.12a}$$

$$v_B^{(0)} = v_I^{(0)} + e^{-z/E^{1/2}}\{(v_L - v_I^{(0)})\cos z/E^{1/2} - (u_L - u_I^{(0)})\sin z/E^{1/2}\}, \tag{9.12b}$$

$$w_B^{(1)} = -\left[\left(\frac{\partial v_L}{\partial x} - \frac{\partial u_L}{\partial y}\right) - \left(\frac{\partial v_I^{(0)}}{\partial x} - \frac{\partial u_I^{(0)}}{\partial y}\right)\right] \times \left[\frac{1}{2} - \frac{\exp(-z/E^{1/2})\sin(z/E^{1/2} + \pi/4)}{2^{1/2}}\right]. \tag{9.12c}$$

The Ekman boundary layer has allowed us to satisfy the proper boundary conditions and we note that as yet the interior flow is still undetermined. Note the characteristic spiral in the hodograph of the boundary layer solutions (9.8). It is perhaps the most kinematically striking feature of the Ekman layer.

Near $z = 1$ another Ekman layer is required. The analysis proceeds precisely as above and we only display the solutions, thus, in the vicinity of $z = 1$,

$$(9.13a)\quad u_B^{(0)} = u_I^{(0)} - \exp[-(1-z)/E^{1/2}]\{(u_T - u_I^{(0)}) \cos[(1-z)/E^{1/2}] + (v_T - v_I^{(0)}) \sin[(1-z)/E^{1/2}]\},$$

$$(9.13b)\quad v_B^{(0)} = v_I^{(0)} + \exp[-(1-z)/E^{1/2}]\{(v_T - v_I^{(0)}) \cos[(1-z)/E^{1/2}] - (u_T - u_I^{(0)}) \sin[(1-z)/E^{1/2}]\};$$

$$(9.14)\quad w_B^{(1)} = \left[\left(\frac{\partial v_T}{\partial x} - \frac{\partial u_T}{\partial y}\right) - \left(\frac{\partial v_I^{(0)}}{\partial x} - \frac{\partial u_I^{(0)}}{\partial y}\right)\right] \times \left[\frac{1}{2} - \frac{\exp[-(1-z)/E^{1/2}]}{2^{1/2}} \sin\left(\frac{1-z}{E^{1/2}} + \pi/4\right)\right].$$

The solutions in the upper Ekman layer arc cxactly the same as in the lower with one vital exception, the change in sign of the vertical velocity. This occurs because an interior flow with vorticity in excess of the vorticity of the boundary will cause a geostrophic pressure gradient, directed toward the vortex center to cause in turn a mass transport in both Ekman layers toward the vortex center. Continuity of mass flux demands that the resulting horizontal mass flux convergence lead to flow out of each boundary layer and consequently upward from the lower layer but downward from the upper layer.

A perusal of (9.12a, b) and (9.13a, b) indicates that the horizontal velocities merge smoothly from their boundary layer to interior values. However, (9.12c) and (9.14) yield two *matching* conditions for the interior flow, i.e. that on $z = 0$ the interior vertical velocity

$$(9.15)\quad w_I(x, y, 0) = E^{1/2} w_B^{(1)}(x, y, \infty) = -\frac{E^{1/2}}{2}\left\{\left(\frac{\partial v_L}{\partial x} - \frac{\partial u_L}{\partial y}\right) - \left(\frac{\partial v_I^{(0)}}{\partial x} - \frac{\partial u_I^{(0)}}{\partial y}\right)\right\},$$

while on $z = 1$

$$(9.16)\quad w_I(x, y, 1) = \frac{E^{1/2}}{2}\left\{\left(\frac{\partial v_T}{\partial x} - \frac{\partial u_T}{\partial y}\right) - \left(\frac{\partial v_I^{(0)}}{\partial x} - \frac{\partial u_I^{(0)}}{\partial y}\right)\right\}.$$

Thus, the *interior* vertical velocity must be $O(E^{1/2})$ on the edge of the Ekman layers and is determined there by the $O(1)$ interior horizontal velocities. Note that (9.15) and (9.16) together with (8.7) imply that $w_I^{(0)}$ is identically zero.

10. **Determination of the interior flow.** If the pressure is eliminated between (8.1a) and (8.1b) we find, for the interior, with the aid of (8.2),

$$(10.1)\quad w_{Iz} = \frac{E}{2}[\zeta_{Izz} + \delta^2\nabla_H^2\zeta_I] + \varepsilon\left[\frac{\partial}{\partial x}(\boldsymbol{q}_I\cdot\nabla)v_I - \frac{\partial}{\partial y}(\boldsymbol{q}_I\cdot\nabla)u_I\right]$$

where $\zeta_I = (\partial v_I/\partial x) - (\partial u_I/\partial y)$. Hence w_{Iz} is no larger than order ε or E, depending which is larger. If $\varepsilon < O(E^{1/2})$, (10.1) implies that the order $E^{1/2}$ vertical velocity pumped into the interior from the Ekman layers must be independent of z. Hence if $\varepsilon < E^{1/2}$, (9.15) and (9.16) yield

$$(10.2)\qquad \zeta_I^{(0)} = \tfrac{1}{2}(\zeta_T + \zeta_L),$$

i.e., the $O(1)$ vertical component of vorticity is equal to the mean of the vorticity of the motion of the bounding surfaces. With (8.5a, b) the problem for the geostrophic velocity becomes, in terms of the pressure,

$$(10.3)\qquad \nabla_H^2 p_I^{(0)} = \tfrac{1}{2}(\zeta_T + \zeta_L).$$

With appropriate lateral boundary conditions, (10.3) may be solved to yield the steady geostrophic flow. If there are rigid lateral boundaries the $O(1)$ geostrophic normal velocity must vanish on those boundaries and the solution is completely determined (unless the region is not simply connected, in which case the circulation around each hole must be determined[3]). As a simple example, suppose $\boldsymbol{q}_L = 0$, $\boldsymbol{q}_T = r\hat{\theta}$; then it is easy to verify that

$$(10.4)\qquad \boldsymbol{q}_{I_H} = r\hat{\theta}/2, \qquad w_I = E^{1/2}/2,$$

where $\hat{\theta}$ is a unit vector in the azimuthal direction. Thus for $\varepsilon < E^{1/2}$ the geostrophic degeneracy has been easily removed by a consideration of the small, viscously controlled, vertical velocity pumped out of the Ekman layer. When $\varepsilon = O(E^{1/2})$ the problem is considerably more complicated, and instead of (10.3), the vertical integral of (10.1) yields

$$(10.5)\quad \nabla_H^2 p_I^{(0)} + (\varepsilon/E^{1/2})(p_{Ix}^{(0)}\nabla_H^2 p_{Iy}^{(0)} - p_{Iy}^{(0)}\nabla_H^2 p_{Ix}^{(0)}) = \tfrac{1}{2}(\zeta_T + \zeta_L)$$

and the equation determining the geostrophic flow is now nonlinear. In either case the flow is still in principle determined by the pumping from the Ekman layer. If $\varepsilon > O(E^{1/2})$ the flow field senses the Ekman layer pumping as a minor effect. The primary control of the geostrophic

[3] This can usually be done by balancing the mass fluxes in the Ekman layer.

flow is inertial rather than viscous although in the problem discussed here ultimately the determination of the flow will require consideration of the Ekman layers. However the structure of the flow field will be constrained primarily by the inertial constraint of vorticity conservation along streamlines, or isobars, i.e. for $\varepsilon/E^{1/2} \gg 1$, (10.5) states that

$$\zeta_I^{(0)} = F(p_I^{(0)})$$

where F is an arbitrary function.

To sum up, when $\varepsilon < E^{1/2}$ the geostrophic degeneracy is removed by the consideration of primarily viscous effects (and the problem is consequently linear), while for $\varepsilon > E^{1/2}$ inertial effects dominate and the problem is fundamentally nonlinear and inviscid in character.

We remark finally that (10.5) (or the simple form (10.3)) has been used as a model for the study of the circulation of large lakes (e.g. Lake Michigan) where, however, the forcing is due to an applied wind stress instead of moving boundaries in which case the curl of the boundary velocity serves here as a model for the curl of the applied wind stress.

LECTURE IV

11. **Conservation of potential vorticity for a shallow layer of fluid.** The results of the last lecture indicated that geostrophic motions which are driven by frictional agents (e.g. tangentially moving boundaries) are most easily studied in that part of parameter space $\varepsilon < E^{1/2}$. We turn our attention now to the alternate situation, $\varepsilon > E^{1/2}$, i.e. inertial control of the geostrophically balanced motion.

As an example consider the motion of a homogeneous, incompressible fluid confined above by a horizontal plane and below by a rigid surface whose distance from the upper plane $H(x, y)$ is variable (see Figure 4).

The fluid is rotating with angular velocity Ω (plus some small, geostrophic departures). Since the motion is, to lowest order, in geostrophic balance the $O(1)$ horizontal velocities will be independent of z (the Taylor–Proudman theorem) and hence the continuity equation, in conjunction

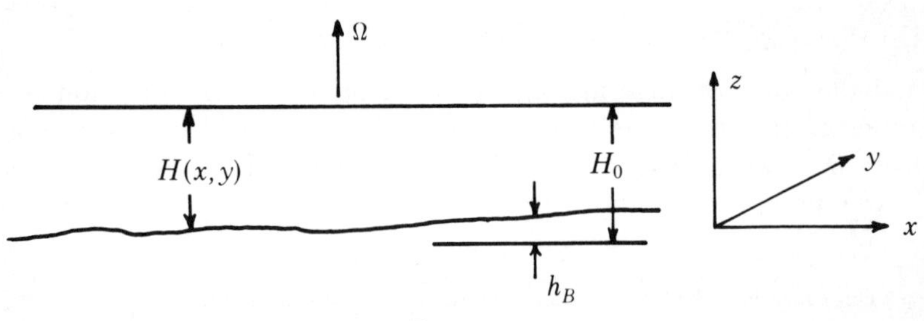

FIGURE 4.

with the kinematic boundary conditions for an inviscid fluid ($\varepsilon \gg E^{1/2}$),

$$w = uh_{Bx} + vh_{By}, \qquad z = h_B, \tag{11.1}$$

yields

$$w = (u_x + v_y)(h_B - z) + uh_{Bx} + vh_{By}. \tag{11.2}$$

Since

$$w(H_0) = 0, \tag{11.3}$$

Equation (11.2) implies that ($H = H_0 - h_B$)

$$(u_x + v_y)H + uH_x + vH_y = (uH)_x + (vH)_y = 0, \tag{11.4}$$

i.e., the conservation of volume flux. Using (11.4) to eliminate the horizontal divergence of velocity, (11.2) becomes

$$w = (uH_x + vH_y)[(z - h_B)/H] + uh_{Bx} + vh_{By}. \tag{11.5}$$

Using the identity

$$Dz/Dt = w, \tag{11.6}$$

it immediately follows that

$$(D/Dt)[(z - h_B)/H] = 0, \tag{11.7}$$

which states that the relative height, or status, of a fluid element in a vertical column is conserved in the flow. Since the $O(1)$ motion is horizontal (the vertical motion being $O(\varepsilon)$ at most) the only $O(1)$ component of the relative vorticity is the vertical component ζ. Since the fluid is of constant density and hence trivially barotropic, the status function $(z - h_B)/H$ fulfills the necessary requirements to play the role of λ in defining the potential vorticity (Lecture I, Equation (4.5)) and hence

$$\rho \Pi = (\zeta + 2\Omega)\hat{k} \cdot \nabla[(z - h_B)/H] = (\zeta + f)/H \tag{11.8}$$

is conserved. This is true for any motion of a shallow layer of homogeneous fluid in which the horizontal velocities are independent of height although we have restricted our attention to quasi-geostrophic motions. Thus if a fluid column is stretched by flowing into a deeper region, the relative vorticity ζ must increase. If ζ is initially zero the stretching (or compression) of a fluid column will produce relative vorticity due to the presence of the planetary vorticity $f = 2\Omega$. Thus since Π is conserved

$$\zeta_t + u\zeta_x + v\zeta_y - (\zeta + f)(uH_x + vH_y)/H = 0. \tag{11.9}$$

Since the Rossby number is assumed small, ζ is negligible as compared with f and hence a consistent approximation to (11.9) is

$$\zeta_t + u\zeta_x + v\zeta_y - f(uH_x + vH_y)/H = 0. \tag{11.10}$$

The system is closed dynamically with the further use of the geostrophic balance valid for small ε, i.e. with

$$(11.11) \qquad u = -\frac{1}{\rho f}\frac{\partial p}{\partial y}, \qquad v = \frac{1}{\rho f}\frac{\partial p}{\partial x}, \qquad \zeta = \frac{1}{\rho f}\nabla_H^2 p.$$

Equation (11.10) may be rewritten entirely in terms of the pressure, viz,

$$(11.12) \qquad \nabla_H^2 p_t + \frac{1}{\rho f}\frac{\partial(p, \nabla_H^2 p)}{\partial(x, y)} + \frac{f}{H}\frac{\partial(p, h_B)}{\partial(x, y)} = 0.$$

We will restrict our attention to those cases (we shall see why in a little while) where the variation of H is sufficiently small so that H may be replaced by its average value, H_0, in (11.12) which yields our final equation

$$(11.13) \qquad \nabla^2\psi_t + [\partial(\psi, \nabla^2\psi + f h_B/H_0)/\partial(x, y)] = 0$$

where we have introduced the geostrophic stream function $\psi \equiv p/\rho f$.

Thus the geostrophic degeneracy is removed, when $\varepsilon > E^{1/2}$, by the application of the theorem of potential vorticity conservation in which all velocities are evaluated geostrophically, leading to the *quasi-geostrophic potential vorticity equation* (11.13). This notion has far wider validity and can be extended to stratified fluids, as we shall see later.

12. **Rossby waves.** The nature of the dynamics of geostrophically balanced flow in the inviscid limit ($\varepsilon > E^{1/2}$) is illustrated by the important phenomenon of wave motion which is made possible by topographic variations. We examine the simple case where there is a linear slope of the bottom inclined upward towards positive y, viz:

$$(12.1) \qquad h_B = \Delta H(y/L),$$

in which case (11.13) becomes

$$(12.2) \qquad \nabla^2\psi_t + [\partial(\psi, \nabla^2\psi)/\partial(x, y)] + \beta\psi_x = 0,$$

where

$$(12.3) \qquad \beta = (f/H_0)(\Delta H/L).$$

It is easily verified that the plane wave

$$(12.4) \qquad \psi = A\cos(kx + ly - \sigma t)$$

is a solution of the *nonlinear equation* (12.2) providing

$$(12.5) \qquad \sigma = -\beta k/(k^2 + l^2).$$

The wave (12.4) with the associated dispersion relation (12.5) is the famous Rossby wave. If the wavelength is of the order of L, the scale over which the bottom varies by an amount ΔH, the resulting frequency of the wave will be $O(f\Delta H/H_0)$. If the flow is to be geostrophically balanced we must obviously insist that $\sigma \ll f$, which in turn requires that $\Delta H/H_0 \ll 1$ which is the condition demanded in passing from (11.12) to (11.13).

The wave is peculiar to rotating flows. The oscillation arises from the fact that an initial excursion of a fluid column across lines of constant y will cause the column to shrink, which will produce relative vorticity, and the velocity field induced by this vorticity, in co-operation with similarly disturbed neighboring columns, ultimately tends to restore the column to its equilibrium position, which it overshoots due to inertia causing the oscillation to persist.

The perceptive reader will at once notice an extraordinary feature of the Rossby wave, i.e. that its phase speed in the x direction

$$c_x = \sigma/k = -\beta/(k^2 + l^2) \tag{12.6}$$

is unidirectional. The wave itself only exists if there is an ambient potential vorticity gradient in the fluid, i.e. a vorticity gradient in the absence of relative motion. In the case at hand the gradient is β, and since the alignment of the direction of the gradient leads to no loss of generality we may note the rule that individual crests in a Rossby wave always move in such a way that an observer moving with a crest (or any fixed wave phase) will see higher ambient potential vorticity on his right.

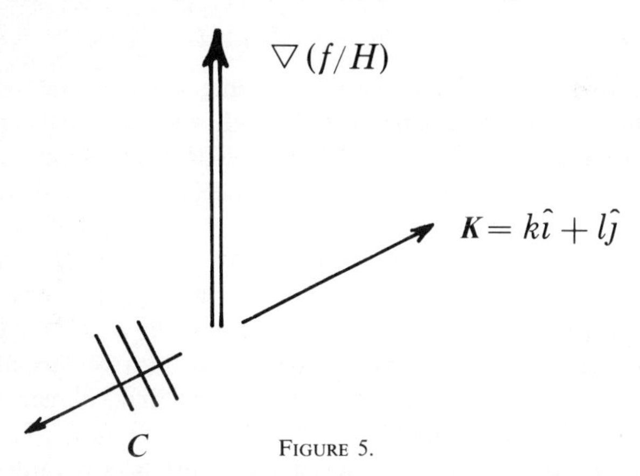

FIGURE 5.

The wave solution (12.4) and dispersion relation (12.5) were studied by Rossby in quite a different context. He reasoned that (as we also did in Lecture II) because large scale motions in the atmosphere were predominantly tangential to the earth's surface, the only significant component of the planetary vorticity, $2\boldsymbol{\Omega}$ was its local normal component $2\Omega \sin \theta$.

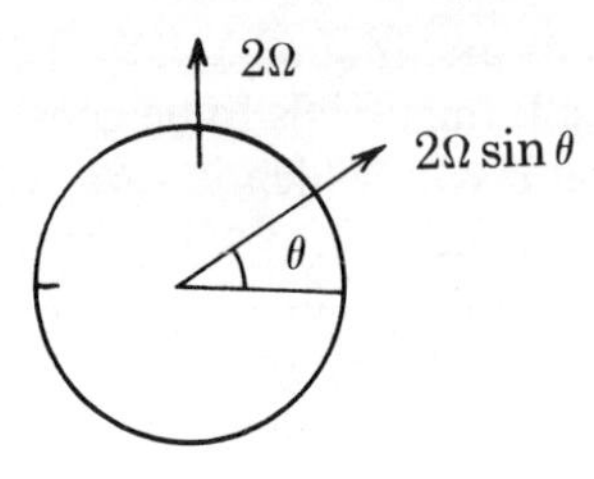

FIGURE 6.

If this is indeed true (as detailed scaling arguments verify) the proper form for the potential vorticity for an atmospheric model which is a homogeneous fluid shell is

$$\rho\Pi = (\zeta + 2\Omega \sin\theta)/(H_0 - h_B), \qquad h_B \ll H_0. \tag{12.7}$$

Furthermore, if the north-south scale of the motion is not too large we may linearize $\sin\theta$ about some central latitude θ_0, viz

$$2\Omega \sin\theta = f_0 + \beta y, \tag{12.8}$$

where

$$f_0 = 2\Omega \sin\theta_0, \qquad \beta = (2\Omega \cos\theta_0)/R, \qquad y = R(\theta - \theta_0), \tag{12.9}$$

and hence aside from an irrelevant constant the potential vorticity is

$$(\zeta + \beta y + f h_B/H_0)/H_0.$$

There is therefore a dynamical equivalence between the variation of the Coriolis parameter, $2\Omega \sin\theta$, with latitude and the linear topographic variation we have used in our model. Rossby considered only the former, we only the latter.[4] Thus the results of our model can be directly transferred to Rossby's simple, but historically ground breaking, theory for large-scale atmospheric waves if we identify β in (12.5) with the northward gradient of the Coriolis parameter and the x and y directions with eastward and northward respectively. Such a model, in which the only effect of the sphericity of the earth is to introduce a linearly varying Coriolis parameter into an otherwise "flat earth" cartesian planar system, is called the β-plane approximation.

Although a single plane wave (12.4) is a solution of the full nonlinear equation (12.2) a superposition or packet of waves is not and the remaining discussion is limited to small disturbance theory, i.e. situations in which the quadratic nonlinearity in (12.2) may be neglected.

---

[4] In more realistic geophysical models both must be included and the topographic contribution to the ambient potential vorticity gradient is rarely constant.

One of the simplest but most illuminating such superpositions occurs in the case of a spatially localized wave packet whose Fourier spectrum is continuous but sharply peaked about the wave vector $\boldsymbol{K} = k\hat{\imath} + l\hat{\jmath}$. In this case the solution of the linear equation

$$(\partial\nabla^2\psi/\partial t) + \beta\psi_x = 0 \tag{12.10}$$

can still be found in the form (12.4), but now A is a *slowly varying* function of x, y, t, i.e. $A_x/A \ll k$, etc. Substitution of (12.4) into (12.10) in that case yields

$$\begin{aligned}&[-\sigma A(k^2 + l^2) - Ak\beta - 2\boldsymbol{K}\cdot\nabla A_t + \sigma\nabla^2 A]\sin(kx + ly - \sigma t)\\ &+[-(k^2 + l^2)A_t + 2\boldsymbol{K}\cdot\nabla A\sigma + \nabla^2 A_t + \beta A_x]\cos(kx + ly - \sigma t) = 0.\end{aligned} \tag{12.11}$$

Since A is a slowly varying function of x, y, and t, to an excellent approximation the coefficients of $\sin(kx + ly - \sigma t)$ and $\cos(kx + ly - \sigma t)$ must vanish separately in (12.11). Retaining only the first order terms in each case we obtain as two conditions for solution

$$\sigma = -\beta k/(k^2 + l^2) \tag{12.12}$$

and, using (12.12),

$$A_t + \frac{\beta(k^2 - l^2)}{(k^2 + l^2)^2}A_x + \frac{2\beta kl}{(k^2 + l^2)^2}A_y = 0. \tag{12.13}$$

Thus, the relation between the frequency and wave number is precisely the same as for a plane wave (after all, locally the packet does appear to be a plane wave) while the propagation of the packet amplitude is determined by (12.13), which in vector form is

$$A_t + \boldsymbol{C}_g\cdot\nabla A = 0 \tag{12.14}$$

where

$$\boldsymbol{C}_g = (\partial\sigma/\partial k)\hat{\imath} + (\partial\sigma/\partial l)\hat{\jmath}, \tag{12.15}$$

i.e. the packet amplitude propagates with the *group velocity*. Note that the group velocity in the x direction can be either positive or negative in distinction to the corresponding phase velocity; so, although individual crests must always propagate to the west, the disturbance packet itself, which really defines the position of the wave disturbance, may propagate either westward or eastward. Naturally the energy of the wave field moves with the packet and hence with the group velocity, so that longitudinally long waves ($k > l$) will propagate their energy to the west while short waves ($k < l$) will propagate their energy to the east. For any given frequency σ, and north-south number l, there are two possible x wave numbers,

$$k = -(\beta/2\sigma) \pm [(\beta^2/4\sigma^2) - l^2]^{1/2}, \tag{12.16}$$

corresponding to the two choices of direction of energy propagation. It is easy to see that the energy preserving reflection of a Rossby wave packet impinging from the right on a rigid north-south (meridional) barrier will produce a reflected wave whose amplitude, frequency, and meridional wave number must be the same (there will be a change in sign of the amplitude). The zonal wave number (k) will be increased by the reflection in order that the reflected packet satisfies the requirement that its energy travels away from the barrier. This curious reflection property has been appealed to in the literature to explain the prevalence of small scale eddy motions near the western oceanic coasts as a consequence of the reflection of energy produced in the ocean by zonally large scale energy inputs by the atmospheric migratory weather systems.

If we put ourselves in a co-ordinate frame moving uniformly through the resting fluid with velocity U to the *west* the plane wave solution (12.4) is still a solution but the frequency suffers a Doppler shift, viz

$$\sigma = Uk - [\beta k/(k^2 + l^2)] \tag{12.17}$$

and hence this dispersion relation is valid for the propagation of a Rossby wave in the presence of a uniform current directed to the *east* with velocity U. In this case if $U \gtrless \beta/(k^2 + l^2)$ the wave crests will propagate to the east (west), and will be held stationary at the critical speed

$$U_c = \beta/(k^2 + l^2), \tag{12.18}$$

which defines the wavelength for a stationary Rossby wave

$$\lambda_c = 2\pi(U_c/\beta)^{1/2}. \tag{12.19}$$

Naturally, the propagation of geostrophic disturbances in the atmosphere and oceans is considerably complicated by the presence of density stratification and the complex atmospheric and oceanic currents in which the disturbances are embedded. The fundamental notion of the description of the propagation in terms of the interaction of the disturbance with a suitably generalized ambient potential vorticity gradient is, however, still the keystone of essentially dissipation-free large scale motions.

Lecture V

The β-effect and steady motions: the oceanic circulation

13. **The wind-driven oceanic circulation: the model.** In Lecture IV, where we discussed a simple example of an inertial closure (i.e. removal

of the geostrophic degeneracy by nonviscous forces), it became apparent that the gradient of the ambient potential vorticity was a feature of great dynamical consequence. It should not be inferred, however, that this feature is limited to fundamentally inviscid motions.

Perhaps one of the most dramatic illustrations of the β-effect is displayed by a simple model of the steady (or time-averaged) oceanic circulation. The simple model we will consider consists of a *homogeneous* fluid in an ocean basin of constant depth D driven by a specified wind stress distribution $\boldsymbol{\tau}(x, y)$ (whose horizontal length scale is L) at its upper surface. Now the ocean is *not* homogeneous but our model may, we hope, be adequate for the discussion of the vertical average of the horizontal motion (the so-called transport) even if it fails to accurately predict the vertical distribution of velocity. We will use the β-plane approximation discussed in the previous lecture by allowing the Coriolis parameter f to be a linear function of y, i.e.

$$f = f_0 + \beta_* y, \tag{13.1}$$

where

$$f_0 = 2\Omega \sin \theta_0, \qquad \beta_* = (2\Omega \cos \theta_0)/R. \tag{13.2}$$

The equations of motion otherwise assumed to be appropriately described for a flat earth are scaled by introducing dimensionless dependent and independent variables, viz:

$$\begin{gathered}
\begin{Bmatrix} u \\ v \\ w \end{Bmatrix} = U \begin{Bmatrix} u' \\ v' \\ \delta w' \end{Bmatrix}, \qquad \begin{Bmatrix} x \\ y \\ z \end{Bmatrix} = L \begin{Bmatrix} x' \\ y' \\ \delta z' \end{Bmatrix}, \\
p = p_0 - \rho g z + \rho U L f_0 p', \\
\boldsymbol{\tau} = \tau_0 \boldsymbol{\tau}', \\
f = f_0(1 + \beta_* L y'/f_0) = f_0 f'.
\end{gathered} \tag{13.3}$$

Naturally our first order of business is to relate the characteristic velocity U to the scale of the stress, τ_0. Since the appropriate Ekman number is small the viscous coupling will occur by means of an Ekman layer near the surface. The vertical shear thus produced will lead to a stress of order

$$\mu U/(\nu f_0)^{1/2},$$

which must balance the applied stress τ_0; hence we will choose the scaling velocity

(13.4) $$U = \tau_0/\rho(\nu f_0)^{1/2}.$$

If $\tau_0 = 1$ dyne/cm^2, and ν (a "turbulent" viscosity) is chosen to be 10^2 cm^2/sec, U turns out to be 1 cm/sec, which is, observationally, perhaps somewhat on the small side. The nondimensional equations for our model become (dropping primes)

(13.5a) $$\varepsilon(uu_x + vu_y + wu_z) - fv = -p_x + (E/2)u_{zz} + \delta^2(E/2)\nabla_H^2 u,$$

(13.5b) $$\varepsilon(uv_x + vv_y + wv_z) + fu = -p_y + (E/2)v_{zz} + \delta^2(E/2)\nabla_H^2 v,$$

(13.5c) $$\delta^2\varepsilon(uw_x + vw_y + ww_z) = -p_z + \delta^2(E/2)w_{zz} + \delta^4(E/2)\nabla_H^2 w$$

(13.5d) $$u_x + v_y + w_z = 0,$$

where

$$\varepsilon = U/f_0L, \qquad E = 2\nu/f_0D^2, \qquad \delta = D/L.$$

The appropriate model boundary conditions in scaled units are

$$(u, v, w) = 0, \qquad z = 0,$$

(13.6) $$u_z\hat{\imath} + v_z\hat{\jmath} = E^{-1/2}(\tau^{(x)}\hat{\imath} + \tau^{(y)}\hat{\jmath}), \qquad z = 1,$$

$$w = 0, \qquad z = 1.$$

Note that in our model we ignore the motion of the free surface of our ocean, a condition whose validity can be demonstrated for small ε. We will discuss the boundary conditions on the lateral boundaries of the oceanic basin a little later.

14. **The interior flow.** In the bulk of the fluid, i.e. outside any boundary layers, the nonlinear and viscous terms in (13.5) are appropriately measured by ε and E respectively, both of which are small. Neglecting those terms and eliminating the pressure between (13.5a) and (13.5b), recalling that

(14.1) $$df/dy = \beta \equiv \beta_* L/f_0,$$

we obtain to lowest order

(14.2) $$\beta v_I = f(\partial w_I/\partial z),$$

where the subscript I reminds us that the variables so subscripted are adequate approximations to the fields in the interior.

Equation (14.2) implies a remarkable constraint between the horizontal and vertical motion. Thus, if there is vortex tube *stretching* $(\partial w_I/\partial z > 0)$, accomplished by differential suction out of the upper and lower Ekman layers, the resulting increase of vorticity experienced by a fluid column is obtained by the column moving northward where the planetary vorticity, which it possesses, is larger. If there is vortex tube squashing ($\partial w_I/\partial z < 0$) the fluid column must move southward.

Now on very general grounds, since our scaling will lead to an $O(1)$ horizontal velocity in the upper Ekman layer, the equation of continuity implies that the vertical velocity pumped out of the upper Ekman layer is $O(E^{1/2})$.[5] If the interior velocity were $O(1)$ the vertical velocity pumped out of the lower Ekman layer would also be $O(E^{1/2})$ which would imply that $\partial w_I/\partial z$ is $O(E^{1/2})$ at most. Hence v_I is at most $O(E^{1/2})$. Now we will see that the presence of meridional bounding coasts will imply that $u_I = O(v_I)$ and hence the interior velocity is entirely of $O(E^{1/2})$, this restriction on the amplitude of the interior motion being entirely due to the constraint imposed by the presence of the potential vorticity gradient. It is important to realize that this constraint requires

$$\varepsilon \ll \beta$$

or equivalently

$$U/\beta_* L^2 \ll 1, \tag{14.3}$$

which holds for the relatively sluggish oceanic circulation but *not* for the more vigorous atmospheric flows.

Furthermore since $\mathbf{q}_{IH}$ is $O(E^{1/2})$ the vertical velocity pumped out of the *lower* Ekman layer is $O(E)$ and, to lowest order, unimportant compared to the $E^{1/2}$ velocity pumped out of the upper Ekman layer.

15. **The upper Ekman layer.** The analysis of the Ekman layer near $z = 1$ proceeds entirely as in Lecture III, §9, except for the change in boundary conditions at $z = 1$ and the absence of an $O(1)$ interior flow. We may therefore skirt the presentation of the detailed analysis and present the solution for the $O(1)$ boundary layer variables viz:

(15.1a)

$$u_B^{(0)} = \frac{1}{2f^{1/2}}\exp(-\eta f^{1/2})[(\tau^{(y)} - \tau^{(x)})\sin\eta f^{1/2} + (\tau^{(y)} + \tau^{(x)})\cos\eta f^{1/2}],$$

(15.1b)

$$v_B^{(0)} = \frac{1}{2f^{1/2}}\exp(-\eta f^{1/2})[(\tau^{(y)} - \tau^{(x)})\cos\eta f^{1/2} - (\tau^{(y)} + \tau^{(x)})\sin\eta f^{1/2}],$$

[5] It is interesting to note that the *dimensional* vertical velocity pumped out of the upper Ekman layer is $O(E^{1/2}U)$ which is *independent* of ν. This is very satisfying since one of the weakest elements of the model is the uncertainty in the size of the turbulent viscosity.

where

(15.2) $$\eta = (1 - z)/E^{1/2}, \qquad f = (1 + (\beta_* L/f_0)y).$$

The horizontal transport in the Ekman layer $\boldsymbol{M}_E$ is given by the relation

(15.3)
$$\begin{aligned} \boldsymbol{M}_E = E^{1/2} \int_0^\infty (u_B^{(0)}\hat{\imath} + v_B^{(0)}\hat{\jmath})\, d\eta &= -(E^{1/2}/2f)\hat{k} \times \boldsymbol{\tau} \\ &= (E^{1/2}/2f)(\hat{\imath}\tau^{(y)} - \hat{\jmath}\tau^{(x)}). \end{aligned}$$

The total mass transport in the Ekman layer is perpendicular (in the northern hemisphere where $f > 0$) and to the right of the direction of the applied wind stress. This is as it must be, for the only force capable of balancing the *net* stress on the Ekman layer as a whole (i.e. on the upper surface) is the net Coriolis force. If the net Coriolis force, i.e. the vertical average of the Coriolis force, is to be in the direction of the applied stress, the net Ekman *velocity* and hence $\boldsymbol{M}_E$ must be at right angles to $\boldsymbol{\tau}$.

The horizontal divergence of the Ekman transport yields the $O(E^{1/2})$ interior velocity sucked into the upper Ekman layer, viz:

(15.4) $$w_I(x, y, 1) = E^{1/2} w_I^{(1)}(x, y, 1) + \cdots = (E^{1/2}/2)\hat{k} \cdot \operatorname{curl} \boldsymbol{\tau}/f.$$

16. **The Sverdrup relation.** Turning our attention once again to the interior flow we scale the interior velocity and pressure

(16.1a) $$\boldsymbol{q}_I = E^{1/2}\boldsymbol{q}_I^{(1)},$$

(16.1b) $$p_I = E^{1/2} p_I^{(1)},$$

which yields to lowest order ($\varepsilon \ll 1$, $E \ll 1$),

(16.2) $$\hat{k} \times \boldsymbol{q}_I^{(1)} = -(\nabla p_I^{(1)}/f),$$

the vertical component of which once more implies that $\boldsymbol{q}_H^{(1)}$ is independent of z. Taking the vertical component of the curl of (16.2) we obtain (14.2) again,

(16.3) $$\beta v_I^{(1)} = f w_{I_z}^{(1)},$$

after use of the continuity equation. With $v_I^{(1)}$ independent of z and $w_I^{(1)}(x, y, 1)$ given by (15.4) while $w_I^{(1)}(x, y, 0)$ is zero we immediately obtain the relation

(16.4a) $$\beta v_I^{(1)} = (1/2) f \hat{k} \cdot \operatorname{curl} (\boldsymbol{\tau}/f),$$

(16.4b) $$w_I^{(1)} = (z/2)\hat{k} \cdot \operatorname{curl} (\boldsymbol{\tau}/f).$$

From the x component of (16.2) we obtain

$$p_I^{(1)} = -\frac{1}{2}\frac{f^2}{\beta}\int_x^1 \hat{k}\cdot\operatorname{curl}(\tau/f)\,dx' + h(y), \tag{16.5a}$$

while from the y component of (16.2),

$$u_I^{(1)} = \frac{1}{2f}\frac{\partial}{\partial y}\left[\frac{f^2}{\beta}\int_x^1 \hat{k}\cdot\operatorname{curl}(\tau/f)\right] - \frac{1}{f}\frac{\partial h}{\partial y}, \tag{16.5b}$$

where $h(y)$ is an *arbitrary* function. The presence of this arbitrary function is a reflection of the fact that in the absence of Ekman suction ($\hat{k}\cdot\operatorname{curl}\tau/f = 0$) the fluid in the interior must conserve potential vorticity. If (14.3) holds, i.e. if the relative vorticity *gradient* is much less than the planetary vorticity gradient, a steady flow must stay on lines of constant y (latitude) to conserve planetary vorticity. Hence in that case v_I would be zero but u_I still undetermined. When the Ekman suction differs from zero v_I is determined but u_I is still free, hence the presence of the unknown function in (16.5b). In a sense, the presence of the strong ambient potential vorticity gradient has reintroduced the problem of geostrophic degeneracy (e.g. (16.2)). The interesting feature in this case is that the removal of the degeneracy must take place in a boundary layer near a lateral wall and not in the interior. For example, once $u_I^{(1)}$ is determined on any meridional plane, $h(y)$ is fixed. Regions (below the Ekman layer) where the hitherto neglected inertial and/or viscous forces are important are located in narrow boundary zones. This is very suggestive, for most oceans are characterized by intense boundary currents on their western sides (e.g. the Gulf Stream in the North Atlantic) where inertial accelerations and turbulent momentum mixing are much larger than typical mid-ocean values.

The northward transport in the interior is

$$M_I^{(y)} = \int_0^1 v_I^{(1)}E^{1/2}\,dz = \frac{E^{1/2}f}{2\beta}\hat{k}\cdot\operatorname{curl}(\tau/f); \tag{16.6}$$

hence the total northward transport is given by

$$M_E^{(y)} + M_I^{(y)} = \tfrac{1}{2}E^{1/2}\hat{k}\cdot\operatorname{curl}\tau, \tag{16.7}$$

which is the famous Sverdrup relation. The Sverdrup relation exemplifies our understanding of mid-oceanic flows. It is, no doubt, true to some extent, but more likely a highly oversimplified view of mid-oceanic dynamics. Recent observations have suggested that small scale Rossby-like waves may play a role in the vorticity balance of the steady circulation.

At the present time we are simply not certain of the extent of their importance.

17. **The western intensification of the oceanic circulation.** Consider an oceanic basin bounded at $x = 0$ and $x = 1$ by meridional solid barriers. Then the total northward transport integrated across the basin is

$$(17.1) \qquad \int_0^1 (M_E^{(y)} + M_I^{(y)})\, dx = \frac{E^{1/2}}{2} \int_0^1 \hat{k} \cdot \operatorname{curl} \boldsymbol{\tau}\, dx.$$

In general $\int_0^1 \hat{k} \cdot \operatorname{curl} \boldsymbol{\tau}\, dx$ is not zero. A simple but representative wind stress pattern would be of the form

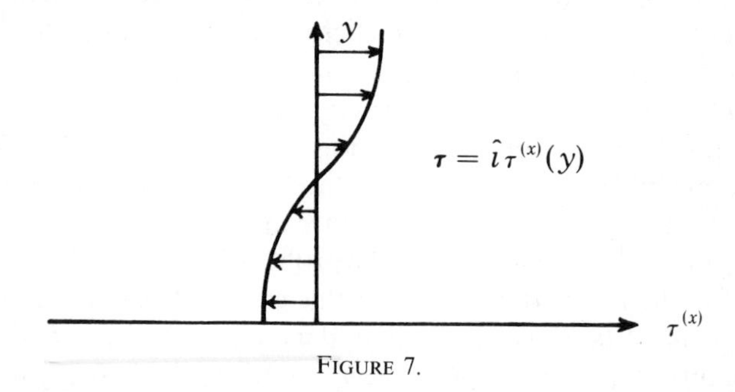

FIGURE 7.

modelling the stresses due to westerly[6] winds in higher latitudes and easterly winds (the Trades) at lower latitudes. In this case

$$(17.2) \qquad \int_0^1 \hat{k} \cdot \operatorname{curl} \boldsymbol{\tau} = -\frac{\partial \tau^{(x)}}{\partial y} < 0.$$

In order to preserve mass there must be a return northward flow in a narrow boundary layer in which forces neglected in deriving (17.1) can no longer be ignored. From a theoretical point of view the first question to be answered is whether the return flow, which must because of its narrowness have relatively high speed, occurs on the eastern or western meridional boundary.

A simple, highly idealized model which answers this question may be derived from the vertical integral of (14.2), in which the vertical velocity pumped out of the lower Ekman layer is not ignored, viz:

$$(17.3) \qquad \frac{\beta v_I}{f} = \frac{E^{1/2}}{2} \hat{k} \cdot \operatorname{curl} \boldsymbol{\tau}/f - \frac{E^{1/2}}{2f^{1/2}} \left\{ f^{1/2} \frac{\partial v_I/f^{1/2}}{\partial x} - f^{1/2} \frac{\partial u_I/f^{1/2}}{\partial y} \right\}.$$

[6] In meteorological parlance a westerly (easterly) wind is a wind blowing *from* the west (east).

In the mid-ocean flow the last term in (17.3) is $O(E)$ because u_I and v_I are $O(E^{1/2})$. However, in a narrow boundary layer on a meridional coast, both v_I and its longitudinal variation may be sufficiently large for the Ekman pumping to be significant. Hence a uniform approximation to (17.3) is

$$\frac{E^{1/2}}{2f^{1/2}}\frac{\partial v_I}{\partial x} + \frac{\beta v_I}{f} = \frac{E^{1/2}}{2}\hat{k}\cdot \operatorname{curl} \tau/f, \tag{17.4}$$

or, using the fact that the total transport $M_T^{(y)}$ is given by

$$M_T^{(y)} = M_I^{(y)} - E^{1/2}\tau^{(x)}/2f, \tag{17.5}$$

we obtain an equation for the total northward transport,

$$\frac{f^{1/2}E^{1/2}}{2}\frac{\partial M_T^{(y)}}{\partial x} + \beta M_T^{(y)} = \frac{E^{1/2}}{2}\hat{k}\cdot \operatorname{curl} \tau, \tag{17.6}$$

where we have explicitly used the fact that in the boundary layer on the meridional wall

$$\partial M_I^{(y)}/\partial x \gg \partial M_E^{(y)}/\partial x, \tag{17.7}$$

i.e. that the return flow is occurring in a current much thinner than the length scale associated with the wind stress which, as we have seen, determines $M_E^{(y)}$. Equation (17.6) must be solved subject to the constraints that (i) over most of the basin the Sverdrup relation holds and (ii) that $\int_0^1 M_T^{(y)}\,dx = 0$.

Using the fact that $f^{1/2}E^{1/2}/2\beta \ll 1$, (17.6) may be easily solved by boundary layer methods to yield

$$M_T^{(y)} = \frac{E^{1/2}\hat{k}\cdot \operatorname{curl} \tau}{\beta} - \frac{1}{f^{1/2}}\int_0^1 \hat{k}\cdot \operatorname{curl} \tau\, dx' \exp(-2\beta x/E^{1/2}f^{1/2}). \tag{17.8}$$

In the case of a wind stress pattern as sketched above, the resulting streamlines of the total transport (note that the *total* horizontal transport must be nondivergent) are sketched below (see Figure 8).

This model indicates that the strong boundary currents closing the streamlines of the total horizontal mass transport should occur on the western side of the ocean. The model presented here is mainly of historical interest. It was the first (due to Stommel) to explain the western intensification of the oceanic circulation and the parallel existence of strong currents such as the Gulf Stream and the Pacific's Kuroshio. We now know that the dynamics in the region of the western boundary current are much more complex than (17.6) would imply.

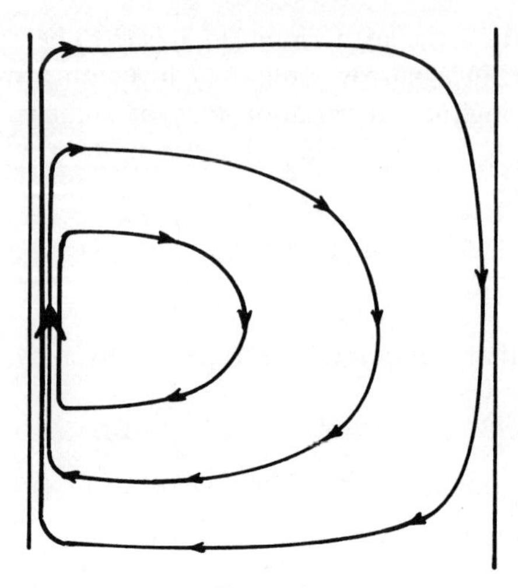

FIGURE 8.

More complicated theories involving effects of lateral friction and nonlinear accelerations have been proposed. The latter especially is more faithful to the observed phenomenon. Nevertheless the important feature revealed by our simple model persists in more complicated theories, namely that the western intensification of the oceanic circulation is a consequence of the presence of a strong ambient potential vorticity gradient.

Incidentally, the fact that the total zonal transport (Ekman + Geostrophic interior) must vanish at the eastern oceanic boundary implies that at $x = 1$,

$$E^{1/2}u_I^{(1)} = -M_E^{(x)}, \tag{17.9}$$

or

$$\partial h/\partial y = \tau^{(y)}(1, y)/2; \tag{17.10}$$

thus

$$u_I^{(1)} = \frac{1}{2\beta}\int_x^1 dx' \frac{\partial}{\partial y}\hat{k}\cdot \operatorname{curl}\boldsymbol{\tau} - \frac{1}{2f}\tau^{(y)}(x, y), \tag{17.11}$$

i.e.

$$E^{1/2}u_I^{(1)} = \int_x^1 dx' \frac{\partial M_T^{(y)}}{\partial y} - M_E^{(x)}. \tag{17.12}$$

Thus once we have determined that the oceanic intensification occurs on the western side of the ocean the interior problem is completely solved.

LECTURE VI

18. **The quasi-geostrophic potential vorticity equation for a stratified fluid**. In Lecture I, I remarked that one of the important characteristics of the atmosphere and the ocean was the presence of a vertical, gravitationally stable, density gradient. Yet we have been able to discuss several interesting phenomena without taking this feature into account. At the very least we have been unable, doing this, to gain any information about the vertical structure of large scale motions. In this lecture we will remove the Taylor–Proudman constraint by introducing the stratification and baroclinicity. We will observe not only the relationship between the homogeneous and stratified models of the same phenomena (like Rossby waves) but will discover the presence of new, and perhaps altogether unexpected, modes of motion.

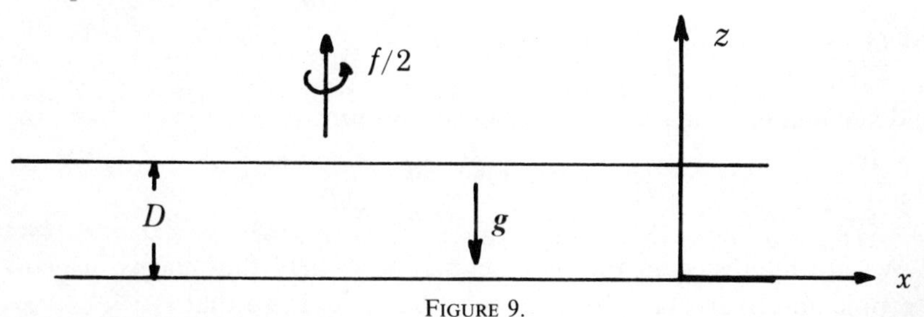

FIGURE 9.

In order to keep our discussion as simple as possible (without sacrificing any dynamical essentials) we consider the motion of a fluid confined between two horizontal planes a distance D apart and rotating with an angular velocity $f/2\hat{k}$, where $\hat{k}$ is a unit vector antiparallel to $\mathbf{g}$ (please see the figure above). In the absence of any relative motion the density field ρ_0 is solely a function of z, the vertical co-ordinate,[7] i.e. $\rho_0 = \rho_0(z)$, and further we assume that

$$(D/\rho_0)(\partial\rho_0/\partial z) \ll 1, \tag{18.1}$$

i.e. that the scale height of the equilibrium density is much greater than D. In the presence of motion the density field will be disturbed so that in general

$$\rho = \rho_0(z) + \delta\rho \tag{18.2}$$

and if the relative motion is small ($\varepsilon \ll 1$) we expect that

$$\delta\rho/\rho_0 \ll 1. \tag{18.3}$$

[7] This implies that the centrifugal potential is negligible compared to the gravitational potential, i.e. that $f^2L^2/gD \ll 1$, where L is a characteristic horizontal measure of the extent of the region.

We will also restrict our attention to a fluid (like water[8]) whose state relation is given by the simple law

$$\rho = \rho_{00}(1 - \alpha(T - T_{00})), \tag{18.4}$$

where ρ_{00} and T_{00} are reference values for ρ and T.

Finally we will examine those motions whose time scales are sufficiently short so that dissipative effects lead to negligible changes of temperature. Hence for a nearly incompressible fluid like water, the first law of thermodynamics reduces, to a good approximation, to the statement of density conservation by individual fluid elements, viz.

$$D\rho/Dt = 0. \tag{18.5}$$

We may thus define the potential vorticity in this case as

$$\Pi = (\boldsymbol{\omega} + f\hat{k}) \cdot \frac{\nabla \rho}{\rho}, \tag{18.6}$$

and keeping only first order terms in the motion,

$$\Pi = \frac{\zeta + f}{\rho_0} \frac{\partial \rho_0}{\partial z} + \frac{f}{\rho_0} \frac{\partial}{\partial z} \delta\rho. \tag{18.7}$$

We are interested in the parameter range where the fluid is in geostrophic and hydrostatic balance, ($\varepsilon \ll 1$, $D/L \ll 1$), so that

$$\zeta = \nabla_H^2 \delta p/\rho_0 f, \qquad \delta\rho = -(1/g)(\partial/\partial z)\delta p, \tag{18.8}$$

where the total pressure is partitioned in the same manner as the density

$$p = p_0(z) + \delta p. \tag{18.9}$$

For the type of motion we will be interested in, the north-south scale of motion is sufficiently small so that the β-plane approximation is adequate, $f = f_0 + \beta_* y$, while the Rossby number, while small, is large enough so that

$$\varepsilon = O(\beta_* L/f_0). \tag{18.10}$$

In which case the conservation of potential vorticity can be written approximately

$$\left(\frac{\partial}{\partial t} + u\frac{\partial}{\partial x} + v\frac{\partial}{\partial y}\right)\left(\frac{\nabla_H^2 \delta p}{\rho_0 f_0} + \beta_* y + \frac{f_0^2}{N^2}\frac{\partial^2}{\partial z^2}\frac{\delta p}{\rho_0 f_0}\right) = 0, \tag{18.11}$$

where N is the so-called Brunt–Väisälä frequency $(-(g/\rho_0)(\partial \rho_0/\partial z))^{1/2}$.

---

[8] The resulting dynamical system is qualitatively similar for a perfect gas such as air.

In obtaining (18.11) we have also used the fact that for an inviscid fluid ($\varepsilon > E^{1/2}$) the vertical velocity in a region bounded below by a horizontal plane is $O(\varepsilon)$ so that

$$D/Dt = \partial/\partial t + u(\partial/\partial x) + v(\partial/\partial y) + O(\varepsilon). \tag{18.12}$$

We introduce the geostrophic stream function

$$\psi = \delta p/\rho_0 f \tag{18.13}$$

for the horizontal motion which allows (18.11) to be rewritten in the neater form

$$\frac{\partial}{\partial t}\left[\nabla_H^2\psi + \frac{f_0^2}{N^2}\frac{\partial^2\psi}{\partial z^2}\right] + \frac{\partial(\psi, \nabla^2\psi + \beta_* y + (f_0^2/N^2)(\partial^2\psi/\partial z^2))}{\partial(x, y)} = 0. \tag{18.14}$$

This quasi-geostrophic form of the potential vorticity equation may be systematically derived from the equations of motion under the following conditions.

(1) $\varepsilon \ll 1$,
(2) $E < \varepsilon$,
(3) $L/R < 1$ (R, the Earth's radius),
(4) $f_0^2L^2/N^2D^2 = O(1)$,
(5) $D/L < 1$,
(6) $(D/\rho_0)(\partial\rho_0/\partial z) < 1$.

The systematic derivation is lengthy!

Finally on $z = 0$ and $z = D$ where w must vanish[9] (18.5) provides us with the necessary boundary condition on δp or equivalently ψ, viz:

$$(\partial/\partial t)(\partial\psi/\partial z) + \partial(\psi, \partial\psi/\partial z)/\partial(x, y) = 0, \qquad z = 0, D. \tag{18.15}$$

19. **Rossby waves in a stratified fluid.** We will now attempt to generalize the analysis of §12 to a stably stratified fluid. Thus we search for wave solutions of (18.14) in the form

$$\psi = AF(z)\cos(kx + ly - \sigma t). \tag{19.1}$$

It may be immediately verified that the nonlinear term in (18.15) is identically zero so that the condition of vanishing w on $z = 0$, D reduces ($\sigma \neq 0$) to

$$dF/dz = 0, \qquad z = 0, D. \tag{19.2}$$

Similarly, all nonlinear terms in (18.14) identically vanish when (19.1) is substituted therein leading to the eigenvalue problem for F,

$$d^2F/dz^2 + \lambda F = 0, \qquad dF/dz = 0, \qquad z = 0, D, \tag{19.3}$$

[9] This requires $\varepsilon > E^{1/2}$.

where

$$\lambda = -(N^2/f_0^2)[\beta k/\sigma + (k^2 + l^2)].$$

The eigenfunctions which are solutions of (19.3) subject to (19.2) are

(19.3a) $\quad F_n(z) = \cos n\pi z/D, \qquad \lambda_n = (n\pi/D)^2, \qquad n = 0, 1, 2 \ldots,$

corresponding to the characteristic *frequencies* of the normal modes

$$\sigma_n = -\frac{\beta_* k}{k^2 + l^2 + (f_0^2/N^2)(\pi^2/D^2)n^2}. \tag{19.3b}$$

We note the following interesting features of the solution. For a given horizontal wavelength the mode with the highest frequency is the $n = 0$ mode. This mode has a horizontal velocity field independent of z (since ψ is independent of z) and has a frequency

$$\sigma_0 = -\beta_* k/(k^2 + l^2),$$

which is identical to the frequency of motion in a homogeneous layer. Thus the first mode ($n = 0$) is *barotropic* in structure, obeys the Taylor–Proudman theorem, and propagates unaffected by the presence of stratification. Hence all our earlier results on the propagation of Rossby waves may be carried over directly to the *barotropic mode* in this stratified problem. To see why, let us calculate the vertical velocity in the Rossby wave. To do this we use (18.15) to calculate w.

Equation (18.5) may be written

$$\frac{\partial}{\partial t}\delta\rho + u\frac{\partial}{\partial x}\delta\rho + v\frac{\partial}{\partial y}\delta\rho + w\left(\frac{\partial}{\partial z}\delta\rho + \frac{\partial}{\partial z}\rho_0\right) = 0. \tag{19.4}$$

Now since the fluid is in hydrostatic balance,

$$\delta\rho = O(\delta p/gD), \tag{19.5a}$$

and is also in geostrophic balance, so that

$$\delta p = O(\rho_0 U f_0 L); \tag{19.5b}$$

hence

$$\delta\rho/\rho_0 = O(Uf_0L/gD) = \varepsilon f_0^2 L^2/gD. \tag{19.6}$$

Therefore

$$\frac{\partial}{\partial z}\delta\rho \Big/ \frac{\partial}{\partial z}\rho_0 = O(\varepsilon f_0^2 L^2/N^2 D^2) \ll 1. \tag{19.7}$$

On the other hand the term involving $w\,\partial\rho_0/\partial z$ cannot be neglected since

$$w\frac{\partial\rho_0}{\partial z}\Big/u\frac{\partial}{\partial x}\delta\rho = O\left(\varepsilon\frac{D}{L}U\frac{\partial\rho_0}{\partial z}\Big/\frac{U}{L}\varepsilon\frac{f_0^2L^2}{gD}\right) = O\left(\frac{N^2D^2}{f_0^2L^2}\right) = O(1).$$

Thus the consistent quasi-geostrophic approximation to (18.15) is

$$\frac{\partial}{\partial t}\frac{\partial\psi}{\partial z} + \frac{\partial(\psi, \partial\psi/\partial z)}{\partial(x, y)} + \frac{N^2}{f_0}w = 0 \tag{19.8}$$

which allows us to calculate w once ψ is known.

Substitution of (19.1) (19.3a, b) into (19.8) yields for each eigenmode

$$w_n = \frac{-A\beta_* k f_0}{k^2 + l^2 + (f_0^2\pi^2n^2/N^2D^2)}\frac{n\pi}{DN^2}\sin\frac{n\pi z}{D}\sin(kx + ly - \sigma t). \tag{19.9}$$

In the barotropic mode, $n = 0$, we see that w is zero for *all* z. That is, the trajectories of fluid elements are flat and consequently lie in planes of constant $\rho_0(z)$. Since the fluid elements in the barotropic mode never venture into regions where the ambient density is different from their own they are oblivious to the fact that the fluid is stratified!

The baroclinic modes, $n = 1, 2, \ldots$, etc., all have *internal nodes*, i.e. the horizontal velocity vanishes on the planes

$$z_j = D(j + \tfrac{1}{2})/n, \qquad j = 0, 1, 2, \ldots, n - 1, \quad n > 0,$$

while the vertical velocity vanishes on the planes

$$z_j = D_j/n, \qquad j = 0, 1, \ldots, n, \quad n > 0.$$

Note that the vertical average of the horizontal velocity is zero except in the barotropic mode (which formed the basis of our study of the homogeneous model) but, naturally, the vertical average of the energy in the internal modes ($n \geqq 1$) is not zero.

We can judge from the dispersion relation (19.3b) that stratification will in general be significant dynamically when

$$N^2 = O(f_0^2L^2/D^2) \tag{19.10}$$

where L is the characteristic wavelength. Restating the criterion another way, the horizontal length scale of the motion must exceed the characteristic scale

$$L_D = DN/f_0 \tag{19.11}$$

if the stratification is to significantly affect the motion. The length scale L_D, which is independent of the motion of the fluid is called the Rossby radius of deformation. For the oceans the characteristic value of

$-(D/\rho_0)(\partial\rho_0/\partial z)$ is 10^{-3} where D is the depth of the ocean (4 km). This leads to values of L_D of O (60–70 km). Motions whose horizontal scales are in excess of these values can be strongly influenced by stratification. On the other hand, the appropriately defined value of N for the atmosphere[10] is 10^{-2} sec^{-1} so that L_D is of O (1,000 km) for motions confined to the troposphere ($D = 10$ km). In short, oceanic internal Rossby waves will have lower frequencies than atmospheric Rossby waves of the same horizontal wavelength.

In fact observations seem to indicate that oceanic meanders are an order of magnitude smaller in scale than their atmospheric counterparts and, furthermore, that the preferred length scales in each case are the order of the Rossby radius of deformation. In the next lecture we will investigate why this scale should yield, at least in order of magnitude, the observed scale of wave fluctuations in the atmosphere and the oceans.

Lecture VII

20. **Baroclinic instability.** One of the most pervasive features of the atmosphere is the presence of large scale "weather" waves. That is, almost any weather map will show large scale horizontal undulations in the pressure and temperature fields migrating from west to east in temperate latitudes. We have already seen in our discussion of Rossby waves that the propagation of large scale features depends critically on the potential vorticity gradient of the ambient state in which the wave is embedded. The major flaw in models of the type we have used to discuss the propagation of Rossby waves is that the model is incapable of explaining the existence of the wave in the first place. Now empirically the weather waves (or cyclone waves, as they are often called) seem to develop ("cyclogenesis") unexpectedly and although there may be preferred regions for the genesis of cyclones their appearance is sufficiently erratic to suggest that they are the result of a fundamental instability of a current, free of waves, that naturally "relaxes" into a wave-like flow. The need, therefore, is a theory to explain the almost spontaneous development of large scale waves, to explain the mechanism of energy transfer which produces the wave from a wave-free mean state, and incidentally to explain a feature touched upon in the last lecture, viz, the apparent preference of waves with length scales of the order of the Rossby radius of deformation.

---

[10] Since the atmosphere is a compressible gas it is not the density but the entropy s which is conserved for nondissipative motions. Hence in the case of the atmosphere (using the perfect gas law) $N^2 = (g/s_0)(ds_0/dz) = (g/T_0)[dT_0/dz + g/c_p]$.

The theoretical answers to these questions, embodied in the theory of baroclinic instability, developed by several workers[11] over the last twenty years, can truly be said to be one of the major triumphs of Geophysical Fluid Dynamics.

The fundamental notion common to all the successful theories is the realization that the presence of unequal solar heating, leading to a temperature field decreasing to the north, provides a store of available potential energy which may be released by a large scale wave. As we shall see the mechanism is subtle and is peculiar to rotating fluids.

21. **The basic state.** We choose the simplest possible basic state embodying the requirements stated above. Let us examine the instability of a rectilinear current in a fluid with a basic stably stratified vertical density gradient which possesses at the same time a horizontal density gradient. To keep matters as elementary as possible we will ignore the β-effect. That is not to say the β-effect is unimportant in the problem of cyclogenesis. Rather, it is felt that it is not an essential ingredient for the development of such waves, a supposition amply verified by laboratory experiments in rotating differentially heated fluids in which cyclone waves are seen to appear and in which the variation of the Coriolis parameter is absent.

Imagine the basic current flowing in the x direction. In the absence of a wave it will be in geostrophic and hydrostatic balance,[12] i.e.

$$u = -(1/\rho_0 f_0)(\partial/\partial y)\delta p, \tag{21.1}$$

$$\delta\rho = -(1/g)(\partial/\partial z)\delta p. \tag{21.2}$$

Eliminating the pressure we obtain

$$\partial u/\partial z = (g/f)(1/\rho_0)(\partial/\partial y)\delta\rho. \tag{21.3}$$

Thus in the presence of a horizontal density (and consequently temperature) gradient, the basic current is obliged to possess a vertical shear (the so-called thermal wind relationship). The sketch below shows the appearance of the density and temperature fields in a stably stratified fluid flowing in the x direction with constant vertical shear confined to the region $0 \leqq z \leqq D$ whose state equation is (18.4) (see Figure 10).

---

[11] Special mention should be made of the pioneering work of J. Bjerknes and Holmboe (1944) and the mathematically complete studies of Charney (1947) and Eady (1949). In the main I will be following Eady's theoretical model.

[12] In fact such a balance is exact within an inviscid approximation.

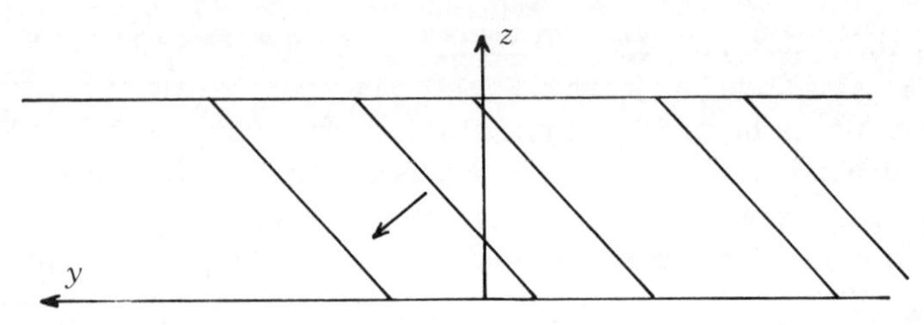

FIGURE 10. Lines of constant density and temperature (arrow shows direction of increasing density and decreasing temperature).

The feature to which I draw your attention is the slope of the isothermal surfaces in the $y - z$ plane. The slope is *not* accompanied by a thermally forced circulation in this plane, a circumstance which can occur only in a rotating fluid in which the resulting northward pressure gradients can be neatly balanced by the Coriolis force produced by the basic current. Indeed in the case where the shear of the basic current is constant and equal to U/D, the slope of the constant density surfaces is given by

$$(\partial z/\partial y)_\rho = (f_0 U/D)/N^2 \tag{21.4}$$

and is directly proportional to the product of the vertical shear and the Coriolis parameter and inversely proportional to N^2, i.e. the gravitationally stable stratification.[13] It seems clear, intuitively, that the basic state, with its sloping isotherms, is in a high potential energy state compared to a state with horizontal isotherms in which the potential energy would be lower. Noting the constraint (21.3) it is also clear that such a redistribution must tend to diminish the vertical shear of the basic state by momentum transports. We now turn our attention to the central questions, i.e. whether a small wave disturbance placed on this basic state will spontaneously grow, what its wavelength will be, and whether the vertical shear of the mean motion is correspondingly reduced.

22. **The wave equation.** To answer these questions we shall seek solutions of the quasi-geostrophic potential vorticity equation (18.14) linearized about the basic state described in (21.1), (21.2) and (21.3). Thus the total geostrophic stream function is written

$$\psi = -Uzy/D + \varphi, \tag{22.1}$$

[13] Note that $f_0 U D^{-1}/N^2 = \varepsilon\delta/S$, where $S = N^2 D^2/f_0^2 L^2 = O(1)$, for the quasi-geostrophic motions under consideration. Since both ε and δ are small the isotherm slopes are very small indeed and very greatly exaggerated in the figure.

where φ is a small amplitude disturbance superimposed on the basic mean state. The linearized form of (18.14) which results when we neglect quadratic terms in φ is (with $\beta_* = 0$)

$$\left(\frac{\partial}{\partial t} + U\frac{z}{D}\frac{\partial}{\partial x}\right)\left(\nabla_H^2\varphi + \frac{f_0^2}{N^2}\frac{\partial^2\varphi}{\partial z^2}\right) = 0, \tag{22.2}$$

while the linearized form of the boundary condition at $z = 0$ and D (18.15) is

$$\frac{\partial}{\partial t}\frac{\partial\varphi}{\partial z} - \frac{U}{D}\frac{\partial\varphi}{\partial x} = 0, \qquad z = 0, \tag{22.3}$$

$$\left(\frac{\partial}{\partial t} + U\frac{\partial}{\partial x}\right)\frac{\partial\varphi}{\partial z} - \frac{U}{D}\frac{\partial\varphi}{\partial x} = 0, \qquad z = D. \tag{22.4}$$

In laboratory experiments the fluid is inevitably confined by walls, usually vertical at, say, $y = 0$ and L on which the y component of velocity must vanish, i.e.

$$\partial\varphi/\partial x = 0, \qquad y = 0, L. \tag{22.5}$$

Such a boundary condition is grossly artificial in the case of the atmosphere. No band of latitude is ever completely isolated from its surroundings. However, the application of (22.5) in the case of the atmosphere can be thought of as an experiment to see whether the process of cyclogenesis is a local one, i.e. one that will produce waves in the latitudes in which they are found without interaction with other latitude zones.

We search for wave solutions of (22.2) in the form

$$\varphi = \operatorname{Re} F(z)\exp(i(kx - \sigma t))\sin n\pi y/L, \tag{22.6}$$

which automatically satisfies (22.5).

The nonsingular general solution of (22.2) yields

$$F(z) = A\cosh\varkappa(\zeta - \tfrac{1}{2}) + B\sinh\varkappa(\zeta - \tfrac{1}{2}), \tag{22.7}$$

where

$$\zeta = z/D, \qquad \varkappa = (ND/f_0)(k^2 + n^2\pi^2/L^2)^{1/2}. \tag{22.8}$$

Substituting (22.7) into (22.3) and (22.4) leads to two homogeneous, linear algebraic equations for A and B, (we have introduced $c \sim \sigma/k$)

$$\begin{aligned} &A[(U - c)\varkappa\sinh\varkappa/2 - U\cosh\varkappa/2] \\ &\quad + B[(U - c)\varkappa\cosh\varkappa/2 - U\sinh\varkappa/2] = 0, \\ &A[c\varkappa\sinh\varkappa/2 - U\cosh\varkappa/2] \\ &\quad + B[-c\varkappa\cosh\varkappa/2 + U\sinh\varkappa/2] = 0, \end{aligned} \tag{22.9}$$

which has nontrivial solutions for A and B only if the determinant of their coefficients vanishes, leading to a quadratic equation for c whose solutions are

$$c = \frac{U}{2} \pm \frac{U}{\varkappa}\left\{(\varkappa/2 - \tanh \varkappa/2)(\varkappa/2 - \coth \varkappa/2)\right\}^{1/2}. \tag{22.10}$$

The phase speed of c of the wave perturbation is a function of the shear and the parameter $\varkappa$. Since $\tanh \varkappa/2 \leqq \varkappa/2$ for all $\varkappa$ we note from (22.10) that c will be strictly real for those values of $\varkappa$ sufficiently large for $\varkappa/2$ to exceed $\coth \varkappa/2$.

On the other hand, for $\varkappa/2 < \coth \varkappa/2$ the radicand is negative and the two roots for c will be complex and the solution corresponding to the root with positive imaginary part, c_i, will grow like e^{kc_it}, where

(22.11)

$$kc_i = U\frac{k}{\varkappa}\{(\coth \varkappa/2 - \varkappa/2)(\varkappa/2 - \tanh \varkappa/2)\}^{1/2}, \qquad \frac{\varkappa}{2} > \coth \varkappa/2.$$

The critical value of $\varkappa$ where c changes from real to complex is determined by

$$\varkappa_c = 2 \coth \varkappa_c/2, \tag{22.12}$$

the solution of which is

$$\varkappa_c = 2.399. \tag{22.13}$$

Hence for instability to occur,

$$ND/f_0 = L_D < (2.399)(k^2 + n^2\pi^2/L^2)^{-1/2}, \tag{22.14}$$

where L_D is the Rossby radius of deformation. Now k may take any real value so that an absolute requisite for instability is

$$L_D < (2.399/\pi)L, \tag{22.15}$$

that is, the north-south scale must certainly be greater than the deformation radius. Considering the artificiality connected with our choice of a finite region in y let us consider the case where $L \gg L_D$ in which case the necessary and sufficient condition for instability (in this problem) is that the zonal wavelength

$$\lambda = 2\pi/k > (2\pi/2.399)L_D = 2.6\ L_D. \tag{22.16}$$

It seems natural to suppose that in an initial wave spectrum that is fairly smooth the wave we shall actually observe is the one that grows to finite amplitude most swiftly.[14] If L is much greater than L_D the wave length with the maximum growth rate is

$$\lambda_{\max} = 3.9\, L_D. \tag{22.17}$$

Thus the most unstable wavelength is of the order of L_D. Waves much smaller than the deformation radius will not grow, and waves very much larger than the deformation radius grow only very slowly. The e folding is of the order of a couple of days, beyond which nonlinear effects hitherto neglected will become important as the wave amplitude increases.

For values of $\varkappa$ such that the wave is only slightly unstable,

$$B = (Aic_i/U\varkappa)/(\varkappa^2/4 - 1) > 0; \tag{22.18}$$

hence to a good approximation

$$\begin{aligned} \varphi = \operatorname{Re} A \cos \varkappa(\zeta - \tfrac{1}{2}) \exp\left[i\left(kx + \frac{c_i}{U}\frac{\varkappa}{(\varkappa^2/4 - 1)}(\zeta - \tfrac{1}{2})\right)\right] \\ \cdot \exp\left(kc_i t + i\frac{U}{2}kt\right) \sin\frac{n\pi y}{L}, \end{aligned} \tag{22.19}$$

and the phase of the wave slopes to the west with height when $c_i > 0$. This feature is one of the most commonly observed characteristics of cyclone waves in the atmosphere.

We are now in the situation where the existence of cyclone waves seems natural since they can be produced by small perturbations on a current that initially lacks them. Furthermore, we have been able to deduce correctly their characteristic scale and structure. What we have not yet done is discuss in detail the physical mechanism for the instability or answer questions relating to the energy flow between the basic current and the wave. These will be our topics for discussion in the next lecture.

LECTURE VIII

23. **More on baroclinic instability.** Unquestionably, the mathematics of the previous lecture has demonstrated that the presence of a horizontal temperature gradient in a stably stratified rotating fluid is a hospitable environment for the development of quasi-geostrophic waves. The results of our analysis showed that the preferred scale of the wave was the order

[14] This assumes of course that the amplitude in the original spectrum is smooth. If a slowly growing wave has a very much larger initial amplitude it may well be the wave actually observed.

of the Rossby radius of deformation.[15] The unstable waves in this simple model propagate in the direction of the flow with its mean speed as they amplify.

It is also equally clear that at first glance the mechanism of instability is not clear at all. Indeed it is rather subtle. It is not a shear flow instability, for the vertical Reynolds stresses which would then be required are not present in the model. The rotational constraint, which makes w of $O(\varepsilon)$, eliminates the vertical velocity as an agent of direct momentum transfer. Furthermore, since the fluid state initially has denser fluid underlying lighter fluid everywhere the instability is not of the classical Benard convective type. As we shall see the mechanism is physically more like the latter while mathematically similar to the former.

24. **The energy equation.** We proceeded directly from the quasi-geostrophic potential vorticity equation to the solution of the perturbation problem in §22. Now let us backtrack for a moment and set down the perturbation equations for the momentum and density appropriate for the wave problem. They are

$$\partial u/\partial t + U(z)\partial u/\partial x - fv = -(1/\rho_0)\partial p/\partial x, \tag{24.1a}$$

$$\partial v/\partial t + U(z)\partial v/\partial x + fu = -(1/\rho_0)\partial p/\partial y, \tag{24.1b}$$

$$\rho g = -\partial p/\partial z, \tag{24.1c}$$

$$\partial \rho/\partial t + U(z)\partial \rho/\partial x + v\partial \bar{\rho}/\partial y + w\partial \rho_0/\partial z = 0, \tag{24.1d}$$

$$\partial u/\partial x + \partial v/\partial y + \partial w/\partial z = 0. \tag{24.1e}$$

In these equations (u, v, w) refer to the small amplitude perturbation velocities and p is the perturbation pressure. The density field is the sum of the basic stable density field $\rho_0(z)$, the horizontally varying density field $\bar{\rho}$ associated with the mean current $U(z)$, while ρ is the perturbation density in the wave. The absence of the term $w(\partial U/\partial z)$ in (24.1a) is a reflection of the fact that w is $O(\varepsilon)$ compared with u and v. On the other hand, $\partial w/\partial z$ is retained in (24.1e) because $(\partial u/\partial x) + (\partial v/\partial y)$, geostrophically evaluated, is $O(\varepsilon)$. If we formed the vorticity equation from (24.1a, b) by eliminating the pressure, the perturbation potential vorticity equation (22.2) would follow naturally.

If (24.1a) and (24.1b) are multiplied by u and v respectively, added and then integrated over the meridional plane $0 \leqq y \leqq L$, $0 \leqq z \leqq D$ and

[15] As a rule of thumb, whenever a quasi-geostrophic motion depends in an essential way on the coupling of the vorticity and thermodynamic fields, its scale is the deformation radius.

over the wavelength in x (a region we shall call R_w) we obtain the mechanical energy equation

$$(24.2)\quad \frac{\partial K}{\partial t} = -\iiint_{R_w}\left[u\frac{\partial p}{\partial x} + v\frac{\partial p}{\partial y}\right] dx\,dy\,dz = \iiint_{R_w} w\frac{\partial p}{\partial z}\,dx\,dy\,dz,$$

where we have repeatedly used the divergence theorem and the fact that v and w vanish on the meridional boundaries of R_w and the fact that the wave properties are periodic in x. The quantity K is the kinetic energy contained in R_w, viz:

$$(24.3)\qquad K = \iiint_{R_w} \rho_0\frac{(u^2 + v^2)}{2}\,dx\,dy\,dz.$$

The vertical velocity does not contribute to the evaluation of K, again due to the relative smallness of w. Use of the hydrostatic relation (24.1c) allows (24.2) to be rewritten

$$(24.4)\qquad \frac{\partial K}{\partial t} = -g\iiint_{R_w} w\rho\,dx\,dydz.$$

Thus if the kinetic energy of the wave field increases with time, as it must for an unstable wave, the vertical velocity must be negatively correlated with the perturbation density so that *on the average* lighter fluid rises while cold fluid sinks. This agrees with the earlier observation we made that the baroclinic system appeared to be unstable to any perturbation motion which would lower its potential energy. Now the needed $w\rho$ correlation is not easy for the wave to obtain. Since in the basic state lighter fluid overlies denser fluid, an element of fluid rising in a nearly vertical trajectory would be heavier than its neighbors which would tend to make $w\rho$ positive. Clearly the mechanism is more complicated than that.

We may evaluate $w\rho$ with the aid of (24.1d) leading to an alternate form of (24.4), namely

$$(24.5)\qquad \frac{\partial}{\partial t}(K + A) = g\iiint_{R_w} v\rho\left(\frac{\partial\bar{\rho}}{\partial y}\bigg/\frac{\partial\rho_0}{\partial z}\right) dx\,dy\,dz,$$

where A is the positive definite quantity defined by

$$(24.6)\qquad A = \frac{g\rho^2}{2(-\partial\rho_0/\partial z)}.$$

We call A the perturbation *available potential energy*. It is a measure of the departure (in this case due to the wave field) of the density field from the basic stable stratification $\rho_0(z)$.[16] In any event, since both K and A are positive definite quantities, quadratic in the perturbation fields, the onset of instability, in which all wave properties grow exponentially, must lead to an exponential increase in $K + A$. In order for this to occur, we must have *on the average*

$$v\rho(\partial\bar{\rho}/\partial y) < 0. \tag{24.7}$$

(Remember that $\partial\rho_0/\partial z$ is less than zero!) Hence if, as in the present case $\partial\bar{\rho}/\partial y > 0$, v and ρ must be so correlated that on the average lighter fluid is carried northward ($y > 0$) and heavier fluid southward. Note that this correlation would tend to tip the lines of constant density in the mean state toward the horizontal. Combining our results, lighter fluid must *rise* but must also be carried horizontally northward. The type of convection which takes place is slanted. In terms of the temperature the waves, on the average, transport heat down the basic temperature gradient. In the atmosphere cyclone waves do just that and play a major role in the heat balance of the earth transporting heat from the tropics to northward latitudes.

We see then, that a fluid element must rise, sloping upward to the north, at the same time being carried downstream by the current where it descends sloping downward to the south. With $U(z) \geqq 0$ for all z, the trajectory is an open rather than a closed convection cell.

It is not to be imagined that every such sloping trajectory contributes to the production of perturbation energy. In fact only a narrow corridor of angles in the meridional plane is energy releasing as we shall see below.

25. **The wedge of instability.** Consider for the moment the one dimensional motion of a solid particle subject to the potential $V(x)$.[17] If we consider small amplitude motions about the equilibrium point x_0, the resulting perturbation equation governing the motion is

$$d^2\zeta/dt^2 + V''(x_0)\zeta = 0, \tag{25.1}$$

where $\zeta = x - x_0$ is the departure from the equilibrium point. Multiplying (25.1) by the particle's velocity $d\zeta/dt$ yields

$$\frac{d}{dt}\left\{\frac{1}{2}\left(\frac{d\zeta}{dt}\right)^2\right\} = -\frac{V''}{2}(x_0)\frac{d\zeta^2}{dt}. \tag{25.2}$$

---

[16] Although there is a substantial amount of potential energy in the basic state in which the isopleths of density are horizontal, none of it is available for the production of kinetic energy.

[17] This is not so gigantic a digression as it may appear—patience!

Consider now an excursion of the particle in a direction away from the equilibrium position, i.e. a motion in which ζ and $d\zeta/dt$ have the *same* sign. Hence $d\zeta^2/dt$ will be positive. This will lead to an increase of the kinetic energy of the particle only if $V''(x_0)$ is negative (i.e. if x_0 is the top of the potential "hill"). Otherwise, if $V''(x_0)$ is positive, the motion will be stable.

Let us apply these notions to our baroclinic problem. First we define η and ζ to be the Lagrangian displacements of a fluid element, which for small amplitude motions are related to v and w by the equations

$$d\eta/dt = (\partial/\partial t + U\partial/\partial x)\eta = v, \tag{25.3}$$

$$d\zeta/dt = (\partial/\partial t + U\partial/\partial x)\zeta = w. \tag{25.4}$$

The equation for conservation of density (24.1d) may then be written

$$(\partial/\partial t + U\partial/\partial x)[\rho + \eta\partial\bar{\rho}/\partial y + \zeta\,\partial\rho_0/\partial z] = 0. \tag{25.5}$$

Since ρ is zero if η and ζ are zero, we may from (25.5) conclude that $\rho = -\eta\partial\bar{\rho}/\partial y - \zeta\partial\rho_0/\partial z$.

Therefore (24.4) may be written in the form

$$\frac{\partial K}{\partial t} = g\iiint_{R_w} \zeta\frac{d\zeta}{dt}\left[1 - \frac{\gamma}{\alpha}\right]\frac{d\rho_0}{dz}\,dx\,dy\,dz \tag{25.6}$$

where

$$\gamma = -\frac{\partial\bar{\rho}}{\partial y}\bigg/\frac{\partial\rho_0}{\partial z} \tag{25.7}$$

is the slope of the lines of constant density in the basic nonwave state, while $\alpha = \zeta/\eta$ is the slope of the fluid trajectory in the $y - z$ plane. We note the similarity between (25.7) and (25.2); namely, as a fluid element leaves its initial position so that $\zeta(d\zeta/dt) > 0$, the change in the kinetic energy of the perturbation wave field will be positive if *on the average* (over the meridional plane)

$$\alpha - \gamma < 0 \tag{25.8}$$

(remembering again that $\partial\rho_0/\partial z < 0$).

Thus only for motions in the wedge γ[18] will perturbation trajectories lead to an increase of perturbation kinetic energy. The reason is clear. If fluid element A slides upward at an angle α to take the position of the heavier element B which in turn takes the position of element A a net decrease of potential energy will occur. Note that the existence of the

[18] Since γ is so small, the angle $\tan^{-1}\gamma$ is approximately γ.

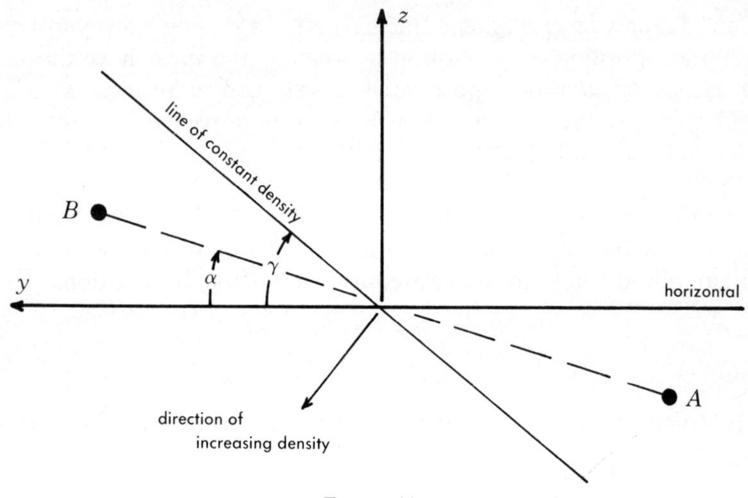

FIGURE 11.

wedge angle $\gamma \neq 0$ can take place only in a rotating fluid in which the sloping density surfaces are, in the basic state, balanced by an increase of the Coriolis force with elevation (the thermal wind relation (25.3)).

From (25.8) we have as a condition for instability that

$$-\frac{\partial \bar{\rho}}{\partial y}\Big/\frac{\partial \rho_0}{\partial z} > \zeta/\eta = \alpha. \tag{25.9}$$

Now for the basic state at hand

$$-\frac{\partial \bar{\rho}}{\partial y}\Big/\frac{\partial \rho_0}{\partial z} = -Uf\Big/\frac{gD}{\rho_0}\frac{\partial \rho_0}{\partial z}, \tag{25.10}$$

while

$$\alpha = (D/\lambda) = \varepsilon\alpha' = (DU/f\lambda^2)\alpha', \tag{25.11}$$

where α' is the scaled slope of $O(1)$ and λ is the wavelength. Then (25.9) with (25.10) and (25.11) yields

$$\frac{Uf}{g(D/\rho_0)|\partial \rho_0/\partial z|} > \frac{UD}{f\lambda^2}\alpha'$$

or

$$\frac{f^2\lambda^2}{g(D^2/\rho_0)|d\rho_0/dz|} = \frac{f^2\lambda^2}{N^2D^2} > \alpha'. \tag{25.12}$$

Since α' is $O(1)$ the condition (25.12) is, aside from $O(1)$ factors, precisely the condition (22.16) we derived from our detailed wave analysis.

Writing (22.7) in the equivalent form

$$F = a \sinh \varkappa\zeta + b \cosh \varkappa\zeta, \tag{25.13}$$

we may use the boundary condition on $z = 0$ to obtain

$$F = a(\sinh \varkappa\zeta - (c/U)\varkappa \cosh \varkappa\zeta) \tag{25.14}$$

so that

$$\begin{aligned}\varphi = \sinh \pi y[(a/2)(\sinh \varkappa\zeta - (c/U)\varkappa \cosh \varkappa\zeta)\, e^{i\theta} \\ + (a^*/2)(\sinh \varkappa\zeta - (c^*/U) \cosh \varkappa\zeta)\, e^{-i\theta*}],\end{aligned} \tag{25.15}$$

where

$$\theta = kx - kct. \tag{25.16}$$

Using the fact that

$$v = \partial\varphi/\partial x \tag{25.17}$$

$$\rho = -(f/g)\rho_0(\partial\varphi/\partial z), \tag{25.18}$$

it follows after a little calculation that

$$\int_x^{x+2\pi/k} v\rho\, dx = -\pi\frac{c_i}{U}\frac{f\rho_0}{gD}|a|^2\varkappa^2 \sin^2 n\pi y\, e^{2kc_it}, \tag{25.19}$$

which is negative for $c_i > 0$ so that $\iiint_{R_w} v\rho\, dx\, dy\, dz$ will certainly be negative.

Similarly

$$\int_x^{x+2\pi/k} w\rho\, dx = -\frac{\pi f^2\rho_0}{gN^2D^2}\sin^2 n\pi y \left\{c_i\left[\left|\frac{dF}{d\zeta}\right|^2 + \varkappa^2|a|^2\right]\right\} e^{2kc_it}, \tag{25.20}$$

which is negative for all $\zeta \neq 0, 1$ (where $w\rho$ vanishes). Thus for the unstable wave the available potential energy of the basic state manifested by the horizontal temperature gradient is converted into wave energy.

LECTURE IX

26. **Slow circulation of a rotating, stratified fluid.** In the preceding lectures we looked at the dynamics of a wave perturbation embedded in a mean current. The meridional (y, z) structure of the basic parallel flow was, in our inviscid formulation, at our disposal to choose. We capitalized on this fact by picking a particularly simple flow to allow us maximum ease in investigating the mechanism of baroclinic instability. We turn our attention now to the question of how the basic state is determined,

and the above remarks indicate that dissipative processes will be important in determining any such parallel flow. Indeed, in Lecture III we saw how Ekman layer suction completely determined slow, axially symmetric flows in a homogeneous fluid.

Our object now is to generalize these results to the case of a stratified fluid. It is beyond the scope[19] of these lectures to consider the problem in a very realistic geophysical context. Instead we shall content ourselves to idealized flows suitable for laboratory study.[20] Nevertheless, similar conditions will apply, even in a somewhat altered form in the consideration of such geophysical problems as the oceanic thermocline and the tropical atmospheric circulations.

27. **The model problem.** Consider the motion of a rotating *stably stratified* fluid in a cylinder of height D, radius R, driven, as in Lecture III, by the differential motion of its upper surface. We again suppose that the fluid is a liquid (e.g. water) whose state relation for our purposes may be approximated as

$$\rho = \rho_0(1 - \alpha(T - T_0)). \tag{2.7b'}$$

In the absence of relative motion, the fluid has the basic temperature field

$$T_s = T_0 + \Delta T\, z/D, \tag{27.1}$$

which assumes that the centrifugal force is sufficiently small so that its tendency to bow the isotherms upward may be ignored ($\Omega^2 R^2/gD \ll 1$).

We introduce nondimensional, scaled variables as follows. We denote the *dimensional* variables henceforth by asterisks and scale according to the following scheme:

$$\begin{aligned} (r_*, z_*) &= D(r, z), \\ (u_*, v_*, w_*) &= \varepsilon\Omega D(u, v, w), \\ T_* &= T_0 + \Delta T z + \varepsilon\frac{\Omega^2 D}{\alpha g}T, \\ p_* &= p_0 - \rho_0 g D z + \tfrac{1}{2}\rho_0 g D\alpha\Delta T z^2 + \varepsilon\rho_0\Omega^2 D^2 p, \end{aligned} \tag{27.2}$$

where $\varepsilon\Omega$ is a measure of the relative angular velocity of the upper surface of the cylinder. The temperature and pressure are scaled in anticipation of the fact that for most of the fluid the motion will be geostrophic and hydrostatic to *lowest* order.

---

[19] This is due primarily to limitations of time rather than any essential difficulty.

[20] Unfortunately, there is still a sore need for experimental work in this area.

The equations of motion for axially symmetric flow become

(27.3a) $$\varepsilon[uu_r + wu_z - v^2/r] - 2v = -(\partial p/\partial r) + E[\nabla^2 u - u/r^2],$$

(27.3b) $$\varepsilon[uv_r + wv_z + uv/r] + 2u = E[\nabla^2 v - v/r^2],$$

(27.3c) $$\varepsilon[uw_r + ww_z] - T = -(\partial p/\partial z) + E\nabla^2 w,$$

(27.3d) $$(1/r)(ru)_r + w_z = 0,$$

(27.3e) $$\varepsilon[uT_r + wT_z] + Sw = (E/\sigma)\nabla^2 T,$$

where

$E = \nu/\Omega D^2$ (the Ekman number),
$S = g\alpha\Delta T/\Omega^2 D$ (the stratification parameter $= N^2/\Omega^2$),
$\sigma = \nu/\varkappa$ (the Prandtl number).

In the equations displayed, the Boussinesq approximation has already been made, i.e. we have assumed that $\alpha\Delta T$ and $\varepsilon\Omega^2 D/g$ are sufficiently small so that density variations may be ignored except where they provide buoyancy forces (which is consistent under the above restrictions). Similarly, the motion to be considered is sufficiently slow and so nearly incompressible that the condition of conservation of mass may be approximated by (27.3d) while the production of internal energy by viscous dissipation and volume expansion have been neglected in the thermodynamic equation, (27.3e).

The boundary conditions are

(27.4)
$$\begin{aligned} (u, v, w) &= 0, && z = 0, \\ (u, v, w) &= (0, v_T(r), 0), && z = 1, \\ (u, v, w) &= 0, && z = r_0 \equiv R/D, \\ \partial T/\partial r &= 0, && r = r_0. \end{aligned}$$

The boundary conditions most easily applied experimentally on $z = 0, 1$ fix the temperature of the horizontal boundaries, i.e.

(27.5) $$T = 0, \qquad z = 0, 1,$$

while for mathematical purposes it is simpler to consider the experimentally difficult (although not wholly artificial) conditions

(27.6) $$\partial T/\partial z = 0, \qquad z = 0, 1.$$

28. **The Ekman boundary layer in a stratified fluid.** Our first order of business is to determine what effect, if any, the presence of stratification has on the structure of the Ekman layer. Setting $\zeta = z/E^{1/2}$, we find for

the Ekman layer near $z = 0$

$$\varepsilon(\tilde{u}\tilde{u}_r + \tilde{w}\tilde{u}_\zeta - \tilde{v}^2/r) - 2\tilde{v} = -\partial\tilde{p}/\partial r + \tilde{u}_{\zeta\zeta} + E[\nabla_H^2\tilde{u} - \tilde{u}/r^2], \tag{28.1}$$

$$\varepsilon(\tilde{u}\tilde{v}_r + \tilde{w}\tilde{v}_\zeta + \tilde{u}\tilde{v}) + 2\tilde{u} = \tilde{v}_{\zeta\zeta} + E[\nabla_H^2\tilde{v} - \tilde{v}/r^2], \tag{28.2}$$

$$\varepsilon E(\tilde{u}\tilde{w}_r + \tilde{w}\tilde{w}_\zeta) - E^{1/2}\tilde{T} = -\partial\tilde{p}/\partial\zeta + E\tilde{w}_{\zeta\zeta} + E^2\nabla_H^2\tilde{w}, \tag{28.3}$$

$$1/r(r\tilde{u})_r + \tilde{w}_\zeta = 0, \tag{28.4}$$

$$\sigma\varepsilon(\tilde{u}\tilde{T}_r + \tilde{w}\tilde{T}_\zeta) + \sigma SE^{1/2}\tilde{w} = \tilde{T}_{\zeta\zeta} + E\nabla_H^2\tilde{T}, \tag{28.5}$$

where a tilde denotes dynamic variables defined for the Ekman layer. In particular the vertical velocity has been written

$$w = E^{1/2}\tilde{w}. \tag{28.6}$$

We observe from (28.5) that as long as $\sigma\varepsilon E^{1/2}$ and σSE are $O(1)$ or less, (in fact since ε and E are small these parameters are generally quite small) the Ekman layer, to lowest order, will be unaffected by the presence of the stratification. This is quite simply because the vertical scale of the Ekman layer is so much smaller than the length scale associated with the stratification that the fluid appears homogeneous on the Ekman layer scale. It follows immediately from this that the interior fields, i.e. the fields external to the Ekman layer, must satisfy the same matching conditions for the vertical velocity as in the homogeneous case, namely

$$w_I(r, 0) = \frac{E^{1/2}}{2}\frac{1}{r}\frac{\partial}{\partial r}rv_I(r, 0) \tag{28.7}$$

and

$$w_I(r, 1) = \frac{E^{1/2}}{2}\frac{1}{r}\frac{\partial}{\partial r}r[v_T(r) - v_I(r, 1)]. \tag{28.8}$$

It furthermore follows from (28.5) that since the *correction* to the interior temperature field in the Ekman layer is of order $\sigma\varepsilon$ or $\sigma E^{1/2}S$, the *interior* temperature field itself must satisfy (27.5) (if that is the relevant boundary condition). If, on the other hand, (27.6) is the appropriate boundary condition, it will apparently apply directly to the interior temperature only if

$$\varepsilon < E^{1/2} \quad \text{and} \quad \sigma S < 1. \tag{28.9}$$

As we shall see these conditions are unnecessarily strong, and we can check after the fact that even if they are not met the *interior* temperature field must satisfy (27.6) if that boundary condition is applied.

29. **The interior flow.** The Ekman layer problem, as we saw, was essentially unchanged by the presence of stratification. This is definitely not true for the interior flow. In the interior we write

$$
\begin{aligned}
v_I &= v_I^{(0)} + \cdots, \\
u_I &= E u_I^{(2)} + \cdots, \\
w_I &= \lambda(E, \sigma S, \varepsilon) w_I^{(1)} + \cdots, \\
p_I &= p_I^{(0)} + \cdots, \\
T_I &= \gamma(E, \sigma S, \varepsilon) T_I^{(0)} + \cdots,
\end{aligned} \tag{29.1}
$$

where the scale functions λ and γ must be determined. We only assume provisionally that they are $O(1)$ or less. With $\varepsilon \ll 1$ there are two main cases of interest. In the first case $\sigma S = O(1)$. Then from (27.3e) we see that λ must be $O(E)$ or less, i.e. the interior vertical velocity is $O(E)$[21] rather than $O(E^{1/2})$ as in the homogeneous case. The stratification so inhibits the vertical velocity that in the interior it falls to this very small value. However, (28.7) and (28.8) imply that any discrepancy between the interior vertical component of vorticity and the vorticity of the bounding surfaces on $z = 0, 1$, will produce an $O(E^{1/2})$ mass flux into the interior. The proper resolution of this difficulty is that for $\sigma S = O(1)$, the interior swirl velocity directly adjacent to horizontal surfaces must possess the vorticity of the surface. In this way, to lowest order, the Ekman layers will be nondivergent. This implies in turn that

$$v_I(r, 0) = C_0/r, \tag{29.2a}$$

$$v_T(r) - v_I(r, 1) = C_1/r. \tag{29.2b}$$

For a simply connected region, C_0 and C_1 must obviously be zero leading to the unexpected result: for a stratified fluid in a simply connected region, the interior velocity must, to lowest order, satisfy the no-slip boundary conditions at $z = 1, 0$. Equivalently there is no $O(1)$ Ekman layer! If the region is not simply connected, e.g. an annulus instead of a cylinder, an $O(1)$ Ekman layer may be present (C_0 and $C_1 \neq 0$), but it must still be nondivergent. It is important to note that since the fluid is not homogeneous the interior velocity may vary with height as it could not in a homogeneous fluid subject to the Taylor–Proudman constraint.

On the other hand, if σS is less than $O(1)$ there will in general be some divergence in the Ekman layer. When σS is decreased to $O(E^{1/2})$ the

[21] This explains why the a priori conditions (26.9) were unnecessarily strong.

interior velocity is once again $O(E^{1/2})$. We will focus our attention, because it turns out to be more illuminating, to the case $\sigma S < 1$. We shall be able to retrieve the behavior of the limiting case $\sigma S = O(1)$ by doing so and indeed obtain a uniform approximation for the interior flow in the range $0 \leqq \sigma S < O(1)$.

In this case, which we will analyze in some detail, we choose $\gamma = 1$, $\lambda = E^{1/2}$ and obtain for the lowest order fields the equations

$$2v_I^{(0)} = \partial p_I^{(0)}/\partial r, \tag{29.3a}$$

$$T_I^{(0)} = \partial p_I^{(0)}/\partial z, \tag{29.3b}$$

$$\partial w_I^{(1)}/\partial z = 0, \tag{29.3c}$$

$$(\sigma S/E^{1/2}) w_I^{(1)} = \nabla^2 T_I^{(0)}. \tag{29.3d}$$

The validity of (29.3d) as an approximation to (27.3e) requires that $\varepsilon < S$ when $E^{1/2} < S < 1$ and $\sigma\varepsilon < E^{1/2}$ when $S < E^{1/2}$. The latter condition is more difficult to meet experimentally. However, in the parameter range $S < E^{1/2}$ the motion field is not essentially coupled to the temperature field and is primarily homogeneous in its dynamics. Hence although the above conditions must formally be met, the set (29.3) provide a useful characterization of the interior flow when $\varepsilon < O(1)$. Indeed when (27.6) applies (29.3d), after the fact turns out to be correct for all $\varepsilon < O(1)$. With (29.3c), (28.7) and (28.8) we find that

$$w_I^{(1)}(r) = \frac{1}{4}\frac{1}{r}\frac{\partial}{\partial r} r v_T - \frac{1}{4}\frac{1}{r}\frac{\partial}{\partial r} r(v_I^{(0)}(r,1) - v_I^{(0)}(r,0)). \tag{29.4}$$

From (29.3a, b) we deduce the thermal wind relation

$$\partial v_I^{(0)}/\partial z = \tfrac{1}{2}(\partial T_I^{(0)}/\partial r), \tag{29.5}$$

which allows us to write $w_I^{(1)}$ entirely in terms of the temperature, viz:

$$w_I^{(1)} = \frac{1}{4}\frac{1}{r}\frac{\partial}{\partial r} r v_T - \frac{1}{8}\frac{1}{r}\frac{\partial}{\partial r} r \frac{\partial}{\partial r}\int_0^1 T_I^{(0)}\, dz. \tag{29.6}$$

So that the problem for the interior temperature becomes

$$\left[\frac{\partial^2}{\partial z^2} + \frac{1}{r}\frac{\partial}{\partial r} r \frac{\partial}{\partial r}\right]\left[T_I^{(0)} + \frac{\sigma S}{8E^{1/2}}\int_0^1 T_I^{(0)}\, dz'\right] = \frac{\sigma S}{4E^{1/2}}\frac{1}{r}\frac{\partial}{\partial r} r v_T. \tag{29.7}$$

Once the temperature has been found, the other interior variables can be immediately determined.

An inspection of (29.7) shows that the interior problem is not completely posed until boundary conditions for $T_I^{(0)}$ on $r = r_0$ are given. This is in contrast to the case of a homogeneous fluid where, as we saw in Lecture III, the flow was completely determined by the Ekman suction. For a stratified fluid, then, not only do the Ekman layers weaken as the stratification increases, but they must share with the side wall boundary layers the role of determining the interior flow. We will leave the determination of this boundary condition and the solution of (29.7) till the next lecture.

LECTURE X

30. **Vertical boundary layers in a rotating, stratified fluid.** We saw in the last lecture that for a stratified, rotating fluid the Ekman layer suction no longer completely determines the inviscid interior flow. Rather, constraints (or boundary conditions) imposed on the interior flow via the side wall boundary layer must also be considered. For a homogeneous, rotating steady flow information is communicated vertically. Any boundary effect is felt uniformly in vertical columns throughout the inviscid part of the fluid. This, after all, is the content of the Taylor–Proudman theorem. Similarly for a substantially stratified *nonrotating* fluid, information is passed laterally through slabs of equal density. When the fluid is both stratified *and* rotating, both constraints are present and information from each direction is required. Such is the situation in the problem at hand.

To analyze the dynamics of the motion in the side wall boundary layer, we split each field into its interior value plus a boundary layer correction, for example, for the temperature

$$T = T_I + T_B \tag{30.1}$$

where T_B vanishes (or is transcendentally small) outside the boundary layer. The boundary layer functions, which are rapidly varying functions of r, correct the interior fields within the boundary layer so that the applied conditions on the sum of interior plus correction fields may be satisfied. Furthermore, we will henceforth assume that ε is sufficiently small so that the nonlinear terms in the boundary layer may be ignored, in which case the correction fields satisfy the same differential equations as the total fields. Since the boundary layer fields are rapidly varying functions of r, a consistent first approximation to (27.3) when $\varepsilon \to 0$ is for the correction fields,

$$-2v_B = -(\partial p_B/\partial r) + [E(\partial^2 u_B/\partial r^2)], \tag{30.2a}$$

$$2u_B = E(\partial^2 v_B/\partial r^2), \tag{30.2b}$$

$$-T_B = -(\partial p_B/\partial z) + E(\partial^2 w_B/\partial r^2), \tag{30.2c}$$

(30.2d) $$(\partial u_B/\partial r) + (\partial w_B/\partial z) = 0,$$

(30.2e) $$Sw_B = (E/\sigma)(\partial^2 T_B/\partial r^2).$$

The bracketed term in (30.2a) may be shown to be negligibly small, that is, the swirl velocity remains in geostrophic balance within the side wall boundary layer. This can easily be seen by eliminating all variables in terms of v_B, say, to obtain

(30.3) $$E^2\{(\partial^6 v_B/\partial r^6) + [(\partial^6 v_B/\partial r^4 \partial z^2)]\} + \sigma S(\partial^2 v_B/\partial r^2) + 4(\partial^2 v_B/\partial z^2) = 0,$$

where the square bracketed term in (30.3) comes directly from the similarly denoted term in (30.2a). Since $(\partial^2/\partial r^2) \gg (\partial^2/\partial z^2)$ for all boundary layer variables, it is clear from (30.3) that the term $\partial^6 v_B/\partial r^4 \partial z^2$ can be consistently neglected, which in turn implies the same is true for the term $E(\partial^2 u_B/\partial r^2)$ in (30.2a).[22]

31. **The boundary condition for the interior problem.** In the case where the side wall is insulated and the flow is axially symmetric, which is the situation at hand, the boundary condition for the interior flow can be determined without the detailed solution of the boundary layer equation (30.3). We will defer a discussion of (30.3) until later.

If we integrate (30.2e) across the width of the boundary layer (a width, remember, we have not yet determined), we find that the net vertical transport in the side wall boundary layer is

(31.1) $$2\pi r_0 \int_0^{r_0} w_B \, dr = 2\pi \frac{E r_0}{\sigma S} \frac{\partial T_B}{\partial r}(r_0)$$

where we have used the fact that all B subscripted variables are zero outside the boundary layer, whose thickness is much less than r_0. On the other hand, from (29.6) the total interior mass flux is

(31.2) $$E^{1/2} 2\pi \int_0^{r_0} w_I^{(1)} r \, dr = 2\pi E^{1/2} \left[\frac{r_0}{4} v_T(r_0) - \frac{r_0}{8} \int_0^1 \frac{\partial T_I^{(0)}}{\partial r}(r_0) \, dz \right].$$

Obviously the *total* vertical mass flux in each horizontal plane must vanish so that

(31.3) $$r_0 \frac{v_T(r_0)}{4} - \frac{r_0}{8} \int_0^1 \frac{\partial T_I^{(0)}}{\partial r}(r_0) \, dz' = -\frac{E^{1/2}}{\sigma S} r_0 \frac{\partial T_B}{\partial r}(r_0).$$

---

[22] More precisely, either the swirl velocity is geostrophically balanced, or if in (30.2a) the square bracketed term is not negligible (30.2a) is uncoupled from the dynamics. For $\sigma S < 1$ the former is true.

We see immediately that $(\partial T_B/\partial r)(r_0)$ must be independent of z. In addition, to satisfy the insulating condition on $r = r_0$,

$$(31.4) \qquad (\partial T_I^{(0)}/\partial r)(r_0) + (\partial T_B/\partial r)(r_0) = 0,$$

which in conjunction with (31.3) yields the desired condition on $\partial T_I^{(0)}/\partial r$, viz:

$$(31.4) \qquad \frac{\partial T_I^{(0)}}{\partial r}(r_0) = \frac{v_T(r_0)}{4} \frac{\sigma S}{E^{1/2}[1 + (\sigma S/8E^{1/2})]}.$$

32. **The solution of the interior problem.** To find the interior temperature field we must solve (29.7) subject to (31.4) and *either*

$$(32.1) \qquad \partial T_I^{(0)}/\partial z = 0, \qquad z = 0, 1,$$

or

$$(32.2) \qquad T_I^{(0)} = 0, \qquad z = 0, 1.$$

If (32.1) applies the solution for $T_I^{(0)}$ is exceedingly simple, viz:

$$(32.3) \qquad T_I^{(0)} = \frac{\sigma S}{4E^{1/2}[1 + (\sigma S/8E^{1/2})]} \int_0^r v_T(r')\, dr'.$$

Using (29.5), (29.3c), (29.4), and (29.3d), we obtain $v_I^{(0)}$ and $w_I^{(1)}$,

$$(32.4) \qquad v_I^{(0)} = \frac{v_T(r)}{2}\left[1 + \frac{\sigma S}{8E^{1/2}} \frac{(2z - 1)}{[1 + (\sigma S/8E^{1/2})]}\right],$$

$$(32.5) \qquad E^{1/2} w_I^{(1)} = \frac{E^{1/2}}{4(1 + \sigma S/8E^{1/2})} \frac{1}{r} \frac{\partial}{\partial r} r v_T.$$

In the limit $\sigma S/E^{1/2} \to 0$ the $O(1)$ swirl velocity is $v_T/2$, i.e. independent of z and equal to the mean of the velocity of the upper and lower plates, while the vertical velocity, of $O(E^{1/2})$, is proportional to the vorticity of the bounding surface. These are precisely the results of Lecture III for the case of a homogeneous fluid. The interior temperature vanishes as $S \to 0$, for in the absence of an ambient stratification, the mechanically driven motions produce no density and temperature variations.

As σS increases, i.e. for a fluid with greater stratification, the azimuthal velocity profile develops a shear (see Figure 12) which is compatible with the interior horizontal temperature gradient developed in the interior.

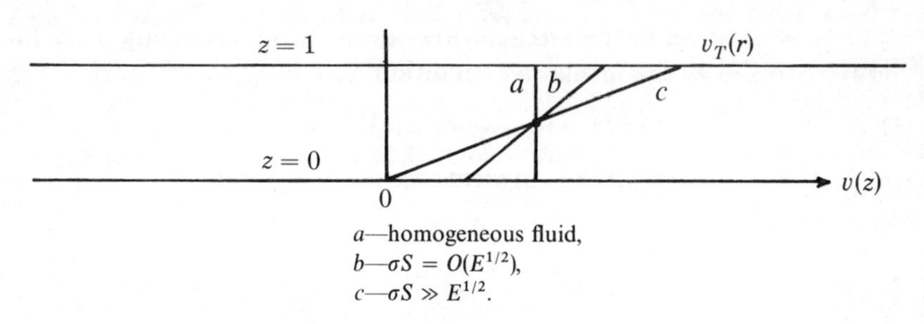

a—homogeneous fluid,
b—$\sigma S = O(E^{1/2})$,
c—$\sigma S \gg E^{1/2}$.

FIGURE 12.

When σS is much greater than $E^{1/2}$, it may be observed from (32.4) that the interior azimuthal velocity itself satisfies the boundary conditions on $z = 0$ and 1 while the interior vertical velocity falls in magnitude from $O(E^{1/2})$ to $(E/\sigma S)$. Both of these results were predicted in the preceding lecture for the case where $\sigma S = O(1)$. We see now that the fluid behaves in this manner as long as σS exceeds $E^{1/2}$. In this sense, then, the fluid behaves as a strongly stratified fluid when the weaker condition $\sigma S > E^{1/2}$ is satisfied.

The role of the side wall boundary layer is now clear. If, say, $v_T(r_0)$ is positive, fluid sucked into the upper Ekman layer is flung out radially to the side wall where it descends in the boundary layer. Since the descending fluid in the boundary layer originated at larger z values and hence higher ambient temperature, the ascending fluid is warmer. The side wall of the cylinder is thermally insulated so this excess heat warms the interior, providing a radial temperature gradient at the edge of the boundary layer given by (31.4). This boundary layer heating produces the tilt in the profile of the swirl velocity as shown above. When $\sigma S/E^{1/2}$ becomes large the velocity profile in the interior obviates the need for the Ekman layers and the mass transport in the meridional plane and the side wall boundary layer falls to $O(E/\sigma S)$.

If the appropriate boundary condition on $z = 0$ and $z = 1$ is (32.2) the solution of (29.7) is more complicated but not qualitatively different. The swirl velocity, for example, in that case is

$$\begin{aligned} v_I^{(0)} &= \frac{v_T}{2} + \frac{\lambda}{1+\lambda} v_T(z - \tfrac{1}{2}) \\ &\quad + \sum_{n=1}^{\infty} \frac{1}{r_0^2} \frac{J_1(k_n r/r_0)}{J_0^2(k_n)} \left[\frac{2\lambda}{1+\lambda} \int_0^{r_0} r' \int_0^{r'} v_T(r'')\, dr''\, J_0(k_n r'/r_0)\, dr' \right] \\ &\quad \times \left[\frac{\sinh (k_n/r_0)(z - \frac{1}{2}) - (\lambda/(1+\lambda))(z - \frac{1}{2}) \sinh (k_n/2r_0)}{\cosh (k_n/2r_0) - (\lambda/(1+\lambda))(r_0/k_n) \sinh (k_n/2r_0)} \right], \end{aligned} \tag{32.6}$$

where $\lambda \equiv \sigma S/8E^{1/2}$, $J_1(k_n) = 0$.

Complicated as (32.6) appears, it can easily be verified that for $\sigma S \gg E^{1/2}$ ($\lambda \gg 1$) the interior velocity satisfies

$$v_I^{(0)}(r, 1) = v_T, \qquad v_I^{(0)}(r, 0) = 0$$

so that once again the Ekman layers are absent in the limit of strong stratification.

33. **The structure of the side wall boundary layer.** Although we have been able to solve the interior problem without considering the detailed nature of the side wall boundary layer, it is important to recognize that this happy state of affairs depended in an essential way on the condition of thermal insulation we imposed on the cylinder side wall. Had we chosen a different boundary condition on $r = r_0$, a detailed consideration of the side wall layer would be necessary for the problem's completion. Furthermore, in some cases, where the fluid may be driven by differential heating of the side wall, motion may never penetrate the interior[23] and the motion only occurs within the vertical boundary layer. Finally it is just interesting to examine the boundary layer structure even in the present case.

The boundary layer equation (30.3) with the square bracketed term deleted is

$$E^2(\partial^6 v_B/\partial r^6) + \sigma S(\partial^2 v_B/\partial r^2) + 4(\partial^2 v_B/\partial z^2) = 0. \tag{33.1}$$

The thickness of the boundary layer depends on *two* parameters E and σS. Consequently the structure of the side wall boundary layer will depend on the relative sizes of σS and E. The following results may be verified by analysis. For all values of $\sigma S < O(1)$ the side wall boundary layer has multiple scales. That is, there are boundary layers within boundary layers. For example, for $\sigma S = 0$ there is a relatively fat region of thickness $E^{1/4}$ in which the swirl velocity is reduced to zero on the side wall. Within this layer there is an $E^{1/3}$ layer in which the vertical and radial velocities are brought to rest and through which the vertical mass transport of the interior is returned. When σS exceeds $E^{2/3}$, the $E^{1/3}$ layer splits into *two* layers, one with thickness $E^{1/2}/(\sigma S)^{1/4}$ (note: this thickness is independent of Ω) and another with thickness $(\sigma S)^{1/2}$. When (σS) becomes comparable with $E^{1/2}$ the $(\sigma S)^{1/2}$ layer merges with the $E^{1/4}$ layer. For $\sigma S \geq O(E^{1/2})$ there are again only two layers, a fat layer whose size is $O(\sqrt{\sigma S})$ and a thin layer with a thickness $E^{1/2}/(\sigma S)^{1/4}$. When $\sigma S = O(1)$ the $\sqrt{\sigma S}$ "layer" becomes indistinguishable from the interior and only a single boundary layer whose size has shrunk to $O(E^{1/2})$ remains.

---

[23] This occurs when $\sigma S < 1$.

The morphology of the multiple boundary layers and the dynamical role of each of them is rather complex, even in this simple situation. Similar changes occur (although even more complex to sort out) in more geophysically realistic situations. Naturally the presence of nonlinearity ($\varepsilon \neq 0$) adds even greater parametric as well as technical difficulties.

References

LECTURES I, II

N. A. Phillips (1963), *Geostrophic motion*, Rev. Geophys. **1** (1963), no. 2, 123–176.

J. G. Charney (1948), *On the scale of atmospheric motions*, Geofys. Publ. Norske Vid.-Akad. Oslo **17** (1948), no. 2. MR **14**, 428.

H. Stommel (1960), *The gulf stream*. Chap. 3, Univ. of California Press, Berkeley, Calif., 1960.

LECTURE III

V. W. Ekman (1905), *On the influence of the Earth's rotation on ocean-currents*, Ark. Mat. Astronom. Fys. **2** (1905), no. 11.

K. Stewartson (1957), *On almost rigid rotations*, J. Fluid Mech. **3** (1957), 17–26. MR **19**, 796.

LECTURE IV

C. G. Rossby, et al. (1937), *Relation between variations in the intensity of the zonal circulation of the atmosphere and the displacements of the semi-permanent centers of action*, J. Marine Res. **2** (1937), 38–55.

H. P. Greenspan (1968), *The theory of rotating fluids*. Chaps. 2, 5, Cambridge Univ. Press, New York, 1968, pp. 85–91, 246–254.

LECTURE V

J. Pedlosky (1967), *An overlooked aspect of the wind-driven oceanic circulation*, J. Fluid Mech. **32** (1967), 809–821.

J. Pedlosky and H. P. Greenspan (1967), *A simple laboratory model for the oceanic circulation*, J. Fluid Mech. **27** (1967), 291–304.

H. Stommel (1960), *The gulf stream*. Chaps. 7, 8, Univ. of California Press, Berkeley, Calif., 1960.

LECTURES VI, VII, VIII

J. Bjerknes and J. Holmboe (1944), *On the theory of cyclones*, J. Meteorol. **1** (1944), 1–22.

J. G. Charney (1947), *The dynamics of long waves in a baroclinic westerly current*, J. Meteorol. **4** (1947), 135–162. MR **9**, 163.

E. T. Eady (1949), *Long waves and cyclone waves*, Tellus **1** (1949), no. 3, 33–52. MR **13**, 86.

J. Pedlosky (1964), *The stability of currents in the atmosphere and the ocean*. I, J. Atmospheric Sci. **21** (1964), 201–219.

LECTURES IX, X

V. Barcilon and J. Pedlosky (1967), *Linear theory of rotating stratified fluid motions*, J. Fluid Mech. **29** (1967), 1–16.

——— (1967), *A unified linear theory of homogeneous and stratified rotating fluids*, J. Fluid Mech. **29** (1967), 609–621.

UNIVERSITY OF CHICAGO

The General Linearised Theory of Wave Propagation

Francis P. Bretherton

1. **Concepts.**

1.1. *What is a wave*? Definition is difficult as the concept is used in different, incompatible senses by different people, and no single statement can embrace all examples of things which seem "wavelike."

Two basic properties are:

(i) Waves involve propagation rather than advection by material particles.

(ii) Waves involve periodic (or quasi-periodic) motion or a combination of such motions.

Not waves, according to this criterion, are, for example, isolated solitary waves, and Kelvin–Helmholtz instability.

In practice we must say a wave is something which satisfies the conditions of validity for one or more of certain general bodies of theory. There are several such—certainly not completely unified and possibly not entirely compatible.

Questions we must ask these theories include:

(a) What propagates (energy? momentum? or influence?)?

(b) What do different dynamical systems have in common?

(c) How are waves generated by nonperiodic influences?

(d) How can they in turn have a systematic nonperiodic effect?

(e) What are the effects of nonlinearities?

(f) What examples can we find in Geophysical Fluid Dynamics?

1.2. *Dispersion relation.* An *elementary wave*

$$\phi(\boldsymbol{x}, t) = A \sin(\boldsymbol{\kappa} \cdot \boldsymbol{x} - \omega t), \qquad \boldsymbol{\kappa} = (k, l, m),$$

AMS 1970 *subject classifications.* Primary 76-02, 76C10, 76Q05.

where A is the amplitude, $\boldsymbol{\kappa}$ is the wave number and ω is the frequency, satisfies a *dispersion relation*

$$\boxed{\omega = \Omega(\boldsymbol{\kappa})}$$

which depends on the medium and is the most important single property of it.

Note. This assumes an unbounded, steady state uniform medium, small amplitude linearised waves, no dissipation. We shall be able to relax each condition *somewhat*, but almost all interesting examples are in some sense contiguous to this case.

1.3. *Wave energy density.* $\mathscr{E} = A^2F(\boldsymbol{\kappa}) > 0$ where $F(\boldsymbol{\kappa})$ depends on the medium. We will postpone discussion till later.

1.4. *Group and phase velocity.* The *group velocity*

$$\boxed{\gamma = \partial\Omega/\partial\boldsymbol{\kappa}}$$

is a vector, components $(\partial\Omega/\partial k, \partial\Omega/\partial l, \partial\Omega/\partial m)$. It is the velocity with which "influence" propagates, and which is much more important than the *Phase speed* c_p, the speed at which a surface of constant phase moves. c_p is *not* a vector.

In direction

$$\|\boldsymbol{\kappa} \qquad c_p = PP'/\tau = \omega/|\boldsymbol{\kappa}| = \omega/(k^2 + l^2 + m^2)^{1/2}.$$

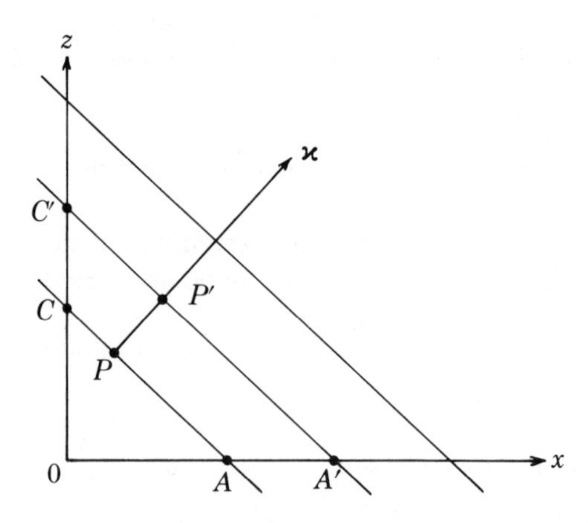

FIGURE 1.

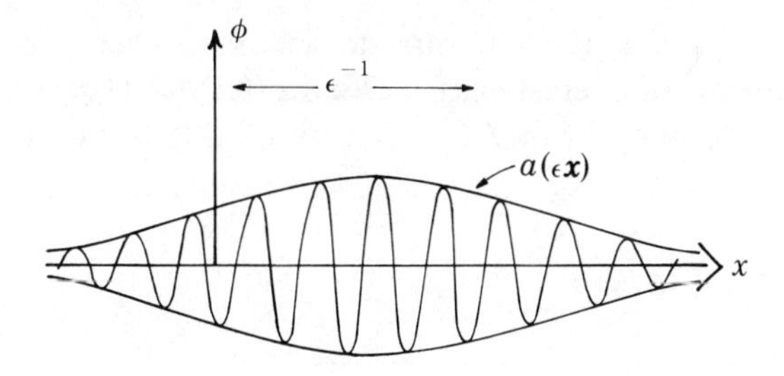

FIGURE 2.

But

$$\|Ox \qquad c_p = AA'/\tau = \omega/k \neq (\omega/|\boldsymbol{\kappa}|)(k/|\boldsymbol{\kappa}|),$$

$$\|Oz \qquad c_p = CC'/\tau = \omega/m \neq (\omega/|\boldsymbol{\kappa}|)(m/|\boldsymbol{\kappa}|).$$

(See Figure 1.)

2. **Wave kinematics.** Linear, Fourier approach.

2.1. *Initial value problem* (in principal trivial). At $t = 0$,

$$\phi(\boldsymbol{x}, 0) = \int_{-\infty}^{\infty} A(\boldsymbol{\kappa}) \exp(i\boldsymbol{\kappa} \cdot \boldsymbol{x})\, d\boldsymbol{\kappa};$$

then assuming $\omega = \Omega(\boldsymbol{\kappa})$ known at $t > 0$,

$$\phi(\boldsymbol{x}, t) = \int_{-\infty}^{\infty} A(\boldsymbol{\kappa}) \exp(i(\boldsymbol{\kappa} \cdot \boldsymbol{x} - \Omega(\boldsymbol{\kappa})t))\, d\boldsymbol{\kappa}.$$

2.2. *Slow modulation on a sine wave* (moves with the group velocity). Suppose

$$\phi(\boldsymbol{x}, 0) = a(\varepsilon \boldsymbol{x}) \exp(i\boldsymbol{\kappa}_0 \cdot \boldsymbol{x}), \qquad \varepsilon \ll 1,$$

where the envelope $a(\varepsilon \boldsymbol{x})$ is infinitely differentiable and the scale is $O(\varepsilon^{-1})$. (See Figure 2.)

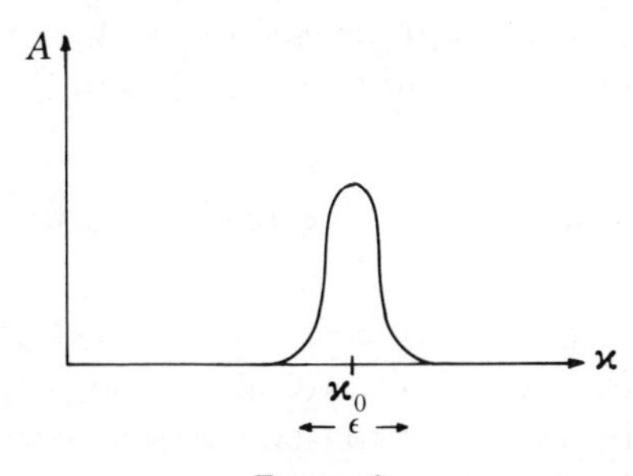

FIGURE 3.

Then $A(\boldsymbol{\kappa}) = \hat{a}((\boldsymbol{\kappa} - \boldsymbol{\kappa}_0)/\varepsilon) \sim 0$ outside $\boldsymbol{\kappa} = \boldsymbol{\kappa}_0 + O(\varepsilon)$ (see Figure 3); $\therefore$ for components of interest $\omega(\boldsymbol{\kappa}) = \Omega(\boldsymbol{\kappa}_0) + (\partial\Omega/\partial\boldsymbol{\kappa}_0)\cdot(\boldsymbol{\kappa} - \boldsymbol{\kappa}_0) + O(\varepsilon^2)$ where $\Omega(\boldsymbol{\kappa}_0)$ is ω_0 and $(\partial\Omega/\partial\boldsymbol{\kappa}_0)$ is $\boldsymbol{\gamma}_0$; $\therefore$ for all t up to but not including $O(\varepsilon^{-1})$,

$$\phi \sim \exp(i(\boldsymbol{\kappa}_0\cdot\boldsymbol{x} - \omega_0 t))\int A(\boldsymbol{\kappa})\exp(i(\boldsymbol{\kappa} - \boldsymbol{\kappa}_0)\cdot(\boldsymbol{x} - \boldsymbol{\gamma}_0 t))(1 + O(\varepsilon))\,d\boldsymbol{\kappa},$$

$$\boxed{\phi(\boldsymbol{x}, t) \sim a(\varepsilon(\boldsymbol{x} - \boldsymbol{\gamma}_0 t))\exp(i(\boldsymbol{x}_0\cdot\boldsymbol{x} - \omega_0 t))};$$

$\therefore$ modulation moves with velocity $\boldsymbol{\gamma}_0$.

Note. (i) If a is complex, it describes phase modulation as well as amplitude.

(ii) Inevitably *variations* in local wave energy density $\mathscr{E}$ (or any other quadratic quantity) also move with velocity $\boldsymbol{\gamma}$.

(iii) If a^2 vanishes outside a compact moving region of dimension of order ε^{-1} wavelengths, V, we have a *wave packet*. Any real quadratic integral over V may be expressed by Parseval's theorem as an integral of $|A(\boldsymbol{\kappa})|^2$, which is independent of time, e.g.

$$\int_V \left|\frac{\partial\phi}{\partial\boldsymbol{x}}\right|^2 d\boldsymbol{x} = (2\pi)^3\int|\boldsymbol{\kappa}|^2|A|^2\,d\boldsymbol{\kappa}.$$

Thus, whatever expression we adopt for the wave energy density $\mathscr{E}$, the *total* wave energy of a packet must be constant, moving with velocity $\boldsymbol{\gamma}_0$. After time $t = O(\varepsilon^{-2})$ (when the packet has moved many diameters), the modulation $a(\boldsymbol{x} - \boldsymbol{\gamma}_0 t)$ may have significantly changed shape because terms $\frac{1}{2}(\partial^2\Omega/\partial\boldsymbol{\kappa}^2)\cdot(\boldsymbol{\kappa} - \boldsymbol{\kappa}_0)^2 t$ in the phase function have become of order unity. The volume V may then be larger, but it still moves with velocity $\boldsymbol{\gamma}_0$ (with fractional error of order ε) and the total energy is still constant.

2.3. *Cauchy Poisson problem* (how a concentrated disturbance breaks up in one dimension). At $t = 0$, $\phi = \delta(x) = (1/2\pi)\int_{-\infty}^{\infty} e^{ikx}\,dk$. For $t > 0$, $\phi = (1/2\pi)\int_{-\infty}^{\infty}\exp(i(kx - \Omega(k)t))\,dk$. (See Figure 4.)

As $t \to \infty$ for fixed $x/t = \gamma_0$, use stationary phase:

$$\boxed{\phi \sim (2\pi(d^2\Omega/dk_0^2)t)^{-1/2}\,e^{-i\pi/4}\exp(i(k_0 x - \omega_0 t)) + O(t^{-3/2})},$$

where k_0 is the root of $(d\Omega/dk)(k) = \gamma_0$, and $\omega_0 = \Omega(k_0)$. Thus disturbance breaks up into a slowly varying wavetrain as shown in Figure 5. At A_1 (moving with speed γ_1) we see a constant wavenumber k_1 (though wave crests move at $\omega_1/k_1 \neq \gamma_1$); wave amplitude a decays like $t^{-1/2}$. At A_2

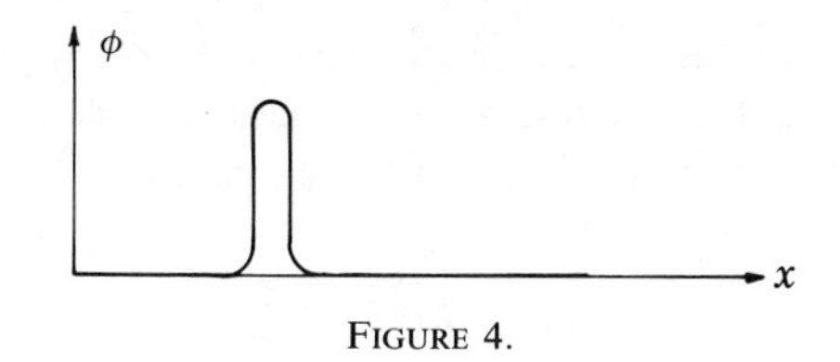

FIGURE 4.

(moving with speed $\gamma_2 \sim \gamma_1 + (d^2\Omega/dk_1^2)\Delta k$), we see $k_2 = k_1 + \Delta k$. Distance $A_1A_2 = x_2 - x_1 = (d^2\Omega/dk_1^2)\Delta kt$. Wave energy in A_1A_2:

$$\sim \frac{F(k_1)}{2\pi(d^2\Omega/dk_1^2)t}(x_2 - x_1) = \frac{1}{2\pi}F(k_1)\,\Delta k$$

which is constant.

2.4. *Steady radiation at large distances from a maintained oscillatory source* [Lighthill, 1965]. At $t = 0$, $\phi(\boldsymbol{x}, 0) = 0$ everywhere. For $t > 0$, $L(\partial/\partial t, \partial/\partial \boldsymbol{x})\phi = f(\boldsymbol{x})\,e^{-i\omega t}$; $f(\boldsymbol{x})$: localised forcing, ω fixed; L: polynomial differential operator, even order, constant coefficients, e.g.

$$L \equiv \frac{\partial^2}{\partial t^2}\left(\frac{\partial^2}{\partial x^2} + \frac{\partial^2}{\partial y^2} + \frac{\partial^2}{\partial z^2}\right) + N^2\left(\frac{\partial^2}{\partial x^2} + \frac{\partial^2}{\partial y^2}\right).$$

Laplace transform in time and let $t \to \infty$. Transients decay to zero. Then

$$\phi(\boldsymbol{x}, t) \to e^{-i\omega t}\int \frac{\hat{f}(\boldsymbol{\kappa})}{L(-i\omega, i\boldsymbol{\kappa})}\,e^{i\boldsymbol{\kappa}\cdot\boldsymbol{x}}\,d\boldsymbol{\kappa}.$$

Here $L(-i\omega, i\boldsymbol{\kappa})$ is a polynomial in ω, k, l, m with real coefficients. $L = 0$ on surfaces

$$\mathscr{S}: \Omega(\boldsymbol{\kappa}) = \omega, \qquad \Omega \text{ may be multiple valued.}$$

Resulting singularities on $\mathscr{S}$ are resolved by replacing ω by $\omega + i\varepsilon$, where ε small, positive.

Note. This method of resolution is a consequence (the only one) of having posed an initial value problem. It is equivalent to "Rayleigh Friction" or a "Radiation condition."

Now let $\boldsymbol{x} \to \infty$ in a given direction, i.e. let $\boldsymbol{x} = r\boldsymbol{n}$, r large and positive. Use a version of multidimensional stationary phase (three dimensional for

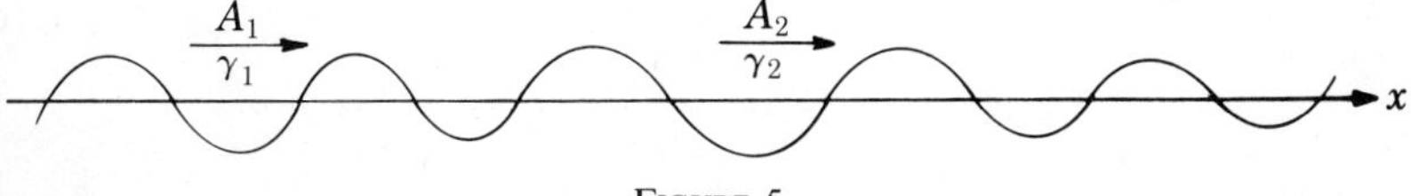

FIGURE 5.

illustration), and obtain asymptotics as $r \to \infty$. Riemann–Lebesgue Lemma $\Rightarrow$ dominant contribution to integral comes from singularities of the integrand, i.e. from a thin region in $\boldsymbol{\kappa}$-space surrounding surface $\mathscr{S}$. Here

$$L(-i\omega, i\boldsymbol{\kappa}) \sim \{\omega - \Omega(\boldsymbol{\kappa})\}\partial L/\partial\omega.$$

Rotate axes $(k, l, m) \to (\lambda, \mu, \nu)$ such that $O\lambda \| \boldsymbol{n}$. Suppose

$$\mathscr{S}: \qquad \Omega(\boldsymbol{\kappa}) = \omega \to \lambda = h(\mu, \nu).$$

Then $\omega - \Omega \sim i\varepsilon - \{\lambda - h(\mu, \nu)\}\boldsymbol{\gamma} \cdot \boldsymbol{n}$ and $\boldsymbol{\kappa} \cdot \boldsymbol{x} = \lambda r$. (See Figure 6.) Singular part of integral

$$= \int -\frac{\hat{f}}{(\partial L/\partial\omega)\boldsymbol{\gamma} \cdot \boldsymbol{n}} \quad \frac{e^{i\lambda r}}{\lambda - h(\mu, \nu) - i(\varepsilon/\boldsymbol{\gamma} \cdot \boldsymbol{n})} \, d\lambda \, d\mu \, d\nu$$

$$= 2\pi i \int_{\mathscr{S}_+} - \frac{\hat{f}}{(\partial L/\partial\omega)\boldsymbol{\gamma} \cdot \boldsymbol{n}} \exp(ih(\mu, \nu)r) \, d\mu \, d\nu$$

where $\mathscr{S}_+$ is that part of $\mathscr{S}$ for which $\boldsymbol{\gamma} \cdot \boldsymbol{n} > 0$.

Now use two-dimensional stationary phase. Dominant contribution comes from a small neighbourhood of P_0: where $\boldsymbol{\gamma}_0 = \partial\Omega/\partial\boldsymbol{\kappa}_0 \| \boldsymbol{n}$, i.e. $\partial h/\partial\mu = \partial h/\partial\nu = 0$. Near P_0:

$$h(\mu, \nu) \sim h_0 + \tfrac{1}{2}\{a(\mu - \mu_0)^2 + b(\nu - \nu_0)^2 + 2c(\mu - \mu_0)(\nu - \nu_0)\},$$

$$\phi \sim \exp(i(\boldsymbol{\kappa}_0 \cdot \boldsymbol{x} - \omega_0 t))\frac{-2\pi i\hat{f}_0}{(\partial L/\partial\omega_0)\boldsymbol{\gamma}_0 \cdot \boldsymbol{n}}$$

$$\times \int \exp(\tfrac{1}{2}i(a\mu'^2 + b\nu'^2 + 2c\mu'\nu')r) \, d\mu' \, d\nu',$$

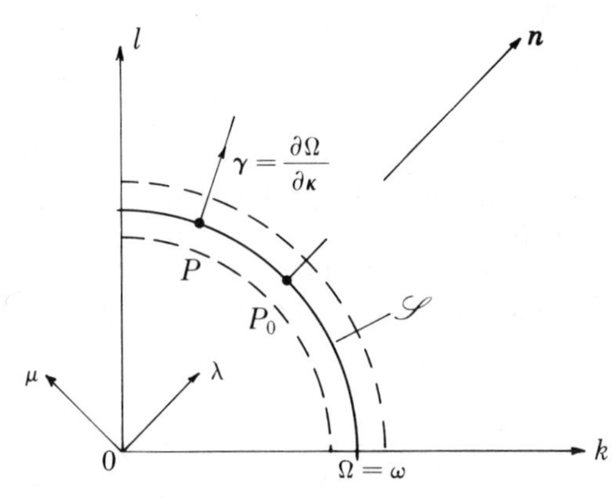

FIGURE 6.

therefore

$$\boxed{\phi(\boldsymbol{x}, t) \sim \frac{(2\pi)^2 \hat{f}_0}{(\partial L/\partial \omega_0)\gamma_0 \cdot \boldsymbol{n}} \cdot \frac{1}{(ab - c^2)^{1/2}} \cdot \frac{1}{r} \exp(i(\boldsymbol{\kappa}_0 \cdot \boldsymbol{x} - \omega t))} \quad \text{as } r \to \infty.$$

Note. (i) Behaviour near $\gamma \cdot \boldsymbol{n} = 0$ needs separate investigation. The contribution is usually negligible.

(ii) Wave number observed at point $\boldsymbol{x}$ is that which would reach there from the vicinity of the forcing if moving *outwards* with its appropriate group velocity.

(iii) Amplitude $\propto r^{-1} \Rightarrow$ Energy density and energy flux $\propto r^{-2}$. In n-dimensions, amplitude $\propto r^{-(n-1)/2}$.

(iv) $ab - c^2$ is Gaussian curvature (*product* of principal curvatures) of the surface $\mathscr{S}$. It is a measure of the dispersion near wavenumber $\boldsymbol{\kappa}_0$. If it vanishes (i.e. $\mathscr{S}$ locally cylindrical) *this* formula for asymptotics breaks down.

(v) $\mathscr{S}_+$ may be multiple valued, and there may be two or more distinct wavetrains superimposed at the point $\boldsymbol{x}$. For certain directions these merge (P_0 approaches branch points of $\mathscr{S}$) and the whole asymptotics needs re-examination.

(vi) The structure of the forcing function $f(\boldsymbol{x})$ is essentially secondary in determining the radiation field. It affects the energy flux in different directions, but that is all.

3. **Internal gravity waves** (in a continuously stratified Boussinesq liquid).

3.1. *The Boussinesq approximation.* Gravitational acceleration: $-g\boldsymbol{n}$ ($\boldsymbol{n}$ unit vector upwards, $\| Oz$). Incompressible liquid: $D\rho/Dt = 0$ (i.e. density ρ constant for a fluid particle). If density slightly nonuniform,

$$\rho = \bar{\rho} + \rho_1 \qquad (\rho_1/\bar{\rho} \ll 1),$$

the buoyancy of a fluid particle (the *net* gravity/unit mass acting on it) is

$$\sigma = g\rho_1/\bar{\rho} \qquad (\ll g).$$

For our purposes, the Boussinesq approximation consists of ignoring variations ρ_1 of inertial density (in terms $\rho(D\boldsymbol{u}/Dt)$), but retaining them in the buoyancy force. It is a consistent first approximation if everywhere

$$|D\boldsymbol{u}/Dt| \ll g \qquad \text{and} \qquad \rho_1/\rho \ll 1.$$

Also set

$$p = (\text{True pressure} + \bar{\rho} g z)/\bar{\rho}.$$

Then the inviscid equations of motion become:
Momentum:

$$D\boldsymbol{u}/Dt + \nabla p + \sigma \boldsymbol{n} = 0.$$

Continuity:

$$\nabla \cdot \boldsymbol{u} = 0.$$

Mass:

$$D\sigma/Dt = 0.$$

3.2. *The basic state.* An equilibrium state is

$$\boldsymbol{u} = (U, 0, 0), \qquad \sigma = -N^2 z,$$

N is the *buoyancy* (Brunt-Väisälä) *frequency*, a measure of the stable density stratification. Although in general $U(z)$, $N^2(z)$, we shall assume them to be uniform.

3.3. *Small perturbations* ($\boldsymbol{u} = \boldsymbol{U} + \boldsymbol{u}'$, $\sigma = -N^2 z + \sigma'$). Linearised equations:

$$(\partial/\partial t + U(\partial/\partial x))\boldsymbol{u}' + \nabla p' + \sigma' \boldsymbol{n} = 0,$$

$$\nabla \cdot \boldsymbol{u}' = 0,$$

$$(\partial/\partial t + U(\partial/\partial x))\sigma' - N^2 w' = 0.$$

If two dimensional,

$$u' = \partial\psi/\partial z, \qquad w' = -\partial\psi/\partial x$$

and

$$L(\psi) \equiv (\partial/\partial t + U(\partial/\partial x))^2(\partial^2\psi/\partial x^2 + \partial^2\psi/\partial z^2) + N^2(\partial^2\psi/\partial x^2) = 0.$$

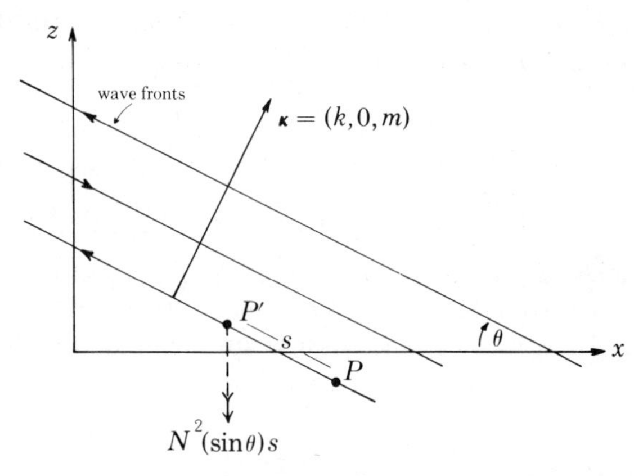

FIGURE 7.

3.4. *Plane waves.* If $\psi = \frac{1}{2}\hat{\psi}\exp(i(kx + mz - \omega t))$ + complex conjugate, we have the dispersion relation:

$$\boxed{\omega = \Omega(\boldsymbol{\kappa}) = Uk \pm N(|k|/(k^2 + m^2)^{1/2})}\,.$$

Note. There are apparently two modes ($\pm$ sign). However, if $+ \to -$ and $\boldsymbol{\kappa} \to -\boldsymbol{\kappa}$, then $\omega \to -\omega$ and all that has happened is $\exp(i(kx + mz - \omega t)) \to$ complex conjugate. Thus without loss of generality we restrict attention to the $+$ sign.

3.5. *In a stationary medium* ($U = 0$). (See Figure 7.) The waves are transverse; velocities $\boldsymbol{u}' \parallel$ wavefronts. If the wave fronts make angle θ with the horizontal

$$\boxed{\omega = N \sin\theta}$$

independent of $|\boldsymbol{\kappa}|$.

An alternative derivation of this equation is: Suppose plane wave solutions exist. Then there can be no pressure or other variations along the plane of the wavefront (AA') and, by continuity, fluid particles must move in this plane. A particle displaced from equilibrium at P through a distance s to P' has risen vertically $s \sin\theta$, and has a buoyancy force $N^2 s \sin\theta$. The component of this acting down the plane is $N^2 s \sin^2\theta$. Thus

$$d^2s/dt^2 + N^2 s \sin^2\theta = 0.$$

Surfaces $\Omega(\boldsymbol{\kappa})$ = constant are straight lines. Thus

$$\boldsymbol{\gamma} = \partial\Omega/\partial\boldsymbol{\kappa} \perp \boldsymbol{\kappa} \qquad (\text{i.e. } \parallel \boldsymbol{u}')$$

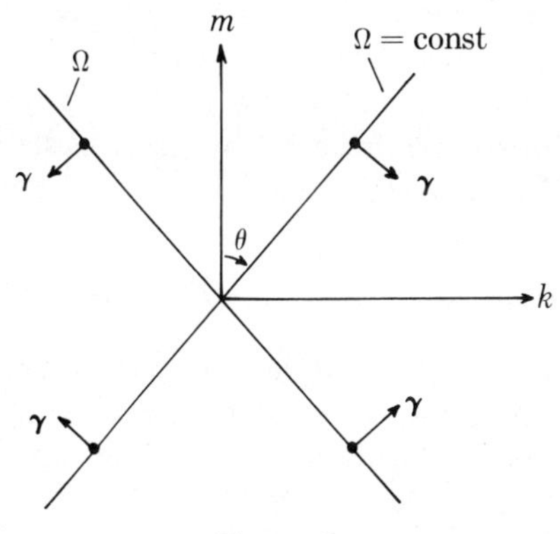

FIGURE 8.

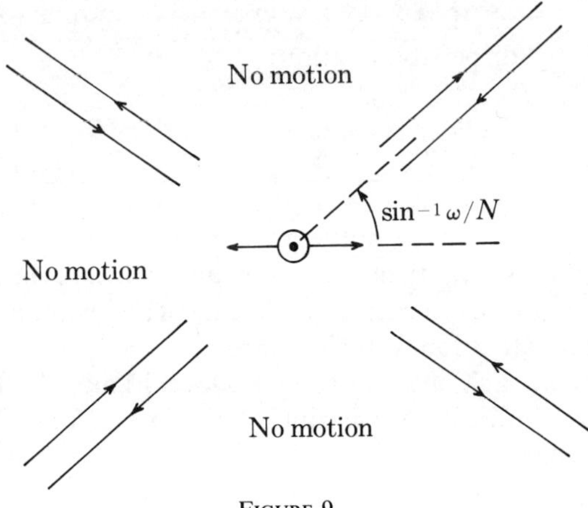

FIGURE 9.

with magnitude $(N \cos \theta)/|\boldsymbol{\kappa}|$, i.e. long waves travel faster. Direction is as indicated by the arrows in the diagram (see Figure 8). Note that for an upward component ($\partial\Omega/\partial m > 0$) we require $m < 0$. As $\theta \to \pi/2$ (particle motions vertical), $\omega \to N, |\boldsymbol{\gamma}| \to 0$. As $\theta \to 0$ (particle motions nearly horizontal), $\omega \to 0$, $|\boldsymbol{\gamma}| \to N/|\boldsymbol{\kappa}|$. For the radiation field far from an oscillatory source, we cannot use the quantitative result of §2.4 because the Gaussian curvature of $\Omega(\boldsymbol{\kappa})$ vanishes (all waves with the same angle θ but different wavelengths travel in the same direction). However, we may expect the pattern of wavefronts: For $\omega < N$, see Figure 9. For $\omega > N$, no waves are possible and no radiation should occur. For good experimental confirmation see Mowbray and Rarity [1967]. For the radiation pattern from an impulsive disturbance see Bretherton [1967].

3.6. *Lee waves* ($\omega = 0$, $U \neq 0$). The term $Uk \equiv \boldsymbol{U} \cdot \boldsymbol{\kappa}$ in the dispersion relation describes advection of waves by the mean flow. The *basic frequency* $\omega - Uk$ (which by convention is always positive) is as in a stationary medium, but for a stationary observer this is Doppler shifted.

For the radiation pattern from a stationary obstacle (e.g. a mountain ridge in a stratified airstream), we have $\omega = 0$ which implies the surface S where $\Omega(\boldsymbol{\kappa}) = \omega$ is:

$$k = 0 \quad or \quad k < 0 \quad \text{and} \quad (k^2 + m^2)^{1/2} = N/U.$$

The direction of increasing Ω is indicated in the diagram (see Figure 10). An observer at a point above the ground level is only interested in the $\frac{1}{4}$ circle $k < 0, m < 0$ and, for these, $\boldsymbol{\kappa} \parallel -\boldsymbol{\gamma}$. The expected pattern of wave crests is shown in Figure 11.

Note. There are no waves upstream.

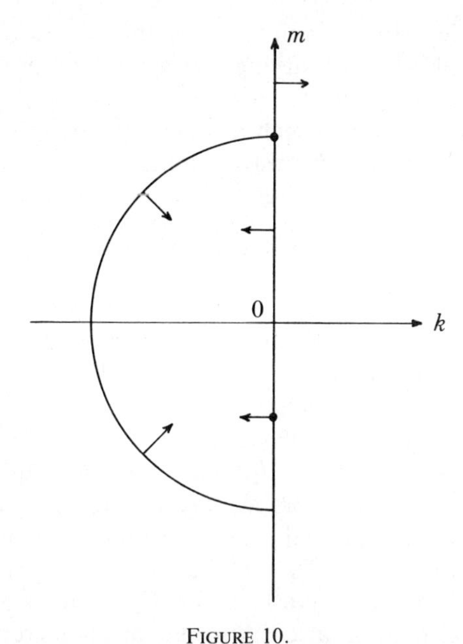

FIGURE 10.

For comparison with an exact computation see Huppert and Miles [1969, Figure 9]. The waves with $k = 0$, $|m| < N/U$ are also possible and could in principle give a horizontally uniform disturbance far upstream of the obstacle (a blocking of the flow by the mountain [Long, 1955]). The status of blocking is controversial, e.g. [Miles, 1969], [Brooke Benjamin, 1970].

4. **Slowly varying wavetrains** (geometrical optics).

4.1. *The local dispersion relation.* The Fourier method is applicable only in a strictly uniform medium, in which sines and cosines are normal

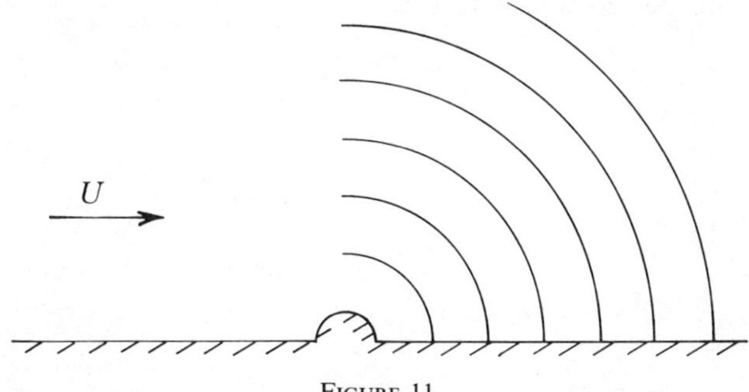

FIGURE 11.

modes of the system. In a nonuniform medium it is quite inappropriate, but the concept of a slowly varying wavetrain may provide a powerful alternative approach.

We suppose a wavetrain which is *locally* periodic (sinusoidal if it is linear), but whose amplitude and wavenumber change gradually over many wavelengths and wave periods (see §2.3 for an example). Thus

$$\phi = a(\boldsymbol{x}, t) \sin \theta(\boldsymbol{x}, t).$$

(See Figure 12.)

The *phase function* θ is constant along wave crests, increasing by 2π over one wavelength or period.

The *local wavenumber* $\boldsymbol{\kappa} = \partial\theta/\partial\boldsymbol{x}$.

The *local frequency* $\omega = -\partial\theta/\partial t$.

Over intervals $|\boldsymbol{\kappa}|^{-1}, \omega^{-1}$: $a^{-1}\Delta a, |\boldsymbol{\kappa}|^{-1}\Delta\boldsymbol{\kappa}, \omega^{-1}\Delta\omega \ll 1$.

Note. (i) $\boldsymbol{\kappa}$ is not a Fourier component; it is the locally dominant value.

(ii) Figure 13 shows what happens as an observer moves from A to B.

No wave crests along $\Gamma_1 = (1/2\pi)_{\Gamma_1}^{B}\int_A \boldsymbol{\kappa}\cdot d\boldsymbol{s} = (1/2\pi)(\theta_B - \theta_A) = (1/2\pi)_{\Gamma_2}^{B}\int_A \boldsymbol{\kappa}\cdot d\boldsymbol{s} =$ no wave crests along Γ_2. Thus $\nabla \times \boldsymbol{\kappa} = 0 \Leftrightarrow$ wave crests have no free ends (as at C). Also rate increase crests along Γ_1

$$= \frac{1}{2\pi}\frac{\partial}{\partial t}\int_A^B \boldsymbol{\kappa}\cdot d\boldsymbol{s} = \frac{1}{2\pi}(\omega_A - \omega_B),$$

i.e. $\partial\boldsymbol{\kappa}/\partial t = -\nabla\omega \Leftrightarrow$ wavecrests are not created or destroyed.

We suppose also that the dynamics of the medium are adequately represented by the *local dispersion relation*

$$\boxed{\omega = \Omega(\boldsymbol{\kappa}; \boldsymbol{x}, t)}\,.$$

Note. (iii) This is an approximation (though a very powerful one), which can have unexpected consequences.

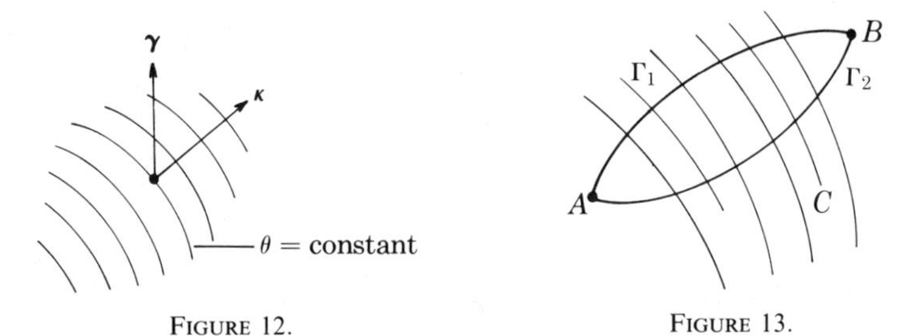

FIGURE 12. FIGURE 13.

(iv) For consistency, the dependence of Ω on $\boldsymbol{x}, t$ must also be on scales large compared to $\boldsymbol{\kappa}^{-1}, \omega^{-1}$. Thus the medium is also slowly varying.

(v) For nonlinear quasi-periodic wavetrains, Ω may also depend on a^2.

4.2. *Linear wave kinematics*. (These ideas have been more or less familiar for a long time from the study of the Hamilton-Jacobi equation, but for an account directed at fluid dynamicists see Whitham [1960].)

An observer moving always with the local group velocity

$$\gamma = \partial\Omega/\partial\boldsymbol{\kappa}$$

of the wavetrain moves along a path in space-time known as a *ray*. From the existence of the phase function θ and the dispersion relation $\Omega(\boldsymbol{\kappa}; \boldsymbol{x}, t)$ alone, we may deduce that changes along rays are given by

$$\boxed{d\boldsymbol{\kappa}/dt = -\partial\Omega/\partial\boldsymbol{x}, \qquad d\omega/dt = \partial\Omega/\partial t}\,.$$

If, say, the medium is uniform in the Ox direction ($\partial\Omega/\partial x = 0$), then the x-component of wavenumber is constant along rays $dk/dt = 0$. If it is constant in time, then so is the frequency ω.

For example: Linear transverse waves on a string of tension $T(t)$ and mass per unit length $m(x)$ satisfy

$$m(\partial^2\eta/\partial t^2) - T(\partial^2\eta/\partial x^2) = 0.$$

For a slowly varying wavetrain the dispersion relation is

$$\omega = (T(t)/m(x))^{1/2}|k| \qquad \text{(by convention } \omega > 0).$$

As waves of frequency ω move into a region of larger m, the wave speed is reduced and the wavenumber increases: $dk/dt = -\partial\Omega/\partial x = \frac{1}{2}(\omega/m)(dm/dx) > 0$. (See Figure 14.) When the tension in a violin string is increased as it is tuned, the pitch of a free oscillation on it rises: $d\omega/dt = \partial\Omega/\partial t = \frac{1}{2}(\omega/T)(dT/dt) > 0$.

Note. (i) The expressions for $d\boldsymbol{\kappa}/dt$, $d\omega/dt$ are not independent, since $\omega, \boldsymbol{\kappa}$ are connected by $\omega = \Omega$. If the medium is nonuniform in one direction only, say Oz, it is in practice most convenient to infer from $\partial\Omega/\partial x =$

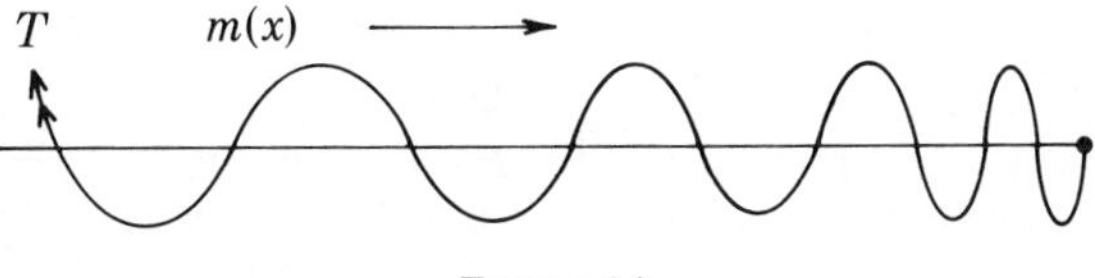

FIGURE 14.

$\partial\Omega/\partial y = \partial\Omega/\partial t = 0$ that k, l, ω are constant along rays, and to use

$$\omega = \Omega(k, l, m; z)$$

to give the local value of m.

(ii) The rays are characteristics of the partial differential equation

$$\partial\theta/\partial t = -\Omega((\partial\theta/\partial \boldsymbol{x}); \boldsymbol{x}, t).$$

For nonsingular Ω, and a given well-behaved $\theta(\boldsymbol{x}, 0)$, the solution is determined in some domain $0 \leqq t < t_1$, but "kinematic shocks" (where $\boldsymbol{\kappa}$ becomes indeterminate) will normally form eventually. (See Figure 15.) Thus in a uniform dispersive medium, rays are straight lines with slope $\boldsymbol{\gamma}(\boldsymbol{\kappa})$. If $\phi(x, 0)$ is wavenumber modulated, some rays will certainly converge and eventually cross (at X). This does not imply a physical discontinuity, only a breakdown of slow variation, and possibly the destruction of wave crests.

PROOF. Use suffix notation ($i = 1, 2, 3$ corresponding to x, y, z). Also remember that although

$$\omega = \Omega(\boldsymbol{\kappa}; \boldsymbol{x}, t) \quad \text{everywhere}$$

and $\partial\omega/\partial x_i$ is the actual gradient in physical space, $\partial\Omega/\partial x_i$ is the partial derivative holding $\boldsymbol{\kappa}$ constant ($= \partial\omega/\partial x_i - (\partial k_j/\partial x_i)(\partial\Omega/\partial k_j)$). Then

$$\begin{aligned}\frac{dk_i}{dt} &= \left(\frac{\partial}{\partial t} + \frac{\partial\Omega}{\partial k_j}\frac{\partial}{\partial x_j}\right)\frac{\partial\theta}{\partial x_i} \\ &= \frac{\partial}{\partial x_i}\left(\frac{\partial\theta}{\partial t}\right) + \frac{\partial}{\partial x_i}\left(\frac{\partial\theta}{\partial x_j}\right)\frac{\partial\Omega}{\partial k_j} \\ &= -\frac{\partial\omega}{\partial x_i} + \frac{\partial k_j}{\partial x_i}\frac{\partial\Omega}{\partial k_j} = -\frac{\partial\Omega}{\partial x_i}.\end{aligned}$$

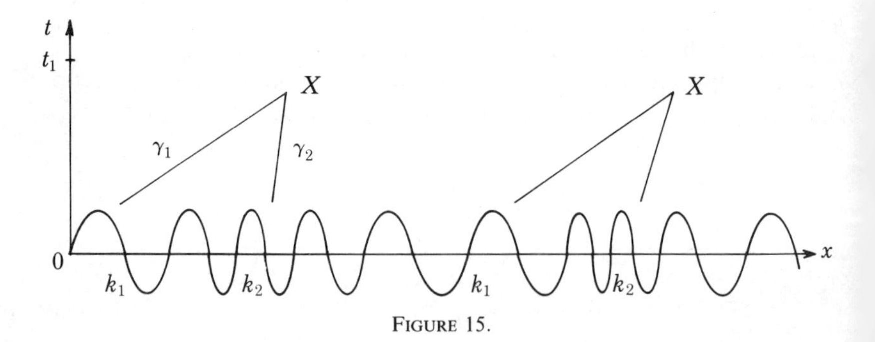

FIGURE 15.

Also

$$\frac{d\omega}{dt} = \frac{\partial\omega}{\partial t} + \gamma_j\frac{\partial\omega}{\partial x_j} = \left(\frac{\partial\Omega}{\partial t} + \frac{\partial k_j}{\partial t}\frac{\partial\Omega}{\partial k_j}\right) + \frac{\partial\Omega}{\partial k_j}\frac{\partial\omega}{\partial x_j}$$

$$= \frac{\partial\Omega}{\partial t} + \frac{\partial\Omega}{\partial k_j}\left(\frac{\partial^2\theta}{\partial t\,\partial x_j} - \frac{\partial^2\theta}{\partial x_j\,\partial t}\right).$$

4.3. *Wave packets.* If the local amplitude $a(\boldsymbol{x}, t)$ vanishes outside a compact volume V over which $\boldsymbol{\kappa}, \omega$ do not vary appreciably, we have a *wave packet* (see §2.2, Note (iii)). Viewed locally, the waves are sinusoidal with wavenumber $\boldsymbol{\kappa}$. Viewed from the scale of variation of the medium V appears as a point, moving with velocity $\boldsymbol{\gamma}$. Its path (a ray) is governed by the equations

$$d\boldsymbol{\kappa}/dt = -\,\partial\Omega/\partial\boldsymbol{x}, \qquad d\boldsymbol{x}/dt = \partial\Omega/\partial\boldsymbol{\kappa}.$$

Note. (i) The similarity to Hamilton's equations for a particle with position $\boldsymbol{x}$ and momentum $\boldsymbol{\kappa}$ moving under a Hamiltonian $\Omega(\boldsymbol{\kappa}, \boldsymbol{x})$.

(ii) The kinematic theory gives no indication of changes in amplitude or energy of a wavetrain or wave packet. Because locally, at least, a wave packet looks like a slow modulation on a carrier wave number $\boldsymbol{\kappa}$, we may infer from §2.2 that the region of nonzero amplitude does indeed move with the group velocity $\boldsymbol{\gamma}$, and that so long as nonuniformities in the medium are imperceptible the total wave energy is constant. However, over distances $\boldsymbol{x}$ and t over which $\boldsymbol{\kappa}$ and ω change substantially, significant changes in total wave energy may occur. Their evaluation is an important dynamical problem.

4.4. *The* WKB *approximation* (see for example Lewis [1965]). A more formal approach to slowly varying wavetrains involves asymptotic expansions in a small parameter ε.

Example: linear transverse waves on a stretched string (cf. §4.2).

$$L(\partial/\partial t, \partial/\partial x)\eta \equiv m(x)\partial^2\eta/\partial t^2 - T(t)\partial^2\eta/\partial x^2 = 0.$$

Try the formal solution

$$\phi = \text{Real part of } \{\hat{\eta}_0(x,t) + \varepsilon\hat{\eta}_1(x,t) + \cdots\}\exp(i\varepsilon^{-1}\theta(x,t)), \quad \varepsilon \ll 1,$$

where $\hat{\eta}_0(x,t)$ is the local amplitude, $\varepsilon\hat{\eta}_1(x,t)$ the first correction, and $i\varepsilon^{-1}\theta(x,t)$ the phase. Over $\Delta x, \Delta t$,

$$\Delta(\varepsilon^{-1}\theta) = \varepsilon^{-1}(\partial\theta/\partial x)\cdot\Delta x + \varepsilon^{-1}(\partial\theta/\partial t)\cdot\Delta t$$

$$= 2\pi \quad \text{when } \Delta x, \Delta t = O(\varepsilon),$$

i.e. wavelength and period are short compared to scales for $\hat{\eta}$. Then

$$\frac{\partial}{\partial t}(\hat{\eta}_0 \exp(i\varepsilon^{-1}\theta)) = \left\{ i\varepsilon_1^{-1}\frac{\partial \theta}{\partial t}\hat{\eta}_0 + \frac{\partial \hat{\eta}_0}{\partial t}\right\} \exp(i\varepsilon^{-1}\theta),$$

where $i\varepsilon^{-1}(\partial\theta/\partial t)\hat{\eta}_0$ is the dominant term. Substituting in the governing equation and equating successive powers of ε:

$$\varepsilon^{-2}: \qquad \{m(\partial\theta/\partial t)^2 - T(\partial\theta/\partial x)^2\}\hat{\eta}_0 = 0,$$

i.e.

$$L(-i\omega, ik) \equiv -m\omega^2 + Tk^2 = 0,$$

or

$$\boxed{\omega = (T(t)/m(x))^{1/2}|k|}$$

where, as in §3.3, we have adopted the convention that the basic frequency, in this case $\omega = -\partial\theta/\partial t$, is positive.

Note. (i) This dispersion relation guarantees the kinematics of §4.2, and determines the rays in terms of $\theta(x, 0)$.

(ii) When it is satisfied,

$$\delta L \equiv \frac{\partial L}{\partial \omega}\delta\omega + \frac{\partial L}{\partial k}\delta k = 0$$

so

$$\gamma = \frac{\partial \omega}{\partial k} = -\frac{\partial L}{\partial k}\bigg/\frac{\partial L}{\partial \omega},$$

a result readily generalised to three dimensions.

$$\varepsilon^{-1}: \qquad -2i\omega m\frac{\partial \hat{\eta}_0}{\partial t} - 2ikT\frac{\partial \hat{\eta}_0}{\partial x} = i\left(m\frac{\partial \omega}{\partial t} - T\frac{\partial k}{\partial x}\right)\hat{\eta}_0 + (m\omega^2 - Tk^2)\hat{\eta},$$

where $-2i\omega m$ is $i\,\partial L/\partial\omega$ and $2ik$ is $-i\,\partial L/\partial k$; these are known from kinematics; $(m\omega^2 - Tk^2)\hat{\eta} = 0$. Thus

$$\boxed{\frac{\partial \hat{\eta}_0}{\partial t} + \gamma\frac{\partial \hat{\eta}_0}{\partial x} = -\frac{1}{2\omega}\left(\frac{\partial \omega}{\partial t} - \frac{T}{m}\frac{\partial k}{\partial x}\right)\hat{\eta}_0},$$

which determines $\hat{\eta}_0$ along rays from its value at $t = 0$ (provided no kinematic shocks have formed and $\partial L/\partial\omega \neq 0$ anywhere).

Note. (iii) This equation is often known as the *transport equation.*

(iv) The true simplicity of the expression for changes in amplitude $|\hat{\eta}_0|$, namely

$$(\partial/\partial t)\mathscr{A} + (\partial/\partial x)(\gamma\mathscr{A}) = 0, \qquad \text{where } \mathscr{A} = \tfrac{1}{2}m\omega|\hat{\eta}_0|^2,$$

is not readily apparent from it.

$\varepsilon^0, \varepsilon^1, \ldots$: A sequence of linear equations of the form

$$(\partial\eta_n/\partial t) + \gamma(\partial\eta_n/\partial x) = f(\eta_n, \eta_{n-1}, \ldots, \eta_0\,; x, t).$$

These determine $\hat{\eta}_1, \hat{\eta}_2, \ldots$ along rays.

Note. (v) This procedure demonstrates how an expression for ϕ of the required form may in principle be found to any desired order. However, it never generates the reflected wave—with phase function found by replacing $\partial\theta/\partial x \to -\partial\theta/\partial x$ but leaving $\partial\theta/\partial t$ unaltered. Yet it is well known that if $m(x)$ is discontinuous partial reflection occurs, and any smooth steplike change in m appears as a discontinuity for sufficiently small wavenumber. For large wavenumbers the reflected amplitude is small, so small that it does not appear to any finite order in the series in powers of ε, but in the exact solution of the differential equation it is still present.

(vi) If the procedure is truncated after terms of order ε^N and the known expression for ϕ substituted in the governing equation, the residual is $O(\varepsilon^{N-2})$ uniformly in some domain in (x, t) space, including the initial line $t = 0$. As $\varepsilon \to 0$, the magnitude of the residual $\to 0$, and, for sufficiently small $\varepsilon < \varepsilon(N)$, it may be reduced by increasing N to $N + 1$. Thus we have an *asymptotic* series in powers of ε. However, in general it is not *convergent* for *any* $\varepsilon > 0$. For if it were, its sum as $N \to \infty$ would be an exact solution of the governing differential equation and would have to include a description of partial reflections.

(vii) A similar approach may be applied for other operators $L(\partial/\partial t, \partial/\partial \mathbf{x}; \mathbf{x}, t)$. However, if the leading derivatives are not all of the same order, it may be necessary to rescale one or other of the independent variables x, y, z, t (thus introducing ε into L explicitly) before inserting the formal expression for ϕ. Otherwise the desired balance of terms in the dynamics of waves with wavelength and period formally of order ε may be upset. Thus, to obtain internal gravity waves in a nonuniform flow $U(z)$, $N^2(z)$, we first define

$$t^* = \varepsilon t, \qquad U^* = \varepsilon^{-1}U.$$

Then

$$\left(\frac{\partial}{\partial t} + U^*(z)\frac{\partial}{\partial x}\right)^2\left(\frac{\partial^2\psi}{\partial x^2} + \frac{\partial^2\psi}{\partial z^2}\right) + \varepsilon^{-2}N^2\frac{\partial^2\psi}{\partial x^2} = \frac{d^2U^*}{dz^2}\left(\frac{\partial}{\partial t} + U^*\frac{\partial}{\partial z}\right)\frac{\partial\psi}{\partial z}.$$

For a situation in which $U^*(z)$ is of order unity, the terms on the left-hand side combine to give the dispersion relation for plane waves with U^*, N uniform (from $O(\varepsilon^{-4})$), and the transport equation (from $O(\varepsilon^{-3})$). The basic vorticity gradient d^2U^*/dz^2 on the right enters as a small correction only at $O(\varepsilon^{-2})$.

However sometimes, when the physical situation is inappropriate, no amount of juggling will give a consistent, slowly varying wavetrain. For example—on an unstratified shear flow

$$(\partial/\partial t + U(z)(\partial/\partial x))(\partial^2\psi/\partial x^2 + \partial^2\psi/\partial z^2) - (d^2U/dz^2)(\partial\psi/\partial x) = 0,$$

the only consistent dispersion relation for locally plane waves is the trivial one, $\omega = Uk$, corresponding to perturbations of such small scale that they are "frozen in" to the fluid.

4.5. *The averaged variational principle* (see Whitham [1965] for the basic idea and its application to nonlinear systems). If the differential equation $L(\phi) = 0$ may be derived from a variational principle of the form

$$\delta \int Q\left(\frac{\partial\phi}{\partial t}, \frac{\partial\phi}{\partial x}, \phi; x, t; \varepsilon\right) dx\, dt = 0,$$

where Q is a homogeneous quadratic form in ϕ and its derivatives, it is instructive to apply the formal expansion procedure directly.

For example: Suppose

$$\delta A \equiv \delta \int \left\{\tfrac{1}{2}m(x)\left(\frac{\partial\eta}{\partial t}\right)^2 - \tfrac{1}{2}T(t)\left(\frac{\partial\eta}{\partial x}\right)^2\right\} dx\, dt = 0$$

for all differentiable $\delta\eta(x, t)$ of compact support (i.e. $\delta\eta = 0$ outside some domain $|x| < x_1, |t| < t_1$). The Euler equation is that for transverse waves on a stretched string, considered in §4.2, §4.4.

Set

$$\eta(x, t) = \sum_0^\infty \varepsilon^n a_n \cos(\varepsilon^{-1}\theta + \alpha_n)$$

where $\theta(x, t)$ and $a_n(x, t)$, $\alpha_n(x, t)$ $(n = 0, 1, \ldots)$ are all real; then

$$\begin{aligned} A \equiv \varepsilon^{-2} &\int \tfrac{1}{4}\left\{m\left(\frac{\partial\theta}{\partial t}\right)^2 - T\left(\frac{\partial\theta}{\partial x}\right)^2\right\} a_0^2\, dx\, dt \\ + \varepsilon^{-1} &\int \tfrac{1}{2}\left\{\left(m\left(\frac{\partial\theta}{\partial t}\right)^2 - T\left(\frac{\partial\theta}{\partial x}\right)^2\right) a_0 a_1 \right. \\ &\qquad \left. + \left(m\frac{\partial\theta}{\partial t}\frac{\partial\alpha_0}{\partial t} - T\frac{\partial\theta}{\partial x}\frac{\partial\alpha_0}{\partial x}\right) a_0^2\right\} dx\, dt + O(\varepsilon^0), \end{aligned}$$

where we have used the result that, for infinitely differentiable $f(x, t)$, $\theta(x, t)$ and $(\partial\theta/\partial t)^2 + (\partial\theta/\partial \mathbf{x})^2 \neq 0$ anywhere,

$$\int f(x, t) \exp(i\varepsilon^{-1}\theta(x, t))\, dx\, dt = o(\varepsilon^N) \quad \text{as } \varepsilon \to 0,$$

for all positive integers N. Strictly speaking, this holds only if $f(x, t)$ vanishes at infinity, but as the expansion for A will be used only after insertion of variations of compact support, this causes no difficulty. Arbitrary smooth variations of compact support $\delta a_0(x, t)$, $\delta\theta(x, t)$, etc., all imply permissible variations $\delta\eta$. Making them and equating coefficients of powers of ε to zero, we obtain a sequence of conditions which are *necessary* if $\eta(x, t)$ is to have a solution in the assumed form. From ε^{-2}, $\delta a_0 : \frac{1}{2}\{m(\partial\theta/\partial t)^2 - T(\partial\theta/\partial x)^2\}a_0 = 0$, which is the dispersion relation

$$\omega = (T(t)/m(x))^{1/2}|k| \qquad \text{if } \omega = -\,\partial\theta/\partial t,\ k = \partial\theta/\partial x.$$

From ε^{-2}, $\delta\theta$:

$$\frac{\partial}{\partial t}\{\tfrac{1}{2}m\omega a_0^2\} + \frac{\partial}{\partial x}\{\tfrac{1}{2}Tka_0^2\} = 0,$$

which is a conservation equation giving changes in a_0 along rays. From ε^{-1}, δa_0:

$$\left(m\frac{\partial\theta}{\partial t}\frac{\partial\alpha_0}{\partial t} - T\frac{\partial\theta}{\partial x}\frac{\partial\alpha_0}{\partial x}\right)a_0 = 0,$$

which determines changes in phase α_0 along rays. From ε^{-1}, $\delta\theta$:

$$\frac{\partial}{\partial t}(m\omega a_0 a_1) + \frac{\partial}{\partial x}(Tka_0 a_1) = \frac{\partial}{\partial t}\left(m\frac{\partial\alpha_0}{\partial t}a_0^2\right) - \frac{\partial}{\partial x}\left(T\frac{\partial\alpha_0}{\partial x}a_0^2\right),$$

which similarly fixes a_1.

Note. Variations δa_1, δa_2, etc. yield the same information as δa_0, but from terms in the expansion of A smaller by powers of ε, ε^2, etc., respectively. Similarly variations $\delta\alpha_1, \delta\alpha_2, \ldots$ yield no more than $\delta\theta$. The sequence from δa_0, $\delta\theta$ fully determines all higher order corrections in the slowly varying wavetrain.

If we define the *averaged Lagrangian*

$$\mathscr{L}(a, \omega, k; m, T) \equiv \tfrac{1}{4}(m\omega^2 - Tk^2)a^2,$$

which is the average value over a wavelength of the quadratic function

$$Q(\partial\phi/\partial t, \partial\phi/\partial x, \phi; m, T)$$

when a strictly sinusoidal wavetrain $\phi = a\sin(kx - \omega t + \alpha)$ is inserted into it, the lowest order terms in the expansion of the variational principle in powers of ε are

$$\boxed{\delta A \equiv \delta \int \mathscr{L}\left(a_0, -\frac{\partial \theta}{\partial t}, \frac{\partial \theta}{\partial x}; m(x), T(t)\right) dx\, dt = 0}\,.$$

This is Whitham's averaged variational principle for slowly varying wavetrains. The dispersion relation and conservation relation derived above, which determine completely the significant parts of the lowest order approximation, are respectively

$$\frac{\delta A}{\delta a_0} = \frac{\partial \mathscr{L}}{\partial a} = 0 \quad \text{and} \quad \frac{\delta A}{\partial \theta} = \frac{\partial}{\partial t}\left(\frac{\partial \mathscr{L}}{\partial \omega}\right) - \frac{\partial}{\partial x}\left(\frac{\partial \mathscr{L}}{\partial k}\right) = 0.$$

Because for linear systems $\mathscr{L}$ is necessarily quadratic in a^2,

$$a(\partial \mathscr{L}/\partial a) = 2\mathscr{L},$$

so the dispersion relation implies $\mathscr{L} = 0$, and

$$\gamma = -\frac{\partial \mathscr{L}}{\partial k}\bigg/\frac{\partial \mathscr{L}}{\partial \omega}.$$

Thus changes in wave amplitude are governed by the conservation equation

$$\boxed{\frac{\partial}{\partial t}\mathscr{A} + \frac{\partial}{\partial x}(\gamma \mathscr{A}) = 0}\,,$$

where $\mathscr{A} = \partial \mathscr{L}/\partial \omega = \frac{1}{2} m \omega T a_0^2$ is known as the *wave action*. Its existence and conservation is common to a wide variety of nondissipative dynamical systems. Its physical significance for a number of fluid dynamical ones will be discussed in §§5 and 6.

5. **Wave energy and wave action** [Bretherton and Garrett, 1968].

5.1. The relationship of wave energy to the total energy of a fluid dynamical system is often more subtle than some elementary accounts would indicate. The simplest approach is to define the wave energy using the linearised equations of motion, leaving the more delicate matter of the energy of the basic state to a separate discussion. It is best described by examples. The method is, after obvious modifications, generally applicable, but the qualifications concerning the interpretation of the formalism are important, and the restrictions which may arise on its consistency must be carefully considered in each case.

5.2. *Wave energy for a stretched string*. Linearised equation:

$$m(x)(\partial^2\eta/\partial t^2) - T(t)(\partial^2\eta/\partial x^2) = f(x, t).$$

Here $f(x, t)$ is an external force applied to material particles, which generates perturbations η of magnitude a. Then

$$\begin{aligned} W &= \int_{x_1}^{x_2} f\frac{\partial\eta}{\partial t}\,dx \\ &= \frac{\partial}{\partial t}\int_{x_1}^{x_2}\left\{\tfrac{1}{2}m\left(\frac{\partial\eta}{\partial t}\right)^2 + \tfrac{1}{2}T\left(\frac{\partial\eta}{\partial x}\right)^2\right\}dx \\ &\quad - T\frac{\partial\eta}{\partial t}\frac{\partial\eta}{\partial x}\bigg|_{x_1}^{x_2} - \int_{x_1}^{x_2}\tfrac{1}{2}\frac{dT}{dt}\left(\frac{\partial\eta}{\partial x}\right)^2 dx. \end{aligned}$$

If m, T are constants, the last term vanishes and in the absence of external forces this is a conservation equation. Also, when $f \neq 0$, the total change in W is indeed the work done by the external forces. Thus we identify

$$\tfrac{1}{2}m(\partial\eta/\partial t)^2 + \tfrac{1}{2}T(\partial\eta/\partial x)^2$$

as the *perturbation energy density*, and $-\ T(\partial\eta/\partial t)(\partial\eta/\partial x)$ as a *perturbation energy flux*.

If we insert a strictly sinusoidal wave

$$\eta = a\cos(kx - \omega t + \alpha)$$

and average over a wave period or wavelength, we have the *wave energy density* $\mathscr{E}$ and the *wave energy flux* $\mathscr{F}$:

$$\mathscr{E} = \tfrac{1}{4}(m\omega^2 + Tk^2)a^2, \qquad \mathscr{F} = \tfrac{1}{2}T\omega k a^2.$$

The governing equation may also be derived from Hamilton's principle

$$\delta\int\{Q + f\eta\}\,dx\,dt = 0,$$

where

$$Q(\partial\eta/\partial t, \partial\eta/\partial x, \eta; x, t) \equiv \tfrac{1}{2}m(x)(\partial\eta/\partial t)^2 - \tfrac{1}{2}T(t)(\partial\eta/\partial x)^2.$$

The Euler equation is

$$\frac{\partial Q}{\partial\eta} - \frac{\partial}{\partial t}\left(\frac{\partial Q}{\partial\eta_t}\right) - \frac{\partial}{\partial x}\left(\frac{\partial Q}{\partial\eta_x}\right) = -f.$$

Now

$$\frac{dQ}{dt} = \frac{\partial Q}{\partial t} + \frac{\partial Q}{\partial \eta_t}\frac{\partial^2 \eta}{\partial t^2} + \frac{\partial Q}{\partial \eta_x}\frac{\partial^2 \eta}{\partial x\,\partial t} + \frac{\partial Q}{\partial \eta}\frac{\partial \eta}{\partial t}$$

$$= \frac{\partial Q}{\partial t} + \frac{\partial}{\partial t}\left(\frac{\partial \eta}{\partial t}\frac{\partial Q}{\partial \eta_t}\right) + \frac{\partial}{\partial x}\left(\frac{\partial \eta}{\partial t}\frac{\partial Q}{\partial \eta_x}\right) - f\frac{\partial \eta}{\partial t};$$

$$\therefore \quad W = \int_{x_1}^{x_2} f\frac{\partial \eta}{\partial t}\,dx$$

$$= \frac{\partial}{\partial t}\int_{x_1}^{x_2}\left(\frac{\partial \eta}{\partial t}\frac{\partial Q}{\partial \eta_t} - Q\right)dx + \frac{\partial \eta}{\partial t}\frac{\partial Q}{\partial \eta_x}\bigg|_{x_1}^{x_2} + \int_{x_1}^{x_2}\frac{\partial Q}{\partial t}\,dx.$$

This expression is term for term identical with the previous one.

Thus if m, T are constants and

$$\mathcal{L} = \tfrac{1}{4}(m\omega^2 - Tk^2)\,a^2$$

is the averaged value of Q (the *averaged Lagrangian*), we have the general formula (applicable in all stationary media):

$$\boxed{\mathcal{E} = \omega(\partial\mathcal{L}/\partial\omega) - \mathcal{L}, \qquad \mathcal{F} = \omega(\partial\mathcal{L}/\partial k)}\,.$$

If $m(x)$, $T(t)$ are not constants, the situation is less clear cut. Perturbation energy is not conserved, there being an interaction per unit length

$$\partial Q/\partial t = -\tfrac{1}{2}(dT/dt)(\partial\eta/\partial x)^2$$

with the basic state. We could equally well define the perturbation energy density as $\frac{1}{2}m(\partial\eta/\partial t)^2$, with interaction $T(\partial\eta/\partial x)(\partial^2\eta/\partial x\,\partial t)$. However, the present arrangement is the only one for which the "interaction" vanishes in a uniform, time independent medium. For a slowly varying medium, there is *local* conservation of perturbation energy, and we *define* the local wave energy density and flux for a quasi-sinusoidal wavetrains by the same expressions for $\mathcal{E}$ and $\mathcal{F}$ in terms of the local values of ω, k, a, m and T. Similarly for $\mathcal{L}$.

Note. (i) The distinguishing feature of Hamilton's principle, as opposed to other variational principles, is that under all permitted variations $\delta\int Q\,dx$ is the virtual work done by the system on its external environment. Only when such a positive identification can be made, are we assured of the physical interpretation of the averaged Lagrangian $\mathcal{L}$ and the wave energy $\omega(\partial\mathcal{L}/\partial\omega) - \mathcal{L}$.

(ii) $\mathcal{E}$ is $O(a^2)$, derived from an equation correct only to $O(a)$. It is *vital* that the state $\eta = 0$ should be one of equilibrium at rest. If an external force F were required (independent of a) to maintain $\eta = 0$, the total external force on a material particle is $F + f$, and to calculate the work done on the system we would need to know changes in η correct to $O(a^2)$. If in the basic state all particles had an $O(1)$ transverse velocity V, we would need to know f correct to $O(a^2)$. The manipulations may still be made, but their interpretation is obscure.

(iii) As shown in §4 from the averaged variational principle, for quasi-sinusoidal wavetrains

$$\mathscr{L} = 0, \qquad \gamma = -\frac{\partial \mathscr{L}}{\partial k} \bigg/ \frac{\partial \mathscr{L}}{\partial \omega}.$$

Thus, using the general formula for stationary media,

$$\boxed{\mathscr{F} = \gamma \mathscr{E}},$$

i.e. the wave energy flux is equal to the wave energy density times the group velocity.

(iv) Also from the general formula the *wave action* is

$$\boxed{\mathscr{A} = \partial \mathscr{L}/\partial \omega = \mathscr{E}/\omega}$$

and we may compute changes in amplitude along rays from the conservation equation for wave action in the form

$$\frac{\partial}{\partial t}\left(\frac{\mathscr{E}}{\omega}\right) + \frac{\partial}{\partial x}\left(\gamma \frac{\mathscr{E}}{\omega}\right) = 0.$$

In a moving medium, the interpretation of $\mathscr{F}$ and $\partial \mathscr{L}/\partial \omega$ need modification, as we shall now see.

5.3. *Wave energy for internal gravity waves.* The linearised equations for a perturbation caused by external forces about a state $\boldsymbol{U}(z) = (U, 0, 0)$, $N^2(z)$ are

$$((\partial/\partial t) + U(\partial/\partial x))\boldsymbol{u}' + w'(d\boldsymbol{U}/dz) + \nabla p' + \sigma' \boldsymbol{n} = \boldsymbol{f},$$

$$\nabla \cdot \boldsymbol{u}' = 0,$$

$$(\partial/\partial t + U(\partial/\partial x)\sigma' - N^2 w' = 0.$$

Multiplying (1) by $\boldsymbol{u}'$, (3) by σ', integrating over a volume V bounded by the surface Σ:

$$W = \int_V \boldsymbol{f} \cdot \boldsymbol{u}' \, dx$$

$$= \frac{\partial}{\partial t} \int_V \left\{ \tfrac{1}{2}|\boldsymbol{u}'|^2 + \tfrac{1}{2}\frac{\sigma'^2}{N^2} \right\} d\boldsymbol{x} + \int_\Sigma \left\{ \left(\tfrac{1}{2}|\boldsymbol{u}'|^2 + \tfrac{1}{2}\frac{\sigma'^2}{N^2} \right) \boldsymbol{U} + p'\boldsymbol{u}' \right\} d\boldsymbol{S}$$

$$+ \int_V u'w' \frac{dU}{dz} \, d\boldsymbol{x}.$$

If $U = 0$, we have conservation of perturbation energy with:
Perturbation energy density $\tfrac{1}{2}|\boldsymbol{u}'|^2 + \tfrac{1}{2}N^{-2}\sigma'^2$,
Perturbation energy flux $p'\boldsymbol{u}'$.
Inserting a sinusoidal wave

$$\psi = a \cos(kx + mz - \omega t + \alpha),$$

we find

$$\boxed{\mathscr{E} = \tfrac{1}{2}(k^2 + m^2)a^2},$$

where the average kinetic and potential energies are equal, and

$$\mathscr{F} = \frac{1}{2} N \frac{|k|m}{k(k^2 + m^2)^{1/2}} (m, 0, -k),$$

which may be shown to equal $\boldsymbol{\gamma}\mathscr{E}$. If $dU/dz \neq 0$, W includes $u'w'(dU/dz)$. This describes the rate of transfer of energy/unit volume to the perturbation from the mean flow via the Reynolds stress. It is no longer obvious whether the perturbation energy flux should be regarded as $(\tfrac{1}{2}|\boldsymbol{u}'|^2 + \tfrac{1}{2}\sigma'^2/N'^2)\boldsymbol{U} + p'\boldsymbol{u}'$ with interaction $u'w'(dU/dz)$, or as $(\tfrac{1}{2}|\boldsymbol{u}'|^2 + \tfrac{1}{2}\sigma'^2/N^2 + u'w')\boldsymbol{U} + p'\boldsymbol{u}'$ with interaction $-U(d/dz)/(u'w')$. However, when $U(z)$ is slowly varying, the first alternative gives a local conservation equation, but unless we choose a frame of reference in which the local value of $\boldsymbol{U}$ vanishes, the conserved quantity is still not perturbation energy. As will be seen later, the most satisfactory procedure is to adopt the above expression for $\mathscr{E}$ under all circumstances—so that it is the wave energy density in the *local basic frame of reference* (moving with velocity $\boldsymbol{U}(z)$). We then define

$$\mathscr{F} = \boldsymbol{\gamma}\mathscr{E}.$$

To exploit conservation of wave action, we need the appropriate form for Hamilton's principle. This is not readily available in the Eulerian formulation used so far, but we shall see that conservation of wave action becomes

$$(\partial/\partial t)(\mathscr{E}/\omega^+) + (\partial/\partial \boldsymbol{x}) \cdot (\gamma(\mathscr{E}/\omega^+)) = 0,$$

where $\omega^+ = \omega - \boldsymbol{U} \cdot \boldsymbol{\kappa}$ is the basic frequency for the wave. This general result can be established only after the complete theory of slowly varying wavetrains has been demonstrated to be consistent for the particular medium under consideration, but it appears to be applicable whenever it is sensible to adopt the concept of a basic frame of reference in which $\boldsymbol{U}$ is negligibly small compared to the phase speed over a local volume of several wavelengths on a time scale of several wave periods.

6. **Radiation stress, wave energy, and the mean flow.**

6.1. Waves of magnitude $O(a)$ cause $O(a^2)$ changes in the mean state of the medium. These are not described by the linearised equations. Yet changes in mean energy are comparable to the wave energy. Waves also exert forces on the medium. The concept of *radiation stress* was discussed in detail for water waves by Longuet-Higgins and Stewart [1964], but the full generality has only recently become apparent. Key stages are finding the most appropriate definition of the mean state, and finding an appropriate formulation of Hamilton's principle. The interrelationships between the waves and the mean flow can then be fully elucidated in a general manner for a large variety of waves in fluid dynamics. It turns out to be important to adopt a Lagrangian viewpoint, following fluid particles rather than looking at velocities at a fixed point. We illustrate using sound in a perfect compressible fluid as an example.

6.2. *The Lagrangian equations for a perfect fluid* (e.g. [Serrin, 1962]).

Kinematics:

$$\boldsymbol{X} = \boldsymbol{X}(\tilde{\boldsymbol{X}}, T)$$

for each fluid particle labelled by $\tilde{\boldsymbol{X}}$, where $\boldsymbol{X}$ is the current position, $\tilde{\boldsymbol{X}}$ is the initial position, and T is the time.

Conservation of mass:

$$R = \tilde{R}(\tilde{\boldsymbol{X}})\{\partial(\boldsymbol{X})/\partial(\tilde{\boldsymbol{X}})\}^{-1},$$

where R is the current density, $\tilde{R}(\tilde{\boldsymbol{X}})$ is the initial density, and $\partial(\boldsymbol{X})/\partial(\tilde{\boldsymbol{X}})$ is the Jacobian.

Conservation of entropy:

$$S = \tilde{S}(\tilde{\boldsymbol{X}}),$$

where S is the entropy per unit mass, and $\tilde{S}(\tilde{\boldsymbol{X}})$ is the initial value.

Equation of state:

$$P = R^2(\partial V/\partial R)_S;$$

$V(R, S)$ = internal energy per unit mass, where P is the pressure.

Hamilton's principle:

$$\delta A \equiv \delta \int \left\{ \frac{1}{2}\left|\frac{\partial \boldsymbol{X}}{\partial T}\right|^2 - V\left(\tilde{R}\,\frac{\partial(\tilde{\boldsymbol{X}})}{\partial(\boldsymbol{X})}, \tilde{S}\right)\right\} \tilde{R}\,d\tilde{\boldsymbol{X}}\,dT = 0$$

for all smooth $\delta \boldsymbol{X}(\tilde{\boldsymbol{X}}, T)$ of compact support. This yields

Equation of motion:

$$\tilde{R}\,\frac{\partial^2 X_i}{\partial T^2} + \sum_{\alpha=1}^{3} \frac{\partial}{\partial \tilde{X}_\alpha}\left\{\frac{\partial(X_{i+1}, X_{i+2})}{\partial(\tilde{X}_{\alpha+1}, \tilde{X}_{\alpha+2})} P\right\} = 0.$$

Note. (i) Conservation of mass and entropy as well as the equation of state are identities, part of the geometrical constraints on the system, and are not violated during the variations permitted under Hamilton's principle.

(ii) Each fluid particle is labelled by its initial position $\tilde{\boldsymbol{X}}$. Yet different particles with the same R, S are dynamically indistinguishable, so redundant information is being carried. This is why Lagrangian equations are inconvenient to use in practice. However, the description of individual particle trajectories (explicit or implied) is crucial for Hamilton's principle, for permitted variations are restricted by the condition $\delta \boldsymbol{X} = 0$ for all $t < t_1$ and $t > t_2$. Merely requiring variations in velocity to vanish there is not enough; there is also a constraint on the integrated time history of the velocity field between times t_1 and t_2. This appears to be the underlying difficulty in Eulerian formulations of Hamilton's principle.

6.3. *The modified Lagrangian description* (for the basic idea see Eckart, [1962]). Now label each particle not by its *initial* position $\tilde{\boldsymbol{X}}$, but by the position $\boldsymbol{x}$ it would have at the relevant time t, had it always moved with a certain *reference flow*. This hypothetical reference flow (still to be identified) is a kinematically possible transition from the initial state, satisfying conservation of mass and entropy but not the equation of motion:

Reference quantity		Solution of	Initial conditions
Trajectory:	$\boldsymbol{x} = \boldsymbol{x}(\tilde{\boldsymbol{X}}, t)$	$dx/dt = \boldsymbol{u}(\boldsymbol{x}, t)$	$\boldsymbol{x}(0) = \tilde{\boldsymbol{X}}$
Density:	$\rho(\boldsymbol{x}, t)$	$\partial\rho/\partial t + (\partial/\partial \boldsymbol{x})\cdot(\rho \boldsymbol{u}) = 0$	$\rho(\boldsymbol{x}, 0) = \tilde{R}(\boldsymbol{x})$
Entropy/unit mass:	$s(\boldsymbol{x}, t)$	$\partial s/\partial t + \boldsymbol{u}\cdot\partial s/\partial \boldsymbol{x} = 0$	$s(\boldsymbol{x}, 0) = \tilde{S}(\boldsymbol{x})$
Pressure:	$p(\boldsymbol{x}, t)$	$p = \rho^2(\partial V/\partial \rho)_s$	$V(\rho, s)$ given

Now use (x, t) as independent variables instead of $(\tilde{X}, T)$.

Note. (i) $t = T$, the dual notation being useful in distinguishing partial derivatives.

(ii) The derivative following a particle is

$$D/Dt \equiv \partial/\partial t + \boldsymbol{u} \cdot \partial/\partial \boldsymbol{x} \equiv (\partial/\partial T).$$

(iii) $\boldsymbol{u}(\boldsymbol{x}, t)$ is still arbitrary. We will make a special choice shortly. Then Hamilton's principle:

$$\delta \int \left\{ \frac{1}{2} \left| \frac{D\boldsymbol{X}}{Dt} \right|^2 - V\left(\rho \frac{\partial(\boldsymbol{x})}{\partial(\boldsymbol{X})}, s \right) \right\} \rho \, d\boldsymbol{x} \, dt$$

for all $\delta \boldsymbol{X}(\boldsymbol{x}, t)$ of compact support, giving

$$\boxed{\rho \frac{D^2 X_i}{Dt^2} + \sum_{\alpha=1}^{3} \frac{\partial}{\partial x_\alpha} \left\{ \frac{\partial(X_{i+1}, X_{i+2})}{\partial(x_{\alpha+1}, x_{\alpha+2})} P \right\} = 0}, \qquad i = 1, 2, 3.$$

Note. (iv) Consider reference volume V, bounded by surface $\mathscr{S}$. Fluid particles referred to actually occupy V', bounded by $\mathscr{S}'$, where $V \leftrightarrow V'$: $\boldsymbol{x} \leftrightarrow \boldsymbol{X}(\boldsymbol{x})$. (See Figure 16.)

Mass fluid concerned $= \int_V \rho \, d\boldsymbol{x}$.

Net force on fluid concerned $= \int_V \rho (D^2 \boldsymbol{X}/Dt^2) \, d\boldsymbol{x}$.

But element of area $\mathscr{S}$:

$$d\mathscr{S}_i' = \sum_\alpha \frac{\partial(X_{i+1}, X_{i+2})}{\partial(x_{\alpha+1}, x_{\alpha+2})} d\mathscr{S}_\alpha.$$

Net pressure force on $\mathscr{S}'$

$$= -\int_{\mathscr{S}'} P \, d\mathscr{S}_i' = -\int_V \sum_\alpha \frac{\partial}{\partial x_\alpha} \left\{ \frac{\partial(X_{i+1}, X_{i+2})}{\partial(x_{\alpha+1}, x_{\alpha+2})} P \right\} d\boldsymbol{x}.$$

6.4. *Slowly varying wavetrains in a moving medium.* Now identify the

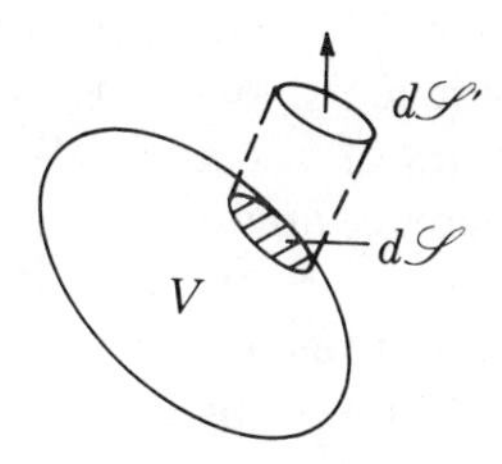

FIGURE 16.

reference flow as the mean state of the medium. Write

$$\boldsymbol{X}(\boldsymbol{x}, t) = \boldsymbol{x} + \boldsymbol{\xi}(\boldsymbol{x}, t),$$

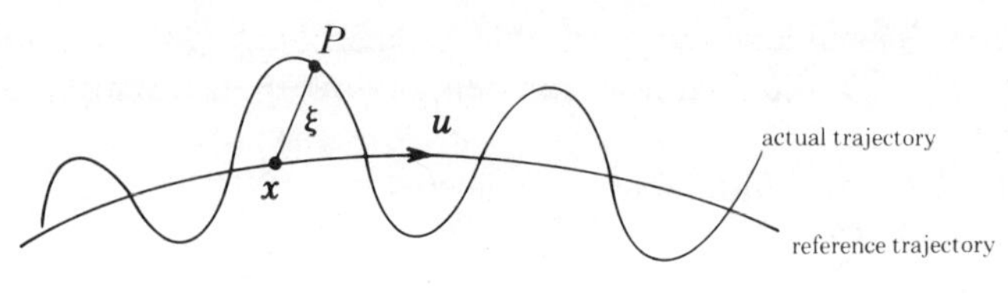

FIGURE 17.

where $\boldsymbol{X}$ is the actual position, $\boldsymbol{x}$ is the reference point, and ξ the particle displacement. (See Figure 17.) Choose the reference trajectories as the running average positions of material particles, found by averaging locally over a large number of wavelengths using a suitably smooth weighting function which vanishes outside a local interval, e.g.

$$\bar{f}(\boldsymbol{x}, t) = \frac{\varepsilon^{1/2}}{C} \int_{t-(\varepsilon)^{1/2}}^{t+(\varepsilon)^{1/2}} f(\boldsymbol{x}, t') \exp\left(-\frac{\varepsilon}{\varepsilon-(t-t')^2}\right) dt';$$

$$C = \int_{-1}^{1} \exp\left(-\frac{1}{1-\lambda^2}\right) d\lambda.$$

Thus we suppose that $\xi(\boldsymbol{x}, t)$ is everywhere small ($O(a)$) and quasi-sinusoidal, and that $O(a^2)$ and higher order corrections forced by nonlinearities consist solely of quasi-sinusoidal harmonics of the fundamental wave frequency, with zero mean value. Formally

$$\xi = a\mathscr{R}\{({}_1\hat{\xi}_0 + \varepsilon\,{}_1\hat{\xi}_1 + \cdots)\exp(i\varepsilon^{-1}\theta)\}$$
$$+ a^2\mathscr{R}\{({}_2\hat{\xi}_0 + \varepsilon\,{}_2\hat{\xi}_1 + \cdots)\exp(2i\varepsilon^{-1}\theta)\} + O(a^2),$$

where $O(a^3)$ terms involve $\exp(3i\varepsilon^{-1}\theta)$ and $\exp(i\varepsilon^{-1}\theta)$, $O(a^4)$ terms involve $\exp(4i\varepsilon^{-1}\theta)$ and $\exp(2i\varepsilon^{-1}\theta)$, etc. Then $\boldsymbol{u}(\boldsymbol{x}, t)$ is the Lagrangian mean (or drift) velocity of fluid particles in the presence of waves, *correct to all orders in* a, ε.

For a given physical wave field and flow pattern, this postulate completely determines the reference velocities $\boldsymbol{u}$ (at least asymptotically as $a, \varepsilon \to 0$) and hence the values of ρ, s, and p. We define the *mean state* as that associated with this reference flow. Conservation of mass and entropy, as well as the equation of state, are exact for the mean flow defined this way, but its equation of motion remains to be determined.

Note. (i) ρ is the mass referred to by unit volume, computed from the mass of those particles whose Lagrangian mean lies within a given region. It is subtly different from the time averaged density following an individual particle, and from the Eulerian mean density, but in practice these differences are usually negligible.

(ii) $s(\boldsymbol{x}, t)$ is precisely the entropy of the particle whose current running mean position is at $\boldsymbol{x}$, usually negligibly different from the Eulerian mean.

(iii) $\boldsymbol{u}(\boldsymbol{x}, t)$ differs from the Eulerian mean normally by $O(a^2)$, a significant amount depending on the local wave structure (the Stokes drift).

(iv) Velocities and accelerations of fluid particles are

$$DX/Dt = \boldsymbol{u} + D\xi/Dt, \qquad D^2X/Dt^2 = D\boldsymbol{u}/Dt + D^2\xi/Dt^2,$$

where $D/Dt = \partial/\partial t + \boldsymbol{u}\cdot\partial/\partial\boldsymbol{x}$, *exactly*.

6.5. *Radiation stress.* We obtain the equations of motion for the mean flow, first by expanding

$$\rho\frac{D^2X_i}{Dt^2} + \frac{\partial}{\partial x_\alpha}\left\{\frac{\partial(X_{i+1}, X_{i+2})}{\partial(x_{\alpha+1}, x_{\alpha+2})}P\right\} = 0$$

in powers of a and ε. Strictly speaking the reference flow should be expanded as well, but no confusion will arise.

We need

$$\frac{\partial(\boldsymbol{X})}{\partial(\boldsymbol{x})} = \frac{1}{6}\varepsilon_{ijk}\varepsilon_{\alpha\beta\gamma}\frac{\partial X_i}{\partial x_\alpha}\frac{\partial X_j}{\partial x_\beta}\frac{\partial X_k}{\partial x_\gamma} \qquad \text{(summation convention)}$$

$$= 1 + \frac{\partial\xi_k}{\partial x_k} + \frac{1}{2}\left(\frac{\partial\xi_i}{\partial x_i}\frac{\partial\xi_j}{\partial x_j} - \frac{\partial\xi_j}{\partial x_i}\frac{\partial\xi_i}{\partial x_j}\right) + \frac{\partial(\xi)}{\partial(\boldsymbol{x})},$$

and

$$\frac{\partial(X_{i+1}, X_{i+2})}{\partial(x_{\alpha+1}, x_{\alpha+2})} = \tfrac{1}{2}\varepsilon_{\alpha\beta\gamma}\varepsilon_{ijk}\frac{\partial X_j}{\partial x_\beta}\frac{\partial X_k}{\partial x_\gamma}$$

$$= \delta_{i\alpha} + \left(\delta_{i\alpha}\frac{\partial\xi_k}{\partial x_k} - \frac{\partial\xi_\alpha}{\partial x_i}\right) + \frac{\partial(\xi_{i+1}, \xi_{i+2})}{\partial(x_{\alpha+1}, x_{\alpha+2})}.$$

Then

$$P = p\left(\frac{\partial(\boldsymbol{x})}{\partial(\boldsymbol{X})}\rho, s\right)$$

$$= p(\rho, s) - \rho\frac{\partial p}{\partial\rho}\frac{\partial\xi_i}{\partial x_i} + \tfrac{1}{2}\rho\frac{\partial p}{\partial\rho}\left\{\left(\frac{\partial\xi_k}{\partial x_k}\right)^2 + \frac{\partial\xi_j}{\partial x_i}\frac{\partial\xi_i}{\partial x_j}\right\} + \tfrac{1}{2}\rho^2\frac{\partial p}{\partial\rho^2}\left(\frac{\partial\xi_i}{\partial x_i}\right)^2$$

$$+ O(\xi^3).$$

Expanding first in the amplitude:

$$O(a^0): \qquad \rho(D\boldsymbol{u}/Dt) + (\partial p/\partial\boldsymbol{x}) = 0,$$

so if $O(a^2)$ forces are neglected, the mean flow must satisfy the Eulerian equation for conservation of momentum.

$$O(a): \quad \rho \frac{D^2}{Dt^2}\,{}_1\xi_i - \rho c^2 \frac{\partial}{\partial x_i}\left(\frac{\partial\,{}_1\xi_k}{\partial x_k}\right) + \frac{\partial}{\partial x_i}(p - \rho c^2)\frac{\partial\,{}_1\xi_k}{\partial x_k} - \frac{\partial p}{\partial x_j}\frac{\partial \xi_j}{\partial x_i} = 0.$$

This is the correctly linearised equation for disturbances on a general flow $\boldsymbol{u}, \rho, s$ where $c^2 = (\partial p/\partial \rho)_s$. When the medium is uniform, it reduces to the wave equation

$$\rho \frac{D^2}{Dt^2}\,{}_1\boldsymbol{\xi}_0 - \rho c^2 \frac{\partial}{\partial \boldsymbol{x}}\left(\frac{\partial}{\partial \boldsymbol{x}}\cdot {}_1\boldsymbol{\xi}_0\right) = 0.$$

Thus we have the local structure for longitudinal waves

$${}_1\boldsymbol{\xi}_0 = a_0\varepsilon(\boldsymbol{\kappa}/|\boldsymbol{\kappa}|)\cos\{\varepsilon^{-1}(\boldsymbol{\kappa}\cdot\boldsymbol{x} - \omega t) + \alpha_0\},$$

and the local dispersion relation

$$\boxed{\omega = \boldsymbol{u}\cdot\boldsymbol{\kappa} + c|\boldsymbol{\kappa}|}.$$

(See Figure 18.)

Note. (i) The amplitude for the displacement ${}_1\hat{\boldsymbol{\xi}}_0$ is taken as $a\varepsilon$, so that the velocities and wave energy are formally $O(a^2\varepsilon^0)$.

$O(a^2)$: There is no contribution from $\rho(D^2X/Dt^2)$. Thus we see at once that the effect of the waves on the mean flow is summed up by a *radiation*

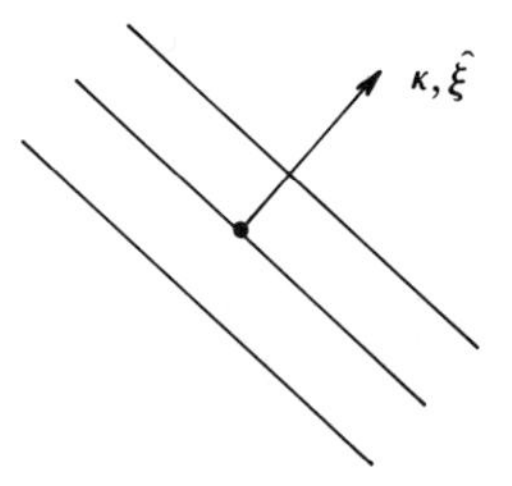

FIGURE 18.

stress tensor $\mathcal{R}_{i\alpha}$, which is the local average of

$$p\delta_{i\alpha} - (\partial(X_{i+1}, X_{i+2})/\partial(x_{\alpha+1}, x_{\alpha+2}))P$$

computed correct to $O(a^2)$ as if the wavetrain were perfectly sinusoidal:

$$\boxed{\mathscr{R}_{ij} = -\tfrac{1}{2}\rho a_0^2 c^2 \left(\kappa_i \kappa_j + \frac{\rho}{c}\frac{\partial c}{\partial \rho} |\boldsymbol{\kappa}|^2 \delta_{ij} \right)},$$

$$\boxed{\rho \frac{Du_i}{Dt} + \frac{\partial p}{\partial x_i} = \frac{\partial}{\partial x_j}\mathscr{R}_{ij}} \qquad \text{correct to } O(a^2 \varepsilon^0).$$

Note. (ii) The computation of $\mathscr{R}_{ij}$ is simplified by noting that any expression like

$$\frac{\partial(\xi_{i+1}, \xi_{i+2})}{\partial(x_{\alpha+1}, x_{\alpha+2})} = \frac{\partial}{\partial x_\beta}\left\{ \frac{1}{2}\varepsilon_{\alpha\beta\gamma}\varepsilon_{ijk}\overline{\xi_j \frac{\partial \xi_k}{\partial x_k}} \right\} = 0,$$

for a homogeneous wavetrain (because $\overline{\xi_j(\partial \xi_k/\partial x_\gamma)}$ is uniform), and hence is negligible for a slowly varying wavetrain.

(iii) (See Figure 19.) The radiation stress is simply the contribution from the waves to the pressure force acting across the interface between two sets of fluid particles from region 1 to region 2. If the normal to the mean position of the interface is $\boldsymbol{n}$, $\mathscr{R}_{ij}n_j$ is not usually $\| \boldsymbol{n}$, because the instantaneous interface is corrugated, and the pressure perturbations are correlated with deviations of the normal. $\mathscr{R}_{ij}$ is only in part associated with the change $\bar{P} - p$ of the average pressure on each particle. From this Lagrangian standpoint, the Reynolds stress does not appear separately identifiable.

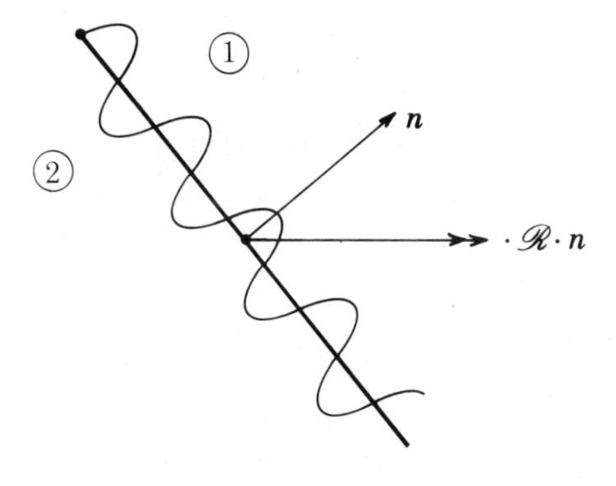

FIGURE 19.

(iv) For sound waves

$$\mathscr{E} = \tfrac{1}{2}\rho a_0^2 c^2 |\boldsymbol{\kappa}|^2$$

so

$$\boxed{\mathscr{R}_{ij} = -\mathscr{E}\left(\frac{\kappa_i \kappa_j}{|\boldsymbol{\kappa}|^2} + \frac{\rho}{c}\frac{\partial c}{\partial \rho}\delta_{ij}\right)}.$$

(v) The particle at $\boldsymbol{x}$ has reference position

$$\boldsymbol{x}' = \boldsymbol{x} - \boldsymbol{\xi}(\boldsymbol{x}') = \boldsymbol{x} - \boldsymbol{\xi}(\boldsymbol{x}) + (\boldsymbol{\xi}\cdot(\partial/\partial \boldsymbol{x}))\boldsymbol{\xi} + O(\xi^3).$$

The Eulerian mean velocity at $\boldsymbol{x}$ is

$$\boldsymbol{u}_E = \overline{\boldsymbol{u}(\boldsymbol{x}') + \frac{D}{Dt}\boldsymbol{\xi}|_{\boldsymbol{x}}},$$

$$\sim \boldsymbol{u}(\boldsymbol{x}) - \overline{\left(\boldsymbol{\xi}\cdot\frac{\partial}{\partial \boldsymbol{x}}\right)\frac{D}{Dt}\boldsymbol{\xi}} = \boldsymbol{u}(\boldsymbol{x}) - \tfrac{1}{2}a^2 c|\boldsymbol{\kappa}|\boldsymbol{\kappa} + O(a^3, a^2\varepsilon),$$

where all gradients of $\boldsymbol{u}$ and other mean quantities are $O(\varepsilon)$ smaller and are ignored. Thus for sound waves the Stokes drift:

$$\boxed{\boldsymbol{u} - \boldsymbol{u}_E = (\mathscr{E}/c)(\boldsymbol{\kappa}|\boldsymbol{\kappa}|)}.$$

6.6. *The averaged variational principle* (see also §4.5). Inserting a slowly varying wavetrain

$$\boldsymbol{\xi} = \tfrac{1}{2}a(\varepsilon\ {}_1\hat{\boldsymbol{\xi}}_0 \exp(i\varepsilon^{-1}\theta) + \text{complex conjugate}) + O(a\varepsilon^2, a^2\varepsilon)$$

into Hamilton's principle

$$\delta A \equiv \delta \int \left\{\frac{1}{2}\left|\frac{D\boldsymbol{X}}{Dt}\right|^2 - V\left(\rho\frac{\partial(\boldsymbol{x})}{\partial(\boldsymbol{X})}, s\right)\right\}\rho\, d\boldsymbol{x}\, dt = 0,$$

after a little manipulation

$$\boxed{\begin{aligned}\delta A = \delta &\int \{\tfrac{1}{2}|\boldsymbol{u}|^2 - V(\rho, s)\}\rho\, d\boldsymbol{x}\, dt \\ &+ \delta \int \mathscr{L}\left(-\frac{\partial\theta}{\partial t}, \frac{\partial\theta}{\partial x}, a, {}_1\hat{\boldsymbol{\xi}}_0; \boldsymbol{u}, \rho, c\right) d\boldsymbol{x}\, dt\end{aligned}} + O(a^4, a^2\varepsilon).$$

Averaged Lagrangian:

$$\mathscr{L}(\omega, \boldsymbol{\kappa}, a, \hat{\xi}; \boldsymbol{u}, \rho, c) = \tfrac{1}{4}\rho\{(\omega - \boldsymbol{u}\cdot\boldsymbol{\kappa})^2\hat{\xi}\cdot\hat{\xi}^* - c^2(\boldsymbol{\kappa}\cdot\hat{\xi})(\boldsymbol{\kappa}\cdot\hat{\xi}^*)\}$$

where * denotes complex conjugate. To apply Hamilton's principle vary everything in sight, and obtain necessary conditions on the lowest order approximation for the slowly varying wavetrain.

$\delta\,{}_1\hat{\xi}_0$: determines the local wave *structure*

$$\partial\mathscr{L}/\partial_1\xi_0 = \tfrac{1}{4}\rho\{(\omega - \boldsymbol{u}\cdot\boldsymbol{\kappa})^2{}_1\hat{\xi}_0^* - c^2(\boldsymbol{\kappa}\cdot{}_1\hat{\xi}_0^*)\boldsymbol{\kappa}\}a^2 = 0.$$

This implies ${}_1\hat{\xi}_0 \| \boldsymbol{\kappa}$, i.e. the waves are longitudinal. The magnitude $|{}_1\hat{\xi}_0|$ is arbitrary; normalise to 1. Then $\partial\mathscr{L}/\partial\omega = \tfrac{1}{2}\rho(\omega - \boldsymbol{u}\cdot\boldsymbol{\kappa})a^2$ $(= \mathscr{A}$ say). This structure is possible only if dispersion relation is satisfied:

$$\omega - \boldsymbol{u}\cdot\boldsymbol{\kappa} = \pm\, c|\boldsymbol{\kappa}|.$$

Without loss of generality we restrict attention to one mode, the positive sign. If dispersion relation is satisfied, $\boldsymbol{\kappa}$ lies on a surface $\mathscr{S}$ in $\boldsymbol{\kappa}$-space. Near $\mathscr{S}$: $\mathscr{L} = \mathscr{A}(\omega - \boldsymbol{u}\cdot\boldsymbol{\kappa} - c|\boldsymbol{\kappa}|)$.

If desired, this information may be used to reformulate the variational principle into a standard form. This is not necessary, but exposes more elegantly the general nature of the results.

Modified Lagrangian:

$$\boxed{\mathscr{L}(\omega, \boldsymbol{\kappa}, \mathscr{A}; \boldsymbol{u}, \rho, c) = \mathscr{A}(\omega - \boldsymbol{u}\cdot\boldsymbol{\kappa} - c|\boldsymbol{\kappa}|)}$$

where variations $\delta\mathscr{A}$ are used instead of δa, and the local structure is assumed known, i.e.

$${}_1\hat{\xi}_0 = \boldsymbol{\kappa}/|\boldsymbol{\kappa}|, \qquad a^2 = 2\mathscr{A}\rho^{-1}/(\omega - \boldsymbol{u}\cdot\boldsymbol{\kappa}).$$

Note. (i) It is somewhat subtle that information derived from a variational principle may be used in this way to modify it.

$\delta\mathscr{A}$: yields the dispersion relation for the mode under consideration

$$\partial\mathscr{L}/\partial\mathscr{A} = \mathscr{L}/\mathscr{A} = \omega - \boldsymbol{u}\cdot\boldsymbol{\kappa} - c|\boldsymbol{\kappa}| = 0.$$

Thus

$$\gamma = -\frac{\partial\mathscr{L}}{\partial\boldsymbol{\kappa}}\Big/\frac{\partial\mathscr{L}}{\partial\omega} = \boldsymbol{u} + c\boldsymbol{\kappa}/|\boldsymbol{\kappa}|.$$

$\delta\theta$: yields conservation of wave action

$$\partial\mathscr{A}/\partial t + (\partial/\partial\boldsymbol{x})\cdot(\gamma\mathscr{A}) = 0.$$

$\delta(\boldsymbol{u}, \rho, s)$: changes in $\boldsymbol{u}$, ρ, s are constrained by the reference flow being a kinematically possible transition from the initial state, conserving particle identity, mass, entropy and the equation of state. This constraint is characterised (for further information see Bretherton [1970]) by setting as permitted variations

$$\delta \boldsymbol{u} = (D/Dt)\Delta \boldsymbol{x} - (\Delta \boldsymbol{x} \cdot (\partial/\partial \boldsymbol{x}))\boldsymbol{u},$$

$$\delta \rho = -\rho(\partial/\partial \boldsymbol{x}) \cdot \Delta \boldsymbol{x} - (\Delta \boldsymbol{x} \cdot (\partial/\partial \boldsymbol{x}))\rho,$$

$$\delta s = -(\Delta \boldsymbol{x} \cdot (\partial/\partial \boldsymbol{x}))s,$$

$$\delta p = (\partial p/\partial \rho)\delta \rho + (\partial p/\partial s)\delta s,$$

where $\Delta \boldsymbol{x}(\boldsymbol{x}, t)$ is smooth and of compact support. $\Delta \boldsymbol{x}$ is the particle displacement under the variation—the fundamental quantity in the Lagrangian statement of Hamilton's principle. It vanishes for large positive or negative t, i.e. the ends of particle trajectories are held fixed.

Note. (ii) We must distinguish between variations Δ following fluid particles, and variations δ at given $\boldsymbol{x}$:

$$\Delta \equiv \delta + \Delta \boldsymbol{x} \cdot (\partial/\partial \boldsymbol{x}).$$

Variations Δ, e.g. $(1/\rho)\Delta\rho = -(\partial/\partial \boldsymbol{x}) \cdot \Delta \boldsymbol{x}$, are those which we might at first have expected.

(iii)

$$\delta\theta = \Delta\theta - \Delta \boldsymbol{x} \cdot (\partial\theta/\partial \boldsymbol{x})$$

and

$$\delta\kappa_i = (\partial/\partial x_i)(\delta\theta).$$

$$\therefore \quad \Delta\kappa_i = (\partial/\partial x_i)(\Delta\theta - \Delta x_j(\partial\theta/\partial x_j)) + \Delta x_j(\partial\kappa_i/\partial x_j)$$
$$= (\partial/\partial x_i)\Delta\theta - \kappa_j(\partial/\partial x_i)\Delta x_j.$$

Thus variations in wavenumber on a given fluid particle are only in part attributable to variations in phase $\Delta\theta$ on that particle. There is also a contribution due to the distortion of relative positions in the original phase gradient. Similarly

$$\Delta\omega = -(\partial/\partial t)\Delta\theta + \boldsymbol{\kappa} \cdot (\partial/\partial t)\Delta \boldsymbol{x}.$$

(iv) If $\Delta \boldsymbol{x} = \boldsymbol{u}\delta t$, δt infinitesimal constant, we obtain the identities

$$\partial \boldsymbol{u}/\partial t = (D/Dt)\boldsymbol{u} - (\boldsymbol{u} \cdot (\partial/\partial \boldsymbol{x}))\boldsymbol{u},$$

$$\partial\rho/\partial t = -\rho(\partial/\partial \boldsymbol{x}) \cdot \boldsymbol{u} - \boldsymbol{u} \cdot (\partial\rho/\partial \boldsymbol{x}),$$

$$\partial s/\partial t = -\boldsymbol{u} \cdot (\partial s/\partial \boldsymbol{x}),$$

showing that any reference flow can be built out of a sequence of such changes in the geometrical configuration of fluid particles, and that these variations do indeed encompass the kinematically possible changes in reference flow.

Inserting these variations:

$$\delta \int \{\tfrac{1}{2}|\boldsymbol{u}|^2 - V(\rho, s)\}\rho \, d\boldsymbol{x} \, dt$$

$$= \int \left[\Delta\{\tfrac{1}{2}\rho|\boldsymbol{u}|^2 - \rho V\} - \Delta\boldsymbol{x} \cdot \frac{\partial}{\partial \boldsymbol{x}} \{\tfrac{1}{2}\rho|\boldsymbol{u}|^2 - \rho V\} \right] d\boldsymbol{x} \, dt$$

$$= \int \left[\rho\boldsymbol{u} \cdot \frac{D}{Dt}\Delta\boldsymbol{x} + \rho^2 \frac{\partial V}{\partial \rho} \frac{\partial}{\partial \boldsymbol{x}} \cdot \Delta\boldsymbol{x} - (\tfrac{1}{2}|\boldsymbol{u}|^2 - V)\rho \frac{\partial}{\partial \boldsymbol{x}} \cdot \Delta\boldsymbol{x} \right.$$

$$\left. - \Delta\boldsymbol{x} \cdot \frac{\partial}{\partial \boldsymbol{x}} \{\tfrac{1}{2}\rho|\boldsymbol{u}|^2 - \rho V\} \right] d\boldsymbol{x} \, dt$$

$$= \int - \left\{ \frac{\partial}{\partial t}\rho u_i + \frac{\partial}{\partial x_j}(\rho u_i u_j + p\delta_{ij}) \right\} \Delta x_i \, d\boldsymbol{x} \, dt.$$

Also

$$\partial \mathscr{L} = \frac{\partial \mathscr{L}}{\partial \omega}\Delta\omega + \frac{\partial \mathscr{L}}{\partial \boldsymbol{\kappa}}\Delta\boldsymbol{\kappa} + \frac{\partial \mathscr{L}}{\partial \mathscr{A}}\Delta\mathscr{A} + \frac{\partial \mathscr{L}}{\partial \boldsymbol{u}}\Delta\boldsymbol{u}$$

$$+ \frac{\partial \mathscr{L}}{\partial \rho}\Delta\rho - \frac{\partial \mathscr{L}}{\partial s}\Delta s - \Delta\boldsymbol{x} \cdot \frac{\partial}{\partial \boldsymbol{x}}\mathscr{L}$$

$$= \frac{\partial}{\partial t}\left\{ - \frac{\partial \mathscr{L}}{\partial \omega}\Delta\theta + \mathscr{P}_i \Delta x_i \right\} + \frac{\partial}{\partial x_j}\left\{ \frac{\partial \mathscr{L}}{\partial \kappa_j}\Delta\theta - (\mathscr{R}_{ij} + \mathscr{L}\delta_{ij})\Delta x_i \right\}$$

$$+ \left\{ \frac{\partial}{\partial t}\left(\frac{\partial \mathscr{L}}{\partial \omega}\right) \cdot \frac{\partial}{\partial \boldsymbol{x}} - \frac{\partial \mathscr{L}}{\partial \boldsymbol{\kappa}} \right\}\Delta\theta + \frac{\partial \mathscr{L}}{\partial \mathscr{A}}\Delta\mathscr{A} + \left\{ - \frac{\partial \mathscr{P}_i}{\partial t} + \frac{\partial}{\partial x_j}\mathscr{R}_{ij} \right\}\Delta x_i,$$

where

$$\mathscr{P}_i = \kappa_i(\partial \mathscr{L}/\partial \omega) + \partial \mathscr{L}/\partial u_i,$$

$$\boxed{\mathscr{R}_{ij} = \kappa_i(\partial \mathscr{L}/\partial \kappa_j) - u_j(\partial \mathscr{L}/\partial u_i) + (\rho[\partial \mathscr{L}/\partial \rho] - \mathscr{L})\delta_{ij}}.$$

The terms $(\partial/\partial t)\{\quad\}$, $(\partial/\partial x)\{\quad\}$ vanish on integration of $\delta\mathscr{L}$ over all space time, and variations $\Delta\theta$, $\Delta\mathscr{A}$ have already been accounted for. Also $\mathscr{P}_i$ vanishes identically, because the only way $\boldsymbol{u}$ enters $\mathscr{L}$ is through the *basic frequency*

$$\omega^+ = \omega - \boldsymbol{u} \cdot \boldsymbol{\kappa}.$$

Thus Hamilton's principle gives the equation of mean motion

$$\boxed{\frac{\partial}{\partial t}(\rho u_i) + \frac{\partial}{\partial x_j}(\rho u_i u_j) + \frac{\partial p}{\partial x_j} = \frac{\partial}{\partial x_j}\mathscr{R}_{ij}} + O(a^4, a^2\varepsilon).$$

Note. (v) The *only* effect of the waves on the mean flow (defined in this way) is the body force $(\partial/\partial \boldsymbol{x}) \cdot \mathscr{R}$. The remaining equations for conservation of mass, entropy, and state all stand unaltered by the presence of waves (they are identities for the reference flow).

(vi) If it did not vanish, $\mathscr{P}$ would be a wave momentum density. *All* the mean momentum of the fluid is accounted for by the mean flow $\rho \boldsymbol{u}$.

(vii) The body force $-(\partial/\partial x_j)\mathscr{R}_{ij}$ is the variational derivative $\delta\mathscr{L}/\Delta xi$, corresponding to virtual displacements of all the fluid particles equally over a volume containing many wavelengths. Wavefronts are advected with fluid particles under this displacement, and the local value $\mathscr{L}$ of the averaged Lagrangian changes. The terms in $\mathscr{R}_{ij}$ may be traced back respectively to:

(a) distortion of the wavefronts under the advection of phase,
(b) changes in $\boldsymbol{u}$,
(c) advection of the wave pattern as a whole,
(d) changes in density associated with virtual dilatations.

6.7. *The energy equation.* The energy equation for the mean flow is

$$\boxed{(\partial/\partial t)\{\tfrac{1}{2}\rho|\boldsymbol{u}|^2 + \rho V\} + (\partial/\partial \boldsymbol{x})\{(\tfrac{1}{2}\rho|\boldsymbol{u}|^2 + \rho V + p)\boldsymbol{u}\} = \boldsymbol{u}\cdot(\partial/\partial \boldsymbol{x})\mathscr{R}},$$

where $(\partial/\partial t)\{\frac{1}{2}\rho|\boldsymbol{u}|^2 + \rho V\}$ is the energy per unit volume,

$$(\partial/\partial \boldsymbol{x})\{(\tfrac{1}{2}\rho|\boldsymbol{u}|^2 + \rho V + p)\boldsymbol{u}\}$$

the energy flux, and $\boldsymbol{u}\cdot(\partial/\partial \boldsymbol{x})\mathscr{R}$ the interaction. Setting $\Delta \boldsymbol{x} = \boldsymbol{u}\,\delta t$, $\Delta\theta = -\omega t\,\Delta t$ $(\omega t = \omega - \boldsymbol{u}\cdot\boldsymbol{\kappa}))$ in the identity for $\delta\mathscr{L}$, and using

$$\mathscr{P}_i = \delta\mathscr{L}/\Delta\theta = \delta\mathscr{L}/\Delta\mathscr{A} = \mathscr{L} = 0,$$

we obtain the changes which actually occur in the real flow

$$0 = \frac{\partial\mathscr{L}}{\partial t} = \frac{\partial}{\partial t}(\omega^{+}\mathscr{L}_{\omega}) + \frac{\partial}{\partial x_j}\left\{-\omega^{+}\frac{\partial\mathscr{L}}{\partial\kappa_j} - u_i\mathscr{R}_{ij}\right\} + u_i\frac{\partial}{\partial x_j}\mathscr{R}_{ij}.$$

But as in §6

$$\omega^{+}(\partial\mathscr{L}/\partial\omega) = \mathscr{E}, \qquad -\omega^{+}(\partial\mathscr{L}/\partial\boldsymbol{\kappa}) = \gamma\mathscr{E} = \mathscr{F},$$

the wave energy density and flux respectively. The wave energy equation is

$$\boxed{\partial\mathscr{E}/\partial t + (\partial/\partial x_j)\mathscr{F}_j = \mathscr{R}_{ij}(\partial u_i/\partial x_j)}.$$

The last term is minus the rate of working by the radiation stress against the mean strain. Adding, the total energy is conserved,

$$\boxed{\frac{\partial}{\partial t}\{\tfrac{1}{2}\rho|\boldsymbol{u}|^2 + \rho V + \mathscr{E}\} + \frac{\partial}{\partial x_j}\{(\tfrac{1}{2}\rho|\boldsymbol{u}|^2 + \rho V + p)u_j + \mathscr{F}_j - \mathscr{R}_{ij}u_i\} = 0}$$

$$+O(a^4, a^2\varepsilon).$$

Note. An extra term $-u_i\mathscr{R}_{ij}$ appears in the total energy flux, in addition to the wave energy flux $\mathscr{F}_j$. This is attributable to the waves, but is probably best left conceptually separate.

6.8 *Kelvin's circulation theorem.* For a circuit Γ' in an inviscid fluid, composed of particles all of the same entropy s, the circulation

$$C' = \int_{\Gamma} \boldsymbol{u}' \cdot d\boldsymbol{l}' = \text{constant},$$

as the circuit moves with the fluid (with velocity $\boldsymbol{u}'$). For slowly varying waves in a mean flow $\boldsymbol{u}$, Γ' may be referred to by the circuit Γ defined by the running average position of the particles. This moves with velocity $\boldsymbol{u}$. Then

$$u_i' = u_i + (D/Dt)\xi_i,$$

$$dl_i' = (\delta_{ij} + \partial\xi_i/\partial x_j)\, dl_j;$$

$$\therefore \quad C' = \int_{\Gamma}\left(u_i + \frac{D}{Dt}\xi_i\right)\left(\delta_{ij} + \frac{\partial\xi_i}{\partial x_j}\right) dl_j \qquad \text{(exactly)}$$

$$= \int_{\Gamma}\left\{u_j + \overline{\frac{\partial\xi_i}{\partial x_j}\frac{D}{Dt}\xi_i}\right\} dl_j \qquad \text{(correct to all orders in } \varepsilon).$$

All linear terms vanish under the averaging procedure. Evaluating locally for sinusoidal sound waves

$$\overline{\frac{\partial\xi_i}{\partial x_j}\frac{D}{Dt}\xi_i} = -\tfrac{1}{2}\kappa_j\omega^+ a^2|\hat{\boldsymbol{\xi}}|^2 = -\kappa_j\mathscr{A}/\rho;$$

therefore

$$\boxed{C' = \int_\Gamma \{\boldsymbol{u} - \boldsymbol{\kappa}\mathscr{A}/\rho\} \cdot d\boldsymbol{l} = \text{constant}}, \quad \text{correct to } O(a^2\varepsilon^0).$$

Note. (i) This shows that when waves propagate into a previously undisturbed region, the *rotational part* of the mean flow is as if a *wave momentum density* $\boldsymbol{\kappa}\mathscr{A}$ had been added to the fluid. The *divergent part* of $\boldsymbol{u}$ cannot be pinned down in this way, presumably because pressure forces can readily redistribute the locally induced compressive effect of the waves.

(ii) In a *weakly dissipative* fluid, the wave action will no longer be conserved, and we may postulate an equation

$$(\partial/\partial t)\mathscr{A} + (\partial/\partial \boldsymbol{x}) \cdot \boldsymbol{\gamma}\mathscr{A} = -D\mathscr{A}$$

where $D(\boldsymbol{\kappa}; \rho, s)$ is a local decay constant. Then using the expression derived in §6.6 for the radiation stress, and assuming the equations of mean motion still hold even though the system is weakly dissipative (this is an intuitively attractive approximation), we have after some manipulation

$$\begin{aligned}\frac{D}{Dt}\int_\Gamma \boldsymbol{u} \cdot d\boldsymbol{l} &= \int_\Gamma \frac{\partial}{\partial x_j}(\mathscr{R}_{ij})\, dl_i \\ &= \frac{D}{Dt}\int_\Gamma (-\boldsymbol{\kappa}\mathscr{A} \cdot d\boldsymbol{l}) + \int_\Gamma D\boldsymbol{\kappa}\mathscr{A} \cdot d\boldsymbol{l}\end{aligned}$$

i.e.

$$\boxed{\frac{D}{Dt}C' = \int_\Gamma D\boldsymbol{\kappa}\mathscr{A} \cdot d\boldsymbol{l}}\,.$$

In the absence of dissipation this reduces to the previous statement of Kelvin's circulation theorem. However, *in the presence of some dissipation* it shows how the radiation stress due to a maintained wave field can systematically increase the circulatory part of the mean flow until the mean velocities become comparable with the phase speed of the waves. Dissipation is essential; otherwise mean velocities are restricted (however long the radiation stress acts) to be $O(a^2)$.

(iii) For sound $\boldsymbol{\kappa}\mathscr{A}/\rho = \boldsymbol{u} - \boldsymbol{u}_E$. Thus in an inviscid fluid, the *Eulerian* mean velocity remains irrotational. If also the fluid is homogeneous and basically at rest, conservation of wave action in a steady state $((\partial/\partial \boldsymbol{x}) \cdot \boldsymbol{\gamma}\mathscr{A} = 0)$ implies $(\partial/\partial \boldsymbol{x}) \cdot (\boldsymbol{\kappa}\mathscr{A}/\rho) = 0$, whereas conservation of mass gives $(\partial/\partial \boldsymbol{x}) \cdot \boldsymbol{u} = 0$. Thus also $(\partial/\partial \boldsymbol{x}) \cdot \boldsymbol{u}_E = 0$ and we recover a result of Eckart, that a steady field of sound in a inviscid fluid gives rise to an $O(a^2)$ Eulerian mean velocity which is precisely zero. With viscosity included, however, substantial mean streaming is still possible.

6.9. *Extension to other wave systems.* This approach is readily applied to a variety of other problems in fluid dynamics, though care is required in formulating precisely the conditions under which the concept of a quasi-sinusoidal wavetrain in a slowly varying medium is consistent.

The inclusion of a frozen-in magnetic field allows Alfvén waves, as well as magneto-acoustic ones, and the Maxwell stresses enter both the radiation stress and the mean flow equations [Dewar, 1970]. The averaged variational principle is readily obtained, and the only effect of the waves on the mean flow is via the radiation stress. The sum of the mean flow energy and the wave energy is conserved. However Kelvin's circulation theorem does not apply for any material circuit (as might be expected) and there is no clear sense in which $\boldsymbol{\kappa}\mathscr{A}$ may be interpreted as a wave momentum (though it is still the wave energy).

For internal gravity waves in a stratified incompressible liquid, more care in the expansion procedure is required. Hamilton's principle is formulated as

$$\delta \int \left[\tfrac{1}{2}\rho \left| \frac{DX}{Dt} \right|^2 - \varepsilon^{-2}\rho g Z + \varepsilon^{-2} P(X) \left\{ \frac{\partial(X)}{\partial(x)} - \frac{\rho}{\tilde{\rho}} \right\} \right] dx\, dt = 0.$$

$P(X)$ is now a Lagrange multiplier, subject to variations δP which vanish at infinity, and has the amplitude expansion

$$P(\boldsymbol{x}, t) = p(\boldsymbol{x}, t) + {}_1\pi(\boldsymbol{x}, t) + {}_2\pi(\boldsymbol{x}, t) + \cdots,$$

where the mean values of ${}_1\pi$, ${}_2\pi$, etc. all vanish. The density $\tilde{\rho}(\boldsymbol{x}, t)$ is constant following fluid particles $(D/Dt)\tilde{\rho} = 0$. The reference flow $\boldsymbol{u}$ is not automatically nondivergent (indeed it cannot consistently be quite so), so that in general $\rho \neq \tilde{\rho}$, but these deviations are not important in practice.

The large parameter ε^{-2} appears explicitly in the action integral, implying that the dominant mechanics are the hydrostatic balance with density uniform in horizontal planes

$$dp_0/dz = -g\rho_0, \qquad \rho_0 = \tilde{\rho}_0(z), \qquad w_0 = 0.$$

Both the mean flow and the waves are in a sense perturbations, the former because $\varepsilon \ll 1$, the latter because $a \ll 1$. The buoyancy frequency is formally

$$\varepsilon^{-1}N = \varepsilon^{-1}\{-(g/\rho_0)(d\rho_0/dz)\}^{1/2},$$

but for consistency the mean flow is constrained to have a time scale of order unity (much longer than a typical wave period which is $O(\varepsilon^{-1})$). This implies that, to lowest order in ε, the mean flow must be horizontal and nondivergent, and that the vertical forces exerted by the waves have negligible effect.

From the terms of order ε^0 we obtain the averaged variational principle, conservation of wave action, and the equations of motion for the mean flow:

$$\frac{\partial}{\partial \boldsymbol{x}} \cdot \boldsymbol{u} = 0, \qquad \boldsymbol{u} \cdot \boldsymbol{n} = 0,$$

$$\rho_0(D/Dt)\boldsymbol{u} + \partial p/\partial \boldsymbol{x} + \lambda \boldsymbol{n} = (\partial/\partial \boldsymbol{x})\mathcal{R},$$

where

$$\mathcal{R}_{ij} = \kappa_i(\partial \mathcal{L}/\partial \kappa_j) - u_j(\partial \mathcal{L}/u_i)$$

and

$$\mathcal{L} = \mathcal{A}\left(\omega - \boldsymbol{u} \cdot \boldsymbol{\kappa} - N\frac{|\boldsymbol{n} \times \boldsymbol{\kappa}|}{|\boldsymbol{\kappa}|}\right).$$

The pressure p here (strictly p_2, the deviation from the hydrostatic value) plays a secondary role, serving only to maintain the horizontal flow nondivergent. The quantity $\lambda(\boldsymbol{x})$, and the vertical components of the radiation stress, have no physical significance. λ is adjusted to keep the vertical accelerations zero (at least to this order in the expansion procedure). The dynamically effective parts of $\mathcal{R}_{ij}$ are equal to the Reynolds stresses. The total energy equation is (correct to this order)

$$\frac{\partial}{\partial t}\{\tfrac{1}{2}\rho_0|\boldsymbol{u}|^2 + \mathcal{E}\} + \frac{\partial}{\partial \boldsymbol{x}} \cdot \{(\tfrac{1}{2}\rho_0|\boldsymbol{u}|^2 + p)\boldsymbol{u} + \gamma\mathcal{E} - \boldsymbol{u} \cdot \mathcal{R}\} = 0.$$

Here $\mathcal{E} = \omega^+\mathcal{A}$ is as derived in §5.3. Note that changes in potential energy of the mean flow do not enter this equation because the mean motion is consistently regarded as horizontal. A similar situation occurs in the passage to the incompressible limit in a gas at low Mach number—changes in internal energy drop out of the incompressible energy equation although they are comparable to the kinetic energy.

Kelvin's circulation theorem applies to all horizontal circuits, and the rotational part of the mean velocity field is augmented when waves arrive by a momentum $\boldsymbol{\kappa}_h\mathcal{A}$ per unit volume, where $\boldsymbol{\kappa}_h$ is the horizontal part of $\boldsymbol{\kappa}$. Thus a wave packet is associated with a patch of vorticity about the vertical axis, and the *impulse* of the patch is $\boldsymbol{\kappa}_h$ times the total wave action contained in it. In this sense every packet of internal gravity waves has horizontal momentum associated with it, that momentum being permanently given up to the mean flow only when the wave packet is absorbed.

Similar dynamical restrictions on the mean state must be imposed in a rotating fluid if we are consistently to talk about slowly varying trains of inertial waves. In a homogeneous liquid, the mean flow must be at low Rossby number, and takes the form of Taylor columns uniform in the direction parallel to the axis of rotation. The radiation stress associated with inertial waves acts effectively only through the couple exerted about this axis. In a rotating stratified fluid, the mean flow in the absence of waves must satisfy the quasi-geostrophic equations of motion, and inertio-gravitational waves may change the potential vorticity. Systematic forcing of large mean velocities may occur in a sustained wavefield, but *only* if the waves are dissipated somewhere.

Irrotational waves in water of finite depth introduce a new complication, the "lateral coordinate" (depth) in which the structure is not quasi-sinusoidal. However, Hamilton's principle may still be used, the averaging procedure over a wave period involving also integration over depth. The averaging Lagrangian yields the lateral structure as a by-product, and the wave energy and radiation stress are just those given by Longuet-Higgins and Stewart [1964]. For consistency the reference flow $\boldsymbol{u}$ must be independent of depth and must satisfy the conventional shallow water equations, except for the horizontal body force exerted by the radiation stress. $\boldsymbol{u}$ is no longer the Lagrangian mean velocity, but the mass transport velocity—the Lagrangian mean velocity averaged over depth.

Finally, it should be noted that all this theory can be extended to quasi-homogeneous quasi-stationary random wave systems [Bretherton, 1969].

References

F. P. Bretherton (1967), *The time-dependent motion due to a cylinder moving in an unbounded rotating or stratified fluid*, J. Fluid Mech. **28** (1967), 545–570.

——— (1969), *Waves and turbulence in stably stratified fluids*, Radio Science **4** (1969), 1279–1288.

——— (1970), *A note on Hamilton's principle for perfect fluids*, J. Fluid Mech. **44** (1970), 19–31.

F. P. Bretherton and C. J. R. Garrett (1968), *Wavetrains in inhomogeneous moving media*, Proc. Roy. Soc. London Ser. A **302** (1968), 529–554.

T. Brooke Benjamin (1970), *Upstream influence*, J. Fluid Mech. **40** (1970), 49–79.

R. L. Dewar (1970), *Interaction between hydromagnetic waves and a time-dependent inhomogeneous medium*, Phys. Fluids **13** (1970), 2710–2720.

C. Eckart (1963), *Some transformations of the hydromagnetic equations*, Phys. Fluids **6** (1963), 1037–1041. MR **27** #5418.

H. E. Huppert and J. W. Miles (1969), *Lee waves in a stratified flow*, J. Fluid Mech. **35** (1969), part 3, 481–496.

R. M. Lewis (1965), *Asymptotic theory of wave propagation*, Arch. Rational Mech. Anal. **20** (1965), 191–250. MR **32** #2023.

M. J. Lighthill (1965), *Group velocity*, J. Inst. Appl. Math. Appl. **1** (1965), 1–28. MR **32** #1930.

R. R. Long (1955), *Some aspects of the flow of stratified fluids*. III, Tellus **7** (1955), 341–357.

M. S. Longuet-Higgins and R. W. Stewart (1964), *Radiation stress in water waves; a physical discussion, with applications*, Deep Sea Res. **11** (1964), 529–562.

J. W. Miles (1969), *Transient motion of a dipole in a rotating flow*, J. Fluid Mech. **39** (1969), 433–442.

D. E. Mowbray and B. S. H. Rarity (1967), *A theoretical and experimental investigation of internal waves of small amplitude in a density stratified liquid*, J. Fluid Mech. **28** (1967), 1–16.

J. B. Serrin (1959), *Mathematical principles of classical fluid mechanics*, Handbuch der Physik 8/1, Springer-Verlag, Berlin, 1959, pp. 125–263. MR **21** #6836b.

G. B. Whitham (1970), *A note on group velocity*, J. Fluid Mech. **9** (1960), 347–352. MR **22** #11709.

——— (1965), *A general approach to linear and non-linear dispersive waves using a Lagrangian*, J. Fluid Mech. **22** (1965), 273–283. MR **31** #6459.

JOHNS HOPKINS UNIVERSITY

Nonlinear Waves

D. J. Benney

1. Concepts and formulations. Much of the recent work in nonlinear waves is in some sense a subtle refinement of a linear theory. Therefore a few remarks about linear waves are in order.

Consider the partial differential equation

$$u_t + \mathscr{L}_{\boldsymbol{x}} u = 0, \tag{1.1}$$

where $\mathscr{L}_{\boldsymbol{x}}$ is a linear operator in which the spatial coordinates $\boldsymbol{x}$ occur only as derivatives. Equation (1.1) has solutions of the form

$$u = a e^{i\boldsymbol{k}.\boldsymbol{x} - i\omega t}, \tag{1.2}$$

where

$$\omega = \omega(\boldsymbol{k}) \tag{1.3}$$

is the dispersion relation. For the most part it will be assumed that ω is real for $\boldsymbol{k}$ real, so that questions of instability do not arise. Here ω is the frequency and $\boldsymbol{k}$ the wave number. The phase velocity $\boldsymbol{c}$ is defined to be

$$\boldsymbol{c} = [\omega(\boldsymbol{k})\boldsymbol{k}]/|\boldsymbol{k}|^2. \tag{1.4}$$

The great mathematical simplification of linear theory is that solutions can be superposed. Also the initial value problem can be solved by using Fourier transforms. In the above example if $u(\boldsymbol{x}, 0) = f(\boldsymbol{x})$ then

$$u(\boldsymbol{x}, t) = \int_{-\infty}^{\infty} a(\boldsymbol{k}) e^{i\boldsymbol{k}.\boldsymbol{x} - i\omega t}\, d\boldsymbol{k}, \tag{1.5}$$

where

$$a(\boldsymbol{k}) = \frac{1}{(2\pi)^n} \int_{-\infty}^{\infty} f(\boldsymbol{x}) e^{-i\boldsymbol{k}.\boldsymbol{x}}\, d\boldsymbol{x}, \tag{1.6}$$

n being the number of spatial dimensions. It is then possible to investigate the long time–far field behavior of this solution by standard asymptotic methods. From such an analysis another velocity arises, the group

AMS 1970 *subject classifications.* Primary 76B15, 76E30, 76B25, 35B20; Secondary 76E05, 42A68.

velocity $\boldsymbol{c}_g$, where

$$\boldsymbol{c}_g = \nabla_{\boldsymbol{k}}\omega. \tag{1.7}$$

The group velocity is the velocity of the envelope of a wave train. For if we consider a sinusoidal wave having a slowly varying complex amplitude

$$u = a(\boldsymbol{X}, T)e^{i\boldsymbol{k}.\boldsymbol{x} - i\omega(\boldsymbol{k})t}, \tag{1.8}$$

where

$$\boldsymbol{X} = \mu\boldsymbol{x}, \qquad T = \mu t, \qquad 0 < \mu \ll 1, \tag{1.9}$$

equation (1.1) takes the form

$$[-i\omega(\boldsymbol{k}) + \mu\partial/\partial T + i\omega(\boldsymbol{k} - i\mu\nabla_{\boldsymbol{X}})]a = 0,$$

or

$$[\partial/\partial T + \nabla_{\boldsymbol{k}}\omega \, . \, \nabla_{\boldsymbol{X}}]a = O(\mu). \tag{1.10}$$

The $O(\mu)$ terms involve higher derivatives of ω and a with respect to $\boldsymbol{k}$ and $\boldsymbol{X}$ respectively, indicating that the envelope will eventually be influenced by dispersion. However to leading order the solution of (1.10) is

$$a = a(\boldsymbol{X} - \boldsymbol{c}_g T), \tag{1.11}$$

and the envelope propagates with the group velocity.

While the preceding example is a very simple one, the essential features persist in more complicated problems. Special mention should be made of the situation where instead of dealing with a partial differential equation with constant coefficients, the coefficients are slowly varying functions of space. This corresponds to linear wave propagation through a slowly varying medium and the problem can be solved by WKB methods.

In a given wave problem of physical interest the magnitude of dimensionless parameters plays a crucial role in both the physics and the mathematics. Apart from these physical parameters there are important geometrical parameters associated with a wave. There is an amplitude a_0, a wave length scale l and often a length h_0 typifying the extent of the region.

To be specific consider waves on the surface of water. Suppose x, z are horizontal coordinates, y measured vertically upwards from a rigid bottom ($y = 0$), t the time, and g the acceleration due to gravity. If h_0 is the undisturbed depth of fluid while $y = h(\boldsymbol{x}, t) = h_0 + \eta(\boldsymbol{x}, t)$ is the disturbed free surface, then the velocity potential satisfies the equation

$$\phi_{xx} + \phi_{yy} + \phi_{zz} = 0, \tag{1.12}$$

with the bottom condition,

$$\phi_y = 0, \qquad y = 0, \tag{1.13}$$

and the free surface conditions,

(1.14) $$h_t + \phi_x h_x + \phi_z h_z = \phi_y, \qquad y = h = h_0 + \eta,$$

(1.15) $$\phi_t + \tfrac{1}{2}(\phi_x^2 + \phi_y^2 + \phi_z^2) + g\eta = 0, \qquad y = h = h_0 + \eta.$$

On going to dimensionless variables by writing

(1.16) $$\begin{aligned} x &= lx', \qquad y = h_0 y', \qquad z = lz', \qquad t = l(gh_0)^{-1/2}t', \\ h &= h_0 h', \qquad \eta = a_0\eta', \qquad \phi = l(gh_0)^{-1/2} g a_0 \phi', \end{aligned}$$

equations (1.12)–(1.15) become (on dropping the primes)

(1.17) $$\phi_{yy} + \beta^2(\phi_{xx} + \phi_{zz}) = 0,$$

(1.18) $$\phi_y = 0, \qquad y = 0,$$

(1.19) $$h_t + \varepsilon(\phi_x h_x + \phi_z h_z) = (\varepsilon/\beta^2)\phi_y, \qquad y = 1 + \varepsilon n,$$

(1.20) $$\phi_t + n + \tfrac{1}{2}[(\varepsilon/\beta^2)\phi_y^2 + \varepsilon(\phi_x^2 + \phi_z^2)] = 0, \qquad y = 1 + \varepsilon n,$$

where $\varepsilon = a_0/h_0$ and $\beta^2 = h_0^2/l^2$ are the geometrical parameters which measure amplitude and wave number.

Various developments are now possible depending on the relative importance of ε and β^2.

(1) Quasi-linear theory, $\beta^2 = O(1)$, $\varepsilon \ll 1$. There is no loss of generality in taking $\beta^2 = 1$, and solutions can then be found by a perturbation expansion

(1.21) $$\phi = \sum_{n=0}^{\infty} \varepsilon^n \phi^{(n)}(x, y, z, t).$$

$\phi^{(0)}$ is the linear theory and it is well known that

(1.22) $$\omega = \pm \sqrt{g|\boldsymbol{k}| \tanh |\boldsymbol{k}| h_0},$$

is the dispersion relation. There are certain modifications needed in the expansion (1.21) due to internal resonances; but these we shall leave until §2.

(2) Shallow water theory, $\varepsilon = O(1)$, $\beta^2 \ll 1$. Again we may suppose $\varepsilon = 1$, and use the expansion

(1.23) $$\phi(x, y, z, t) = \sum_{n=0}^{\infty} \beta^{2n}\phi^{(n)}(x, y, z, t) = f - \frac{\beta^2 y^2}{2}(f_{xx} + f_{zz}) + O(\beta^4),$$

where $f = f(x, z, t)$, and equations (1.17) and (1.18) have been satisfied. Equations (1.19) and (1.20) yield two equations for h and f, namely,

(1.24) $$h_t + f_x h_x + f_z h_z + h(f_{xx} + f_{zz}) = O(\beta^2),$$

(1.25) $$f_t + h - 1 + \tfrac{1}{2}(f_x^2 + f_z^2) = O(\beta^2).$$

Differentiation of this last equation with respect to x and z yields

$$\begin{aligned} h_t + (uh)_x + (wh)_z &= O(\beta^2), \\ u_t + uu_x + wu_z + h_x &= O(\beta^2), \\ w_t + uw_x + ww_z + h_z &= O(\beta^2) \end{aligned} \tag{1.26}$$

where $u = f_x$ and $w = f_z$ are the horizontal velocities to leading order. Neglecting the $O(\beta^2)$ terms, these are the familiar equations of shallow water theory.

The above two developments are distinct. From the quasi-linear theory there are energy exchanges possible between certain bands of wave numbers; but there is no indication of breaking. On the basis of shallow water theory all waves will break, yet such deformations must inevitably lead to $\beta^2 = O(1)$ and all the nonlinear dispersive terms will become important. The question as to when and how a wave will break remains an intriguing but unresolved theoretical problem.

Finally if $\varepsilon = O(\beta^2) \ll 1$ so that weak dispersion and nonlinearity are in balance, the equations (1.20) give the Boussinesq equations to leading order (or the Korteweg de Vries equation for unidirectional waves), and finite amplitude waves of permanent form can be found. These are the well-known solitary and cnoidal waves.

The preceding discussion is somewhat superficial; but the important point to realize is that two quite distinct theoretical developments are possible in water waves depending on the magnitude of the geometrical parameters. With minor modifications these ideas apply to many other physical systems.

2. Resonances and instabilities. In this section we investigate the properties of quasi-linear wave trains, whose amplitudes vary slowly in both space and time. (In §5 the theory will be generalized to fully nonlinear waves.) The essential idea is to make use of the asymptotic methods familiar in the study of weakly coupled oscillators.

For a set of N interacting discrete waves in a weakly nonlinear conservative physical system, the equations governing the amplitudes are well known. Thus if the wave numbers are $\boldsymbol{k}_l$; $l = 1, 2, \ldots, N$; with complex Fourier amplitudes $A(\boldsymbol{k}_l t)$, these equations are

$$\begin{aligned} \frac{dA_l}{dt} + i\omega_l A_l &= i\varepsilon \sum_{m,n} \alpha_{lmn} A_m^* A_n^* \\ &\quad + i\varepsilon^2 \left[\sum_p \beta_{lp} A_l A_p A_p^* + \sum_{q,r,s} \gamma_{lqrs} A_q^* A_r^* A_s^* \right] + O(\varepsilon^3), \end{aligned} \tag{2.1}$$

where $\omega_l = \omega(\boldsymbol{k}_l) = -\omega_{-l}$ is the linear dispersion relation and the notational contraction $A_l = A(\boldsymbol{k}_l, t) = A^*_{-l}$ has been used. The coupling constants $\alpha_{lmn}, \beta_{lp}, \gamma_{lqrs}$ are real for a conservative system and $\alpha_{lmn}, \gamma_{lqrs}$ are nonzero only for those nontrivial combinations of wave numbers such that $\boldsymbol{k}_l + \boldsymbol{k}_m + \boldsymbol{k}_n = 0$, and $\boldsymbol{k}_l + \boldsymbol{k}_q + \boldsymbol{k}_r + \boldsymbol{k}_s = 0$, respectively. The constants β_{lp} correspond to the direct modal transfer terms.

Equation (2.1) is readily modified to describe interacting quasi-linear wave trains. As in §1 we write

$$\boldsymbol{X} = \mu \boldsymbol{x}, \qquad T = \mu t, \qquad 0 < \mu \ll 1, \tag{2.2}$$

and use the multiple scaling procedure. Equation (2.1) becomes

$$\begin{aligned} \mu\left(\frac{\partial a_l}{\partial T} + \sum_r \frac{\partial \omega_l}{\partial k_{lr}} \frac{\partial a_l}{\partial X_r}\right) &= \frac{i\mu^2}{2} \sum_{r,s} \frac{\partial^2 \omega_l}{\partial k_{lr} \partial k_{ls}} \frac{\partial^2 a_l}{\partial X_r \partial X_s} \\ &\quad + i\varepsilon \sum_{m,n} \alpha_{lmn} a_m^* a_n^* + i\varepsilon^2 \sum_p \beta_{lp} a_l a_p a_p^* \\ &\quad + i\varepsilon^2 \sum_{q,r,s} \gamma_{lqrs} a_q^* a_r^* a_s^* + O(\varepsilon^3, \varepsilon\mu, \mu^3). \end{aligned} \tag{2.3}$$

Here the fast time scale t has been removed by writing

$$A_l = a_l(\boldsymbol{k}_l, \boldsymbol{X}, T) e^{-i\omega_l t} = a_l e^{-i\omega_l t}, \tag{2.4}$$

and the only nonzero $\alpha_{lmn}, \gamma_{lqrs}$, are those for which the triad and quartet resonance conditions are satisfied, namely

$$\omega_l + \omega_m + \omega_n = 0, \qquad \boldsymbol{k}_l + \boldsymbol{k}_m + \boldsymbol{k}_n = 0, \tag{2.5}$$

$$\omega_l + \omega_q + \omega_r + \omega_s = 0, \qquad \boldsymbol{k}_l + \boldsymbol{k}_q + \boldsymbol{k}_r + \boldsymbol{k}_s = 0. \tag{2.6}$$

Equation (2.3) is fundamental for the determination of properties of interacting weakly nonlinear wave trains. There are several possibilities depending on the relative magnitude of the small parameters μ and ε. Crudely speaking if $\mu \ll \varepsilon$ discrete wave interactions are most relevant, while if $\mu \gg \varepsilon$ the system is dominated by linear envelopes propagating with the appropriate group velocities. A few special cases illustrate the possibilities.

(a) Suppose $\mu = \varepsilon^2$, and that none of the wave numbers concerned form a resonant triad or quartet. The leading terms in (2.3) give

$$\partial a_l / \partial T + \boldsymbol{c}_{gl}, \nabla_{\boldsymbol{X}} a_l = i \sum_p \beta_{lp} a_l a_p a_p^*, \tag{2.7}$$

and the general solution is

$$a_l = f_l(\boldsymbol{X} - \boldsymbol{c}_{gl}T) \times \exp\left[i \int_0^T \sum_p \beta_{lp} |f_p(\boldsymbol{X} - \boldsymbol{c}_{gl}T + (\boldsymbol{c}_{gl} - \boldsymbol{c}_{gp})S)|^2 \, dS \right], \tag{2.8}$$

where

$$a_l(\boldsymbol{X}, 0) = f_l(\boldsymbol{X}).$$

If each f_l is independent of $\boldsymbol{X}$ then the frequencies are merely dependent on amplitudes as in weakly coupled oscillators.

On considering only one mode, the solution (2.8) gives

$$a_l = f_l(\boldsymbol{X} - \boldsymbol{c}_{gl}T) \exp[i\beta_{ll}|f_l(\boldsymbol{X} - \boldsymbol{c}_{gl}T)|^2 T], \tag{2.9}$$

so that to an observer moving with the group velocity, a_l is locally periodic in time. Note that if $f_l(\boldsymbol{X}) = b + c \exp[i\boldsymbol{K} \cdot \boldsymbol{X}]$ so that initially there are two modes, then subsequently all higher side band harmonics will be generated. This clearly points to the dangers of a discrete analysis when the wave numbers are close together.

(b) With three modes $l = 1, 2, 3$ which form a resonant triad and the balance $\mu = \varepsilon$, the leading order terms in (2.3) yield

$$\begin{aligned} \partial a_1/\partial T + \boldsymbol{c}_{g1} \cdot \nabla_{\boldsymbol{X}} a_1 &= i\alpha_{123} a_2^* a_3^*, \\ \partial a_2/\partial T + \boldsymbol{c}_{g2} \cdot \nabla_{\boldsymbol{X}} a_2 &= i\alpha_{231} a_3^* a_1^*, \\ \partial a_3/\partial T + \boldsymbol{c}_{g3} \cdot \nabla_{\boldsymbol{X}} a_3 &= i\alpha_{312} a_1^* a_2^*. \end{aligned} \tag{2.10}$$

No general solution to the initial value problem posed by the equations (2.10) is known; but if the initial values are $\boldsymbol{X}$ independent, an exact solution in which the amplitudes are periodic functions of time has been found [**1**].

Resonances have been discussed by many authors [**2**], [**3**] and they play an important role in weak interaction theory.

(c) A single finite amplitude wave can be unstable to nearby wave numbers [**4**], [**5**], [**6**]; that is side band instabilities are possible. This phenomenon can be investigated by taking $\mu = \varepsilon$ and considering one mode. Equation (2.3) then gives

$$\frac{\partial a}{\partial T} + \boldsymbol{c}_g \cdot \nabla_{\boldsymbol{X}} a = i\varepsilon \left[\frac{1}{2} \sum_{r,s} \frac{\partial^2 \omega}{\partial k_r \partial k_s} \frac{\partial^2 a}{\partial X_r \partial X_s} + \beta a^2 a^* \right], \tag{2.11}$$

where for convenience the subscript l has been omitted. On changing to a coordinate system $\boldsymbol{X}'$ moving with the group velocity and to a slower time scale T' by writing

(2.12) $$\mathbf{X}' = \mathbf{X} - \mathbf{c}_g T, \qquad T' = \varepsilon T, \qquad \delta_{rs} = \frac{1}{2}\frac{\partial^2 \omega}{\partial k_r \partial k_s},$$

equation (2.11) becomes

(2.13) $$\frac{\partial a}{\partial T'} = i\left[\sum_{r,s} \delta_{rs} \frac{\partial^2 a}{\partial X_r \partial X_s} + \beta a^2 a^*\right].$$

The general solution of equation (2.13) is not known. (Certain special solutions can easily be found.) One very simple solution is the nonlinear discrete wave

(2.14) $$a^{(0)}(T') = a^{(0)}(0) e^{i\beta|a^{(0)}(0)|^2 T'}.$$

The linear stability of this solution can be studied by writing

(2.15) $$a(\mathbf{X}', T') = a^{(0)}(T') + b^{(1)}(\mathbf{X}', T') e^{i\beta|a^{(0)}(0)|^2 T'}.$$

The linearized equation for $b^{(1)}$ is found to be

(2.16) $$\left[\frac{\partial^2}{\partial T'^2} + \sum_{r,s} \delta_{rs} \frac{\partial^2}{\partial X'_r \partial X'_s}\left(\sum_{p,q} \delta_{pq} \frac{\partial^2}{\partial X'_p \partial X'_q} + 2\beta|a^{(0)}(0)|^2\right)\right] b^{(1)} = 0,$$

and so for a side band perturbation $b^{(1)} = b(T') e^{i\mathbf{K}' \cdot \mathbf{X}'}$,

(2.17) $$\left[\frac{d^2}{dT'^2} + \sum_{r,s} \delta_{rs} K'_r K'_s \left(\sum_{p,q} \delta_{pq} K'_p K'_q - 2\beta|a^{(0)}(0)|^2\right)\right] b = 0.$$

The above equation may have oscillatory or exponential solutions, that is some side bands may be stable or unstable depending on the sign of the quadratic form $\beta_{rs} K'_r K'_s$ and the magnitude of the amplitude.

The details of the problem in water waves are more complicated as a small large scale mean elevation is induced and plays an important part in the stability calculation. It is found that a periodic wave always has instability in some direction, except in the pathological case when $|\mathbf{k}|h_0 = .38$. Of course it must be realized that this type of instability does not imply any persistence of exponential growth; rather it is to be interpreted as the slow development of some incoherence in the wave. Experiments confirm this point.

The type of instability discussed here is a special case of a more general problem to which we will return in §5.

3. Random waves. There is much interest in the statistical initial value problem for a system of waves [7], [**8**], [9], [**10**]. Assuming the system to be spatially homogeneous, one usually attempts to describe the time evolution of the system by deriving a hierarchy of equations for the statistical moments, or equivalently the Fourier space cumulants. Unfortunately the equation for the time rate of change of a cumulant of given order depends on cumulants of higher order, and one is faced with the well-known closure difficulty fundamental to nonlinear random processes. In fully nonlinear problems no entirely satisfactory closure scheme has been derived. However, for a quasi-linear system a consistent set of closure equations can be found, and we shall sketch the analysis for a simple model equation.

Consider the equation

$$u_t + \mathscr{L}_{\boldsymbol{x}} u = \varepsilon \mathscr{N}_{\boldsymbol{x}} u^2, \tag{3.1}$$

where $u(\boldsymbol{x}, t)$ is a stationary random function of position, t is the time variable, and ε is a small parameter to denote the weak nonlinear coupling. $\mathscr{L}_{\boldsymbol{x}}$ and $\mathscr{N}_{\boldsymbol{x}}$ are quite general operators. Our interest lies in the solution to the statistical initial value problem in an unbounded spatial region, and so it is convenient to Fourier transform (3.1) to obtain

$$\begin{aligned} &\left(\frac{\partial}{\partial t} + i\omega(\boldsymbol{k})\right) A(\boldsymbol{k}, t) \\ &\quad = \varepsilon \int_{-\infty}^{\infty} H(\boldsymbol{k}_1, \boldsymbol{k}_2) A(\boldsymbol{k}_1, t) A(\boldsymbol{k}_2, t) \delta(\boldsymbol{k}_1 + \boldsymbol{k}_2 - \boldsymbol{k})\, d\boldsymbol{k}_1\, d\boldsymbol{k}_2, \end{aligned} \tag{3.2}$$

where

$$\begin{aligned} u(\boldsymbol{x}, t) &= \int_{-\infty}^{\infty} A(\boldsymbol{k}, t) e^{i\boldsymbol{k}.\boldsymbol{x}}\, d\boldsymbol{k}, \\ A(\boldsymbol{k}, t) &= \frac{1}{(2\pi)^{:}} \int_{-\infty}^{\infty} u(\boldsymbol{x}, t) e^{-i\boldsymbol{k}.\boldsymbol{x}}\, d\boldsymbol{x}. \end{aligned} \tag{3.3}$$

The linearized form of equation (3.1) has sinusoidal wave solutions $u = \exp[i(\boldsymbol{k} \cdot \boldsymbol{x} - \omega(\boldsymbol{k})t]$, where $\omega = \omega(k) = -\omega(-k)$ is the dispersion relation. It will be assumed that the waves are dispersive so that $\omega(k) \neq \boldsymbol{\alpha} \cdot \boldsymbol{k}$ ($\boldsymbol{\alpha}$ constant). The latter case has recently been investigated [**11**]. It will also be assumed that the mean value property $\langle u(\boldsymbol{x}, t)\rangle = 0$ is consistent with the physical model, corresponding to the requirement that $H(\boldsymbol{k}, -\boldsymbol{k}) = 0$. There is no loss of generality in taking a symmetric function $H(\boldsymbol{k}_1, \boldsymbol{k}_2)$. Note that if $u(\boldsymbol{x}, t)$ is real then $A(-\boldsymbol{k}, t) = A^*(\boldsymbol{k}, t)$, and $H(-k_1, -k_2) = H^*(k_1, k_2)$.

In order to proceed with a naive perturbation approach we write

$$A(\boldsymbol{k}, t) = a(\boldsymbol{k}, t) e^{-i\omega(\boldsymbol{k})t}, \tag{3.4}$$

so that (3.2) becomes

$$\frac{\partial a}{\partial t}(\boldsymbol{k}, t) = \varepsilon \int_{-\infty}^{\infty} H(\boldsymbol{k}_1, \boldsymbol{k}_2) a(\boldsymbol{k}_1, t) a(\boldsymbol{k}_2, t) e^{i(\omega(\boldsymbol{k}) - \omega(\boldsymbol{k}_1) - \omega(\boldsymbol{k}_2))t} \times \delta(\boldsymbol{k}_1 + \boldsymbol{k}_2 - \boldsymbol{k})\, d\boldsymbol{k}_1\, d\boldsymbol{k}_2. \tag{3.5}$$

This equation is rewritten in a condensed notation to be

$$\frac{\partial a_l}{\partial t} = \varepsilon \int H_{mn} a_m a_n e^{i\omega_{l,mn} t} \delta_{l,mn}\, d\boldsymbol{k}_{mn}, \tag{3.6}$$

where

$$\begin{aligned} a_l = a(\boldsymbol{k}_l, t), \qquad H_{mn} &= H(\boldsymbol{k}_m, \boldsymbol{k}_n), \qquad d\boldsymbol{k}_{m_1 m_2 \dots m_r} = \prod_{p=1}^{r} d\boldsymbol{k}_{m_p}, \\ \delta_{l_1 \dots l_r, m_1 \dots m_s} &= \delta\left(\sum_{\alpha=1}^{r} \boldsymbol{k}_{l_\alpha} - \sum_{\beta=1}^{s} \boldsymbol{k}_{m_\beta} \right), \\ \omega_{l_1 \dots l_r, m_1 \dots m_s} &= \sum_{\alpha=1}^{r} \omega(\boldsymbol{k}_{l_\alpha}) - \sum_{\beta=1}^{s} \omega(\boldsymbol{k}_{m_\beta}), \end{aligned} \tag{3.7}$$

and the limits on all integrations are from $-\infty$ to $+\infty$.

On taking a formal perturbation solution for a_l by writing

$$a_l = a_{0l} + \varepsilon a_{1l} + \varepsilon^2 a_{2l} + \dots, \tag{3.8}$$

it is clear that a_{0l} is constant as for linear theory while a_{1l} and a_{2l} are given by

$$\begin{aligned} a_{1l} &= \int H_{mn} a_{0mn} \Delta(\omega_{l,mn}) \delta_{l,mn}\, d\boldsymbol{k}_{mn}, \\ a_{2l} &= \int H_{mnpq} a_{0mpq} E(\omega_{l,mpq}; \omega_{l,mn}) \delta_{l,mn} \delta_{n,pq}\, d\boldsymbol{k}_{mnpq}. \end{aligned} \tag{3.9}$$

Here the following additional notational contradictions have been used:

$$\begin{aligned} a_{0mnp} \dots &= a_{0m} a_{0n} a_{0p} \dots, \qquad H_{mnpq} \dots = H_{mn} H_{pq} \dots, \\ \Delta(x) &= \int_0^t e^{ixt}\, dt = \frac{e^{ixt} - 1}{ix}, \\ E(x; y) &= \int_0^t \Delta(x - y) e^{iyt}\, dt = \frac{\Delta(x) - \Delta(y)}{i(x - y)}. \end{aligned} \tag{3.10}$$

The aim of the perturbation procedure is to keep the physical space cumulants well ordered in time. The nth order physical space cumulant is defined by

$$R^{(n)}(\boldsymbol{r}, \boldsymbol{r}', \ldots, \boldsymbol{r}^{(n-2)}, t) = \langle u(\boldsymbol{x}, t)u(\boldsymbol{x} + \boldsymbol{r}, t) \ldots u(\boldsymbol{x} + \boldsymbol{r}^{(n-2)}, t)\rangle$$

$$\text{(3.11)} \qquad - \sum_{\alpha,\beta} \langle u(\boldsymbol{x}, t) \ldots u(\boldsymbol{x} + \boldsymbol{r}^{(\alpha)}, t)\rangle$$

$$\times \langle u(\boldsymbol{x} + \boldsymbol{r}^{(\beta)}, t) \ldots u(\boldsymbol{x} + \boldsymbol{r}^{(n-2)}, t)\rangle,$$

and is related to the nth order Fourier space cumulant by the equation

$$R^{(n)}(\boldsymbol{r}, \boldsymbol{r}', \ldots, \boldsymbol{r}^{(n-2)}, t)$$

$$= \int Q^{(n)}(\boldsymbol{k}_l, \boldsymbol{k}_{l'}, \ldots, \boldsymbol{k}_{l^{(n-2)}}, t)\; e^{i\boldsymbol{k}_l.\boldsymbol{r} + \ldots i\boldsymbol{k}_{l^{(n-2)}}.\boldsymbol{r}^{(n-2)}}\, d\boldsymbol{k}_{ll'\ldots l^{(n-2)}},$$

$$\text{(3.12)} \qquad = \int q^{(n)}_{ll'\ldots l^{(n-2)}} e^{-i\omega_{ll'\ldots l^{(n-2)}}t}\; e^{i\boldsymbol{k}_l.\boldsymbol{r} + \ldots i\boldsymbol{k}_{l^{(n-2)}}.\boldsymbol{r}^{(n-2)}}\, d\boldsymbol{k}_{ll'\ldots l^{(n-2)}},$$

where $\boldsymbol{k}_l + \boldsymbol{k}_{l'} + \ldots \boldsymbol{k}_l(n-1) = 0$, and where

$$\text{(3.13)} \qquad \delta_{ll'\ldots l^{(n-1)}} q^{(n)}_{ll'\ldots l^{(n-2)}} = \langle a_{ll'\ldots l^{(n-1)}}\rangle - \sum_{\alpha,\beta} \langle a_{l\ldots l^{(\alpha)}}\rangle \langle a_{l^{(\beta)}\ldots l^{(n-1)}}\rangle.$$

For $n = 2, 3$ the cumulants and correlations are identical, $\boldsymbol{q}^{(2)}_l = Q^{(2)}_l$ being the energy spectrum.

On using the assumed expansion for the a_l (equation (3.8)), a similar expansion can be derived for each $\boldsymbol{q}^{(n)}$, namely,

$$\text{(3.14)} \qquad q^{(n)} = q^{(n)}_0 + \varepsilon q^{(n)}_1 + \varepsilon^2 q^{(n)}_2 + \ldots .$$

For example,

$$\delta_{ll'} q^{(2)}_{1l} = \mathscr{P}_{ll'} \langle a_{1l} a_{0l'}\rangle, \qquad \delta_{ll'l''} q^{(3)}_{1ll'} = \mathscr{P}_{ll'l''} \langle a_{1l} a_{0l'} a_{0l''}\rangle,$$

$$\text{(3.15)} \qquad \delta_{ll'l''l'''} q^{(4)}_{1ll'l''} = \mathscr{P}_{ll'l''l'''} \{\langle a_{1l} a_{0l'} a_{0l''} a_{0l'''}\rangle - \langle a_{1l} a_{0l'}\rangle \langle a_{0l''} a_{0l'''}\rangle$$

$$- \langle a_{1l} a_{0l''}\rangle \langle a_{0l'} a_{0l'''}\rangle - \langle a_{1l} a_{0l'''}\rangle \langle a_{0l'} a_{0l''}\rangle\}.$$

The symbol $\mathscr{P}_{ll'\ldots l^{(n)}}$ is used to imply a cyclic summation over $ll' \ldots l^{(n)}$.

It is to be anticipated that the ordering of each $R^{(n)}$ will involve certain nonuniformities as $t \to \infty$. Indeed it is precisely these nonuniformities which enable us to obtain the desired asymptotic closures. The multiple scale method will be used to collect the secular terms.

To be specific only the first time closure will be considered, and the details will be given only for the energy spectrum $q^{(2)}_l$. For a more elaborate discussion the reader is referred to [**10**].

A straightforward calculation shows that

$$q_{1l}^{(2)} = \mathscr{P}_{ll'} \int H_{mn} q_{0mn}^{(3)} \Delta(\omega_{l,mn}) \delta_{l,mn}\, d\boldsymbol{k}_{mn},$$

$$\begin{aligned} q_{2l}^{(2)} = \mathscr{P}_{ll'} \Bigg[& \int H_{mnpq} q_{0mnp}^{(4)} \Delta(\omega_{l,mn}) \Delta(\omega_{-l,pq}) \delta_{l,mn} \delta_{lpq}\, d\boldsymbol{k}_{mnpq} \\ & + \int H_{mn-m-n} q_{0m}^{(2)} q_{0n}^{(2)} \Delta(\omega_{l,mn}) \Delta(\omega_{mn,l}) \delta_{l,mn}\, d\boldsymbol{k}_{mn} \\ & + 2 \int H_{mnpq} q_{0mpq}^{(4)} E(\omega_{l,mpq}; \omega_{l,mn}) \delta_{l,mn} \delta_{n,pq}\, d\boldsymbol{k}_{mnpq} \\ & + 4 \int H_{mnl-m} q_{0l}^{(2)} q_{0m}^{(2)} E(\omega_{l,mn}; 0) \delta_{l,mn}\, d\boldsymbol{k}_{mn} \Bigg]. \end{aligned} \tag{3.16}$$

In order to obtain a uniform representation it is necessary to interpret the $q_0^{(n)}$ to be slowly varying functions of time, so that

$$q_0^{(n)} \to q_0^{(n)} + \varepsilon_1 q_0^{(n)} + \varepsilon^2 ({}_2 q_0^{(n)} - t \partial q_0^{(n)} / \partial T_2) + O(\varepsilon^3), \tag{3.17}$$

where only even order time scales prove to be necessary. The first such time scale is $T_2 = \varepsilon^2 t$. It is assumed that at $t = 0$ all the spectral functions are smooth to a sufficiently high order in ε and that ${}_r q_0^{(n)}$ are the higher order free statistical properties.

The entire closure procedure now depends on the asymptotic properties of certain integrals. To order ε^2 only the time dependent functions $\Delta(x)$, $\Delta(x)\Delta(y)$, $\Delta(x)\Delta(-x)$, $E(x; y)$ and $E(x; 0)$ need be considered.

By taking a path in the complex x plane indented at $x = 0$ it is readily shown that

$$\lim_{t \to \infty} \int f(x) \Delta(x)\, dx = \pi f(0) + iP \int \frac{f(x)}{x}\, dx, \tag{3.18}$$

where P denotes the Cauchy principal value of the integral. Thus in operational form we may write

$$\Delta(x) \sim \tilde{\Delta}(x) = \pi \delta(x) + iP(1/x). \tag{3.19}$$

Also as $\Delta(0) = t$, and as

$$E(x; 0) = \int_0^t \Delta(x)\, dt = \left(t + i \frac{\partial}{\partial x} \right) \Delta(x), \tag{3.20}$$

integration by parts gives

$$E(x; 0) \sim \tilde{\Delta}(x)(t - i\partial/\partial x). \tag{3.21}$$

To find the asymptotic form of $\Delta(x)\Delta(y)$ one can avoid products with the identity

$$\Delta(x)\Delta(y) \equiv i[y^{-1}\Delta(x) + x^{-1}\Delta(y) - (x^{-1} + y^{-1})\Delta(x + y)], \tag{3.22}$$

and use (3.15) to obtain

$$\begin{aligned} \Delta(x)\Delta(y) \sim\ & i\pi\delta(x)P_y(1/y) + i\pi\delta(y)P_x(1/x) - 2P(1/xy) \\ & + P_x(1/x)P_y(1/(y + x)) + P_y(1/y)P_x(1/(x + y)). \end{aligned} \tag{3.23}$$

By virtue of the Poincaré–Bertrand formula, namely,

$$\begin{aligned} P_x(1/x)P_y(1/(y + x)) &+ P_y(1/y)P_x(1/(x + y)) \\ &= \pi^2\delta(x)\delta(y) + P(1/xy), \end{aligned} \tag{3.24}$$

equation (3.19) takes the expected form

$$\Delta(x)\Delta(y) \sim \tilde{\Delta}(x)\tilde{\Delta}(y). \tag{3.25}$$

Similar exercises show that

$$\begin{aligned} E(x; y) &\sim \tilde{\Delta}(x)\tilde{\Delta}(y), \\ \Delta(x)\Delta(-x) &\sim 2\pi t\delta(x) + 2P(1/x)\partial/\partial x. \end{aligned} \tag{3.26}$$

Armed with these asymptotics it is now a simple matter to write down the first time closure for the energy. From equations (3.12) and (3.13) one finds that

$$\begin{aligned} \frac{\partial q_{0l}^{(2)}}{\partial T_2} = \mathscr{P}_{ll'}\Bigg[& \int 2\pi\delta(\omega_{l,mn})H_{mn-m-n}q_{0m}^{(2)}q_{0n}^{(2)}\delta_{l,mn}\,d\boldsymbol{k}_{mn} \\ & + 4\int \tilde{\Delta}(\omega_{l,mn})H_{mnl-m}q_{0l}^{(2)}q_{0m}^{(2)}\delta_{l,mn}\,d\boldsymbol{k}_{mn}\Bigg], \end{aligned}$$

a closed nonlinear integro-differential equation for the evolution of the energy spectrum. (By using a long space scale, so that only local spatial homogeneity is required, an equation not unlike the Boltzmann equation can be derived.) Similar equations can be found for the evolution of each $q_0^{(n)}$, at this and higher time scales. There are some interesting issues concerning retraceability of the solution and the form of the higher closures which we shall not try to discuss at this time [**10**]. It is also possible to re-examine the entire analysis for diffusion operators [**12**].

4. Waves in shear flows. In this lecture we will discuss one of the many fascinating aspects of hydrodynamic stability. The question is whether finite amplitude spatially periodic steady flows can exist which almost everywhere are perturbations from a given parallel flow [**13**].

Consider a steady parallel flow with velocity components $(\bar{u}(y), 0, 0)$ referred to rectangular axes (x, y, z). Any $\bar{u}(y)$ is acceptable as a solution of the Euler equations; but only if $\bar{u}(y)$ is parabolic will it also be a solution of the Navier–Stokes equations. However, as is well known, flows which are quasi-parallel may be treated as parallel, provided the length scale of the disturbance is small compared to the scale of the flow variations.

We ask for a two dimensional neutral wave having a real wave speed c. In a frame moving with the wave we may write the total streamfunction Ψ as

$$\Psi(x, y) = \int^{y} (\bar{u}(y) - c)\, dy + \varepsilon\psi(x, y), \tag{4.1}$$

where ε is a measure of the disturbance amplitude and ψ is the perturbation streamfunction. The equations for Ψ and ψ are

$$\Psi_y \Delta\Psi_x - \Psi_x \Delta\Psi_y = \nu\Delta\Delta\Psi, \tag{4.2}$$

$$(\bar{u} - c)\Delta\psi_x - \bar{u}_{yy}\psi_x + \varepsilon(\psi_y \Delta\psi_x - \psi_x \Delta\psi_y) = \nu\Delta\Delta\psi. \tag{4.3}$$

Here ν is an inverse Reynolds number, all quantities having been suitably nondimensionalized.

Our interest centers on solutions to equation (4.3) when the nonlinear parameter ε and the viscous parameter ν are both small, and so the Rayleigh equation

$$(\bar{u} - c)\Delta\psi_x - \bar{u}_{yy}\psi_x = 0, \tag{4.4}$$

is the dominant balance almost everywhere. There are really two regions where (4.4) must fail. The first is where real boundary conditions are applied so that some type of viscous boundary layer would be needed, and the second is near the point (or points) where $\bar{u} = c$ $(y = y_c)$, the so called critical layer. Here attention will be restricted to the second case.

Examination of equation (4.3) for $y - y_c$ small shows that there are two boundary layer scales $y - y_c \sim \varepsilon^{1/2}$, and $y - y_c \sim \nu^{1/3}$. Accordingly two analytical developments are possible depending on whether $\varepsilon^{1/2} \ll \nu^{1/3}$ or $\varepsilon^{1/2} \gg \nu^{1/3}$. The first case corresponds to the Orr–Sommerfeld balance and the analysis proceeds along the lines of a quasi-linear theory.

The alternative is to consider a nonlinear critical layer and write

$$\Psi = \tilde{\Psi}, \qquad y - y_c = \varepsilon^{1/2} Y, \qquad \lambda = \nu\varepsilon^{-3/2} \ll 1. \tag{4.5}$$

The inner problem is then

$$\begin{aligned} &\tilde{\Psi}_Y \tilde{\Psi}_{YYx} - \tilde{\Psi}_x \tilde{\Psi}_{YYY} + \varepsilon(\tilde{\Psi}_Y \tilde{\Psi}_{xxx} - \tilde{\Psi}_x \tilde{\Psi}_{Yxx}) \\ &\quad = \lambda\varepsilon(\tilde{\Psi}_{YYYY} + 2\varepsilon\tilde{\Psi}_{YYxx} + \varepsilon^2\tilde{\Psi}_{xxxx}), \end{aligned} \tag{4.6}$$

and it is possible to match $\tilde{\Psi}$ to Ψ (the outer solution) by standard methods. The expansion for $\tilde{\Psi}$ is of the form

$$\tilde{\Psi} = \varepsilon\tilde{\Psi}^{(0)} + \varepsilon^{3/2}\log\varepsilon\tilde{\Psi}^{(1)} + \varepsilon^{3/2}\tilde{\Psi}^{(2)} + \dots, \tag{4.7}$$

with

$$\tilde{\Psi}^{(0)} = (\bar{u}_c'/2)Y^2 + \cos\alpha x, \qquad \tilde{\Psi}^{(1)} = (\bar{u}_c''/2\bar{u}_c')Y\cos\alpha x. \tag{4.8}$$

$\tilde{\Psi}^{(2)}$ involves all the harmonics and $\partial^2\tilde{\Psi}^{(2)}/\partial Y^2$ has jump discontinuities across the edge of the cats-eyes. At these regions separating closed from open streamlines thin viscous layers $O(\lambda^{1/2})$ are needed, and the analysis reduces to a Wiener–Hopf problem.

One interesting consequence is that there is no phase change across the critical layer and so a new set of modes is found. It should be stressed that only neutral modes have been investigated, and whether such modes will be observed is an open question; but at least there should be some tendency toward the nonlinear configuration before the initiation of secondary instabilities. For oblique waves strong local shears $O(\lambda^{-1/2})$ develop in the thin viscous layers and breakdown must be expected.

While the nonlinear theory may be little more than a mathematical exercise it does have some appealing features, e.g., high local shears, the initiation of turbulence due to free stream disturbances, and the importance of three dimensional disturbances. Of course it also points very clearly to the severe amplitude restrictions in applying the viscous theory ($\lambda \gg 1$).

Recently Kelly and Maslowe [**14**] have applied the nonlinear theory to stratified shear flows. Analytically the problem is tractable at small values of the Richardson number. Their calculations are interesting because the analysis offers a very plausible explanation for the initiation of Clear Air Turbulence (CAT). Here we sketch some of the ideas.

For the two dimensional motion of a Boussinesq fluid the streamfunction Ψ and temperature T satisfy the equations

$$\Delta\Psi_t + \partial(\Delta\Psi, \Psi)/\partial(x, y) = -g\beta T_x + v\Delta\Delta\Psi, \tag{4.9}$$

$$T_t + \partial(T, \Psi)/\partial(x, y) = (v/P)\Delta T, \tag{4.10}$$

where v is an inverse Reynolds number, P the Prandtl number, g the gravitational acceleration, x and y the horizontal and vertical coordinates, t the time and β the constant occurring in the assumed equation of state $\rho = \rho_0(1 - \beta(T - T_0))$.

Given a steady flow $\Psi = \int^y (\bar{u}(y) - c)\,dy + \varepsilon\psi$, $T = \bar{T}(y) + \varepsilon T$ the equations for ψ, T are

$$(\bar{u} - c)\Delta\psi_x - \bar{u}_{yy}\psi_x + g\beta T_x + \varepsilon\partial(\Delta\psi, \psi)/\partial(x, y) = v\Delta\Delta\psi, \tag{4.11}$$

(4.12) $$(\bar{u} - c)T_x - \bar{T}_y\psi_x + \varepsilon\partial(T, \psi)/\partial(x, y) = (\nu/P)\Delta T.$$

The corresponding Rayleigh equation has solutions ψ which behave like $(y - y_c)^{1/2 \pm \gamma}$ near $u(y) = c$. Here $\gamma = (1/4 - J)^{1/2}$ where the Richardson number J is given by

(4.13) $$J = g\beta\bar{T}'_c/\bar{u}'^2_c.$$

For a nonlinear balance we write

(4.14) $$y - y_c = \varepsilon^p Y, \qquad \Psi = \varepsilon^{2p}\bar{u}'^2_c\tilde{\Psi}, \qquad T = \bar{T}_c + \varepsilon^p\bar{T}'_c\tilde{T},$$

where $p = (3/2 + \gamma)^{-1}$ if $J < 1/4$, and $p = 2/3$ if $J > 1/4$. In either case the inner problem takes the form

(4.15) $$\begin{aligned}\tilde{\psi}_Y\tilde{\psi}_{YYx} - \tilde{\psi}_x\tilde{\psi}_{YYY} + J\tilde{T}_x + \varepsilon^{2p}(\tilde{\psi}_Y\tilde{\psi}_{xxx} - \tilde{\Psi}_x\tilde{\Psi}_{Yxx})\\ = \lambda(\tilde{\Psi}_{YYYY} + 2\varepsilon^{2p}\tilde{\Psi}_{YYxx} + \varepsilon^{4p}\tilde{\Psi}_{xxxx}),\end{aligned}$$

(4.16) $$\tilde{\Psi}_Y\tilde{T}_x - \tilde{\Psi}_x\tilde{T}_Y = (\lambda/P)(\tilde{T}_{YY} + \varepsilon^{2p}T_{xx}),$$

where $\lambda = \nu/\varepsilon^{3p} \ll 1$.

To leading order the solutions of (4.15), (4.16) are

(4.17) $$\tilde{T} = H(\tilde{\Psi}),$$

(4.18) $$\tilde{\Psi}_{YY} = JYH'(\tilde{\Psi}) + G(\tilde{\Psi}),$$

where G and H are determined in principle by using the viscous secularity and matching conditions. Within closed streamlines G and H must take constant values; and on the critical streamlines which separate the open and closed streamlines, thin viscous and thermal boundary layers are needed. The details of the flow must be determined by numerical methods, but some very definite qualitative features emerge.

In the linear theory it is well known that $J > \frac{1}{4}$ everywhere is a sufficient condition for stability. In the nonlinear theory there is no such critical Richardson number and waves are expected to be possible for a broad range of values of J. The critical layer problem is fully nonlinear and it is in this region where the most dramatic effects are to be expected. Of particular significance is the existence of thin shear layers which surround closed streamline regions. For these shear layers the local Richardson number is very small $O(\lambda^{1/2})$, and the layers are natural candidates for the development of local small scale secondary instabilities. Such instabilities may lead to patches of turbulence which will be convected and intensify or decay depending on the local environment, without necessarily destroying the coherence of the large scale wave. It is believed that the stability or instability of the large scale wave is relatively unimportant compared to the local spatial configuration it creates and in this spirit the assumption of a permanent large scale wave is probably a reasonable one.

5. Multi-phase modes. Finally I wish to discuss certain aspects of fully nonlinear waves. This work, which is continuing, has been done in collaboration with M. J. Ablowitz [**15**]. The investigation extends some of the interesting results obtained by Whitham [**16**], [**17**], [**18**], [**19**] and Luke [**20**].

Whitham's approach is to start with a known fully nonlinear permanent wave having a single periodicity and to derive new equations which govern the evolution of wave properties (amplitude, wave number and frequency), assuming that these properties vary slowly. The present contribution extends Whitham's results to fully interacting waves having many periodicities and shows how the properties of such nonlinear modes evolve due to large scale variations. Such modes will be called multi-phase modes. At this time we restrict attention to a study of the nonlinear Klein–Gordon equation; but the method is general and results from other theories can be recovered as special cases.

The essential features of the method are illustrated by considering the two phase modes. The one dimensional Klein–Gordon equation is

$$u_{tt} - u_{xx} + V'(u) = 0, \tag{5.1}$$

and on supposing there are two independent phases

$$\begin{gathered} \theta_1 = \Theta_1(X, T)/\mu, \quad \theta_2 = \Theta_2(X, T)/\mu, \quad X = \mu x, \\ T = \mu t, \quad 0 < \mu \ll 1, \end{gathered} \tag{5.2}$$

equation (5.1) takes the form

$$\begin{aligned} m_1 u_{\theta_1\theta_1} + 2\lambda u_{\theta_1\theta_2} + m_2 u_{\theta_2\theta_2} + V'(u) \\ = \mu[M_1 u_{\theta_1} + M_2 u_{\theta_2} + 2(\omega_1 \partial/\partial T + k_1 \partial/\partial X) u_{\theta_1} \\ + 2(\omega_2 \partial/\partial T + k_2 \partial/\partial X) u_{\theta_2}] + \mu^2[u_{XX} - u_{TT}], \end{aligned} \tag{5.3}$$

where

$$\begin{aligned} &\omega_j = -\,\partial\Theta_j/\partial T, \qquad k_j = \partial\Theta_j/\partial X, \qquad j = 1, 2, \\ &m_j = \omega_j^2 - k_j^2, \qquad M_j = \partial\omega_j/\partial T + \partial k_j/\partial X, \qquad j = 1, 2, \\ &\lambda = \omega_1\omega_2 - k_1 k_2, \end{aligned} \tag{5.4}$$

$k_j, \omega_j, j = 1, 2$ being the local wave numbers and frequencies. The existence of two independent phase functions implies conservation of each wave number, that is

$$\partial k_j/\partial T + \partial\omega_j/\partial X = 0, \qquad j = 1, 2. \tag{5.5}$$

The solution $u(\theta_1, \theta_2, X, T)$ is to be periodic with period 2π in θ_1 and θ_2 and this requirement determines the expansions for m_j and M_j. If one writes

$$
\begin{aligned}
u &= f + \mu u^{(1)} + \mu^2 u^{(2)} + \cdots, \\
m_1 &= g + \mu m_1^{(1)} + \mu^2 m_1^{(2)} + \cdots, \\
m_2 &= h + \mu m_2^{(1)} + \mu^2 m_2^{(2)} + \cdots, \\
M_1 &= G + \mu M_1^{(1)} + \mu^2 M_1^{(2)} + \cdots, \\
M_2 &= H + \mu M_2^{(1)} + \mu^2 M_2^{(2)} + \cdots,
\end{aligned} \tag{5.6}
$$

a sequence of problems results, namely,

$$
g f_{\theta_1\theta_1} + 2\lambda f_{\theta_1\theta_2} + h f_{\theta_2\theta_2} + V'(f) = 0, \tag{5.7}
$$

$$
L u^{(n)} = F^{(n)}, \qquad n \geqq 1, \tag{5.8}
$$

where $F^{(n)}$ depends on $f, u^{(1)}, \ldots, u^{(n-1)}$. For example

$$
\begin{aligned}
F^{(1)} = &-m_1^{(1)} f_{\theta_1\theta_1} - m_2^{(1)} f_{\theta_2\theta_2} + 2(\omega_1 f_{\theta_1 T} + k_1 f_{\theta_1 X}) \\
&+ 2(\omega_2 f_{\theta_2 T} + k_2 f_{\theta_2 X}) + G f_{\theta_1} + H f_{\theta_2}.
\end{aligned} \tag{5.9}
$$

Given a doubly periodic function f satisfying (5.7) it is necessary to choose the slowly varying functions $m_j^{(n)}, M_j^{(n-1)}$ to ensure that each $u^{(n)}$ is periodic in θ_1 and θ_2. Using appropriate amplitude normalizations and applying Green's Identity the entire expansion can be found. For details the reader is referred to [**15**].

To leading order the equations of slow variation (which may be hyperbolic or elliptic) provide the two phase generalization of Whitham's equations, namely,

$$
\partial k_1/\partial T + \partial \omega_1/\partial X = 0, \qquad \partial k_2/\partial T + \partial \omega_2/\partial X = 0, \tag{5.10}
$$

$$
\begin{aligned}
&\int_0^{2\pi}\int_0^{2\pi}\left[\left(\frac{\partial \omega_1}{\partial T} + \frac{\partial k_1}{\partial X}\right) f_{\theta_1}^2 + \left(\frac{\partial \omega_2}{\partial T} + \frac{\partial k_2}{\partial X}\right) f_{\theta_1} f_{\theta_2}\right. \\
&\qquad \left. + 2 f_{\theta_1}\left\{\left(\omega_1 \frac{\partial}{\partial T} + k_1 \frac{\partial}{\partial X}\right) f_{\theta_1} + \left(\omega_2 \frac{\partial}{\partial T} + k_2 \frac{\partial}{\partial X}\right) f_{\theta_2}\right\}\right] d\theta_1\, d\theta_2 = 0, \\
&\int_0^{2\pi}\int_0^{2\pi}\left[\left(\frac{\partial \omega_2}{\partial T} + \frac{\partial k_2}{\partial X}\right) f_{\theta_2}^2 + \left(\frac{\partial \omega_1}{\partial T} + \frac{\partial k_1}{\partial X}\right) f_{\theta_1} f_{\theta_2}\right. \\
&\qquad \left. + 2 f_{\theta_2}\left\{\left(\omega_1 \frac{\partial}{\partial T} + k_1 \frac{\partial}{\partial X}\right) f_{\theta_1} + \left(\omega_2 \frac{\partial}{\partial T} + k_2 \frac{\partial}{\partial X}\right) f_{\theta_2}\right\}\right] d\theta_1\, d\theta_2 = 0,
\end{aligned} \tag{5.11}
$$

$$
\begin{aligned}
E_1 &= \frac{1}{2\pi}\int_0^{2\pi}\left[\frac{g}{2} f_{\theta_1}^2 - \frac{h}{2} f_{\theta_2}^2 + V(f)\right] d\theta_2, \\
E_2 &= \frac{1}{2\pi}\int_0^{2\pi}\left[\frac{h}{2} f_{\theta_2}^2 - \frac{g}{2} f_{\theta_1}^2 + V(f)\right] d\theta_1.
\end{aligned} \tag{5.12}
$$

The higher order terms are dispersive, somewhat like a shallow water expansion so that no shock structure can be predicted without the addition of some dissipation. The generalization to N independent phases is routine. Quasi-linear theory can be recovered by choosing $V'(u)$ to be almost linear. Nonlinear wave propagation through a slowly varying medium can be dealt with by the above method. There are many other questions, e.g., stability of solutions, the connection to the averaged Lagrangian, nonlinear resonance phenomena etc., which we shall not try to discuss at this time.

References

1. F. P. Bretherton, *Resonant interactions between waves. The case of discrete oscillations*, J. Fluid Mech. **20** (1964), 457–479. MR **30** #3672.

2. O. M. Phillips, *On the dynamics of unsteady gravity waves of finite amplitude*. I. *The elementary interactions*, J. Fluid Mech. **9** (1960), 193–217. MR **22** #6266.

3. D. J. Benney, *Non-linear gravity wave interactions*, J. Fluid Mech. **14** (1962), 577–584. MR **28** #1839.

4. T. B. Benjamin, Proc. Roy. Soc. Ser. A **299** (1967), 59–75.

5. D. J. Benney and A. C. Newell, *The propagation of nonlinear wave envelopes*, J. Math. and Phys. **46** (1967), 133–139. MR **39** #2397.

6. D. J. Benney and G. J. Roskes, *Wave instabilities*, Studies in Appl. Math. **48** (1969), 377–385.

7. K. Hasselmann, *On the non-linear energy transfer in a gravity-wave spectrum*. I. *General theory*, J. Fluid Mech. **12** (1962), 481–500. MR **24** #B2243.

8. M. M. Litvak, Avco-Everett Research Report #92, 1960.

9. D. J. Benney and P. G. Saffman, *Nonlinear interaction of random waves in a dispersive medium*, Proc. Roy. Soc. Ser. A **239** (1966), 301–320.

10. D. J. Benney and A. C Newell, *Random wave closures*, Studies in Appl. Math. **48** (1969), 29–53.

11. A. C. Newell and P. J. Aucoin (to appear).

12. D. J. Benney and C. G. Lange, *The asymptotics of nonlinear diffusion*, Studies in Appl. Math. **49** (1970), 1–19.

13. D. J. Benney and R. F. Bergeron, Jr., *A new class of nonlinear waves in parallel flows*, Studies in Appl. Math. **48** (1969), 181–204.

14. R. E. Kelly and S. A. Maslowe, *The nonlinear critical layer in a slightly stratified shear flow*, Studies in Appl. Math. **49**(1970), 301–326.

15. M. J. Ablowitz and D. J. Benney, *The evolution of multi-phase modes for nonlinear dispersive waves*, Studies in Appl. Math. **49** (1970), 225–238.

16. G. B. Whitham, *Non-linear dispersive waves*, Proc. Roy. Soc. Ser. A **283** (1965), 238–261. MR **31** #996.

17. ———, *A general approach to linear and non-linear dispersive waves using a Lagrangian*, J. Fluid Mech. **22** (1965), 273–283. MR **31** #6459.

18. ———, *Non-linear dispersion of water waves*, J. Fluid Mech. **27** (1967), 399–412. MR **34** #8711.

19. ———, Proc. Roy. Soc. Ser. A **299** (1967), 6–25.

20. ———, *A perturbation method for nonlinear dispersive wave problems*, Proc. Roy. Soc. Ser. A **292** (1966), 403–412. MR **33** #3491.

Massachusetts Institute of Technology

Rotating and Stratified Flows

L. N. Howard

Lecture I

1. These lectures are about the general structure of the initial value problem for flows which are markedly affected by both a basic rotation and a basic density stratification in a gravitational field. We think of the fluid as being enclosed in a container of fixed shape which is rotating about a fixed axis with constant angular velocity. The fluid itself is compressible (with a more or less arbitrary equation of state), viscous, and heat conducting; but the coefficients of viscosity and thermal conductivity, expressed in a suitable dimensionless form, are taken to be small. This fluid is also subjected to a conservative force per unit mass, due partially to the centrifugal acceleration and partially to a gravitational field which is regarded as fixed in the system rotating with the container, but is otherwise essentially arbitrary. We shall be concerned with motions (and variations in the thermodynamic state of the fluid) which are small deviations from a basic state of:

(a) rigid rotation with the container, and

(b) statically stable but otherwise arbitrary density stratification.

Thus the mathematical model on which our considerations will be based is obtained by linearizing the full equations about such a basic state. Such motions may be excited by initial perturbations or may be driven by external agencies—stresses, differential velocities, heat fluxes or temperatures prescribed on parts of the boundary, body forces or heat source distributions, precession or other accelerations of the container, etc. In these lectures we shall concentrate on motions excited by initial conditions; thus we shall be dealing with the initial value problem for a linear homogeneous system of partial differential equations, subject to homogeneous boundary conditions.

In formulating the mathematical model, it seems most convenient to take

AMS 1970 *subject classifications*. Primary 76U05, 76V05; Secondary 76N05, 76–02.

the basic state about which we linearize to be a solution of the basic equations in the absence of viscosity and heat conduction. This is easily seen to imply (the velocity field being zero in the rotating system) that the basic density, pressure, temperature, and entropy fields are constant along the surfaces of constant potential of the combined centrifugal and gravity field. This basic state remains a solution of the momentum (and continuity) equations when viscosity is introduced (with a model of the Navier–Stokes type) but is *not* in general a solution of the heat equation in the presence of heat conduction. Thus the linearized heat equation is not homogeneous but contains an apparent heat source term. In cases of interest this term is normally small (because the heat conductivity is) and may be treated in the same manner as externally imposed heat sources. We shall suppose this done, but it should be remembered then that what we describe mathematically as a "pure initial value problem" without external forcing really corresponds to one with an external heat source (or sink) distribution so selected as to just cancel the apparent source and make the basic state a solution of the heat equation.

2. To put our mathematical model in a dimensionless form, we introduce a length L, characterizing the size of the container, and use it and the angular velocity Ω of the container to construct dimensionless length and time variables. As a velocity scale (in the rotating system) we use $\varepsilon\Omega L$, where the small parameter ε ("Rossby number") is characteristic of the magnitude of the initial perturbations relative to the basic rigid rotation. The potential of the gravitational plus centrifugal accelerations $(gz - \frac{1}{2}\Omega^2(x^2 + y^2)$ in a laboratory case, for instance) we call $L^2\Omega^2\varphi$, and we write the basic density field as $\tilde{\rho}\rho_0(\varphi)$, where $\tilde{\rho}$ is a characteristic density of the fluid; the *perturbation* density is then called $\varepsilon\tilde{\rho}\rho$. We use a dynamical scale for the pressure, writing the basic pressure field as $\tilde{\rho}\Omega^2L^2p_0(\varphi)$ and the perturbation as $\varepsilon\tilde{\rho}\Omega^2L^2p$. The hydrostatic equilibrium of the basic state then gives

$$p_0'(\varphi) + \rho_0 = 0. \tag{1}$$

Thus we may regard the basic state of stratification as given by $p_0(\varphi)$; ρ_0 then follows from (1) and the other thermodynamic variables from equations of state characterizing the fluid. We introduce a temperature scale $\tilde{T}$ and some convenient entropy scale $\tilde{s}$ (for example the gas constant R in the ideal gas law $p = R\rho T$, though we do not assume this special equation of state) to construct the dimensionless basic temperature and (specific) entropy fields T_0 and s_0, scaling the perturbation quantities with $\varepsilon\tilde{T}$ and $\varepsilon\tilde{s}$. Let us write for the dimensionless equations of state:

$$\rho_0 = f(p_0, s_0), \qquad T_0 = g(p_0, s_0). \tag{2}$$

For the perturbation we then have the linearized equations

(3) $$\rho = f_p p + f_s s$$

and

(4) $$T = g_p p + g_s s,$$

where the coefficients f_p etc. are evaluated at the basic state p_0, s_0 and so are functions only of φ. With a little thermodynamics one can show that these coefficients are given in a physically more familiar form by: $f_p = \Omega^2 L^2/c^2$, where $c^2 = (\partial p_*/\partial \rho_*)_s$ is the square of the adiabatic sound speed (asterisks indicate dimensional variables); $f_s = -\alpha\tilde{T}\rho_0 T_0 \tilde{s}/c_p$, where $\alpha = -(1/\rho_*)(\partial \rho_*/\partial T_*)_p$ is the thermal expansion coefficient at constant pressure and $c_p = T_*(\partial s_*/\partial T_*)_p$ is the specific heat at constant pressure; $g_p = (\alpha/c_p)\Omega^2 L^2 T_0/\rho_0$; and $g_s = \tilde{s}T_0/c_p$. For an ideal gas with constant specific heats (and with $\tilde{s} = R$) these reduce to

$$f_p = \frac{\Omega^2 L^2}{\gamma R\tilde{T}}\frac{1}{T_0}, \qquad f_s = -\frac{\gamma - 1}{\gamma}\rho_0,$$

$$g_p = \frac{\gamma - 1}{\gamma}\frac{\Omega^2 L^2}{R\tilde{T}}\frac{1}{\rho_0}, \qquad g_s = \frac{\gamma - 1}{\gamma}T_0,$$

where $\gamma = c_p/c_v$.

We shall suppose that the viscous stress tensor is given by a formula of the Navier–Stokes type, namely

$$\tau_{*ij} = \mu_*(\partial u_{*i}/\partial x_{*j} + \partial u_{*j}/\partial x_{*i}) + \lambda_*\delta_{ij}\nabla_* \cdot \boldsymbol{u}_*,$$

where the viscosity coefficients λ and μ are functions of the thermodynamic state, and hence for the basic state are functions of φ. To put this in dimensionless form we set

(5) $$\mu_* = \tilde{\rho}\Omega L^2 E \mu_0.$$

Here $\mu_0(\varphi)$ is taken to be $O(1)$; for definiteness we might take $\mu_0(\tilde{\varphi}) = 1$, if $\tilde{\varphi}$ is the place where $\rho_0 = 1$. The Ekman number E is thus defined, and we are here concerned with cases of small E. Similarly for the heat conduction vector we assume the usual Coulomb form $-k_*\nabla_* T_*$, and set

(5′) $$k_* = \tilde{\rho}\tilde{s}\Omega L^2(E/\sigma)k_0,$$

where again we take, say, $\tilde{s}k_0(\tilde{\varphi}) = c_p(\tilde{\varphi})$, defining thereby the Prandtl number σ. The linearized momentum equation then becomes, supposing that the angular velocity vector is $\Omega\boldsymbol{k}$:

(6) $$\rho_0(\boldsymbol{u}_t + 2\boldsymbol{k} \times \boldsymbol{u}) + \nabla p + \rho\nabla\varphi = E\nabla \cdot \tau,$$

where

$$\tau_{ij} = \mu_0(\partial u_i/\partial x_j + \partial u_j/\partial x_i) + \lambda_0\delta_{ij}\nabla \cdot \boldsymbol{u}.$$

For the linearized heat equation we have

$$\rho_0 T_0(s_t + \boldsymbol{u} \cdot \nabla s_0) = E/\sigma \nabla \cdot \boldsymbol{K} \tag{7}$$

where $\boldsymbol{K} = k_0 \nabla T + k \nabla T_0$. The second term here is necessary if the heat conductivity is not constant and the basic state is not isothermal. k is the dimensionless perturbation of the conductivity due to the perturbation in the thermodynamic variables. Finally the linearized continuity equation is

$$\rho_t + \nabla \cdot (\rho_0 \boldsymbol{u}) = 0. \tag{8}$$

Equations (3), (4), (6), (7) and (8) are the basic equations of our linearized mathematical model. The *nondissipative model* is obtained by setting $E = 0$; it is evidently *relevant* to the problem for small E, but since $E \to 0$ is a singular perturbation we must anticipate some nonuniformities in this "relevance." In particular there will normally be "boundary layers" in which, because of large gradients, the dissipative terms are not in fact all small compared to the others. Likewise we may anticipate nonuniformity in time. Evidently a time dependent solution (or part of the solution) varying on a time scale of order E^{-1} cannot be computed correctly without inclusion of the dissipative terms, and we should be warned by the homogeneous spin-up problem that because of the boundary layers such phenomena may also occur on a shorter time scale. Still, we expect the nondissipative model to give an accurate description of the solution to the initial value problem *away* from boundary layers and for times of order 1, and we need to understand it first in any case. We proceed now to this, setting $E = 0$ for the moment.

3. Let us call the flow region R and its boundary B. Various types of boundary conditions are of interest, but we shall assume that one of them is always the impenetrability of the container wall:

$$\boldsymbol{u} \cdot \boldsymbol{n} = 0 \quad \text{on } B. \tag{9}$$

This is in fact the only boundary condition which is relevant to the nondissipative case; the others for the full problem may be zero tangential velocity or zero stress, and zero perturbation temperature or zero heat flux, or combinations of these, but for the moment we need not specify this more closely.

Now it is easy to show from the basic equations in the nondissipative case that

$$\frac{\partial}{\partial t} \tfrac{1}{2}\{\rho_0 |\boldsymbol{u}|^2 + \rho_0^{-1} f_p p^2 - (s_0')^{-1} f_s s^2\} + \nabla \cdot (p\boldsymbol{u}) = 0. \tag{10}$$

Integrating this over R and using (9) we obtain the temporal conservation of the "energy" integral

$$\mathscr{E} = \frac{1}{2}\int_R [\rho_0|\boldsymbol{u}|^2 + \rho_0^{-1}f_p p^2 - (s_0')^{-1}f_s s^2]\,dV. \tag{11}$$

We shall assume that $f_s(s_0')^{-1} \leqq 0$; since most fluids expand on being heated at constant pressure ($\alpha > 0$) this usually means $s_0'(\varphi) > 0$ (entropy increases with altitude). This is the condition of static stability of the basic state and it evidently insures that $\mathscr{E}$ is a *positive definite* quadratic form. The existence of this temporally conserved positive definite form $\mathscr{E}$ implies in the usual way that any (possibly complex) *normal mode* solutions (i.e. with a time dependence of the form $e^{i\omega t}$) must have real frequencies ω, and that any two such normal modes of different frequencies are orthogonal in the sense of the hermitian inner product corresponding to $\mathscr{E}$:

$$\langle \boldsymbol{u}_1, p_1, s_1 | \boldsymbol{u}_2, p_2, s_2\rangle = \int_R \{\rho_0 \bar{\boldsymbol{u}}_1 \cdot \boldsymbol{u}_2 + \rho_0^{-1}f_p \bar{p}_1 p_2 - (s_0')^{-1}f_s \bar{s}_1 s_2\}\,dV. \tag{12}$$

These observations enable us to envisage the following familiar method for solving the initial value problem: first find all the normal modes; their spatial parts (hopefully) provide a system of functions forming an orthogonal basis for the space of initial data. Then represent the initial data in terms of this basis, i.e. express (by suitable projections) the initial conditions as a superposition of the initial values of the normal modes. Finally, the solution to the initial value problem is the same superposition of the normal modes, each evolving in time at its own proper frequency.

This all sounds very much the same as the usual elementary examples of heat conduction (and as in that case can be explicitly carried out in terms of formulas only in very special circumstances), but the present problem has one feature that is less familiar. This is the fact that the frequency *zero* is here often not only present but highly degenerate. The corresponding subspace of the initial conditions may be of infinite dimension, and the required projection of the given data onto it cannot be obtained by calculating a coefficient or two. We call the normal modes of zero frequency "geostrophic normal modes"; it turns out to be possible to describe them all fairly explicitly, even in the present generality. We shall now do this and then show how to find the "geostrophic part" of the solution to the initial value problem.

4. With $E = \partial/\partial t = 0$, equation (7) says simply that $s_0'\boldsymbol{u}\cdot\nabla\varphi = 0$; we assume $s_0' \neq 0$ (i.e. the flow really *is* stratified) and so see that any geostrophic flow must be "horizontal," i.e. directed along equipotential surfaces. (Note that in this sense vectors perpendicular to $\boldsymbol{k}$ are not usually

horizontal.) Using this, the cross product of $\nabla\varphi$ with (6) shows that $-2\varphi_z\boldsymbol{u} + \nabla\varphi \times \nabla p = 0$, while the $\boldsymbol{k}$ component of (6) gives $p_z + \rho\varphi_z = 0$. Thus $\boldsymbol{u}$ and ρ can be expressed in terms of p by:

$$\boldsymbol{u} = \frac{1}{2\varphi_z}\nabla\varphi \times \nabla p, \tag{13}$$

$$\rho = -p_z/\varphi_z. \tag{14}$$

(If $\varphi_z \equiv 0$, i.e. the rotation vector is everywhere horizontal, the problem must be reconsidered. It turns out then to be in many ways quite similar to the case of homogeneous incompressible flow, but we cannot go into this here. The most interesting example of this is when there is no gravity at all, so that φ comes entirely from the centrifugal acceleration.) Knowing p and ρ (in terms of it), s and T are then also determined by (3) and (4), and one finds that in fact (13) and (14) give solutions of (6) and (7) (with $E = \partial/\partial t = 0$) for *any* p. However (8) will be satisfied also only if

$$\nabla p \cdot \nabla\varphi \times \nabla\varphi_z = 0. \tag{15}$$

This condition is automatic if φ and φ_z are functionally related—we call this the "free" case, and a region (R or a part R_F of it) throughout which $\nabla\varphi \times \nabla\varphi_z \equiv 0$ will be called a *free region*. The part of the boundary B which is adjacent to a free region we call B_F, and we further subdivide B_F into the part B_{hF} (if any) which is *horizontal*, and the part B_{nF} which is not, i.e. on which $\boldsymbol{n} \times \nabla\varphi \neq 0$. (13) shows that the boundary condition $\boldsymbol{u} \cdot \boldsymbol{n} = 0$ is satisfied for any p on B_{hF}, but that on B_{nF} p must be constant along the intersections of equipotential surfaces with B_{nF}, i.e. along horizontal curves on B_{nF}. Subject to this restriction, *any* p gives a geostrophic normal mode in a free region. In a nonfree part of R, where $\nabla\varphi \times \nabla\varphi_z \neq 0$, (15) shows that p must be a function of φ and φ_z, i.e. constant along the integral curves of the vector field $\nabla\varphi \times \nabla\varphi_z$; we call these integral curves "geostrophic curves." With such a p, (13) becomes:

$$\boldsymbol{u} = \frac{1}{2\varphi_z}\frac{\partial p(\varphi, \varphi_z)}{\partial\varphi_z}\nabla\varphi \times \nabla\varphi_z, \tag{16}$$

showing that the streamlines of any geostrophic normal mode in a nonfree region must be geostrophic curves. Now the geostrophic curves may intersect the boundary or they may not; in the latter event, except perhaps in some very pathological case (I can think of no examples), they will form closed curves lying entirely in the region. We thus distinguish two kinds of nonfree regions: *guided* regions, R_G, covered by closed geostrophic curves not intersecting B_G, and *blocked* regions, R_B, covered by geostrophic curves which *do* cross B_B. Along a given geostrophic curve in a blocked region the velocity must be a constant multiple of $\nabla\varphi \times \nabla\varphi_z$, and so we see from $\boldsymbol{u} \cdot \boldsymbol{n} = 0$ that in fact in such a region $\boldsymbol{u} \equiv 0$, and $p = p(\varphi)$. It is easily checked that then also $\rho = -p'(\varphi)$, so the only geostrophic normal modes in a blocked region are not really flows at all, but merely perturbations to a new state of hydrostatic equilibrium characterized by the arbitrary function of one variable $p(\varphi)$. In a guided region, on the other hand, p may be any function of the *two* variables φ and φ_z, and the geostrophic normal modes are then given by (16), (14), and the equations of state. The boundary condition $\boldsymbol{u} \cdot \boldsymbol{n} = 0$ is here always satisfied. This completes the description of the geostrophic normal modes; they are evidently very numerous, particularly in the free case. It is of interest to note that in laboratory experiments we have $\varphi_* = gz - \frac{1}{2}\Omega^2(x^2 + y^2)$ and consequently we always have the *free* case, whereas for geophysical or astrophysical flows the guided or blocked cases are typical.

5. To determine the geostrophic part of the solution to the initial value problem, we must determine the projection, orthogonal in the sense of the inner product (12), of an *arbitrary* initial flow $(\boldsymbol{u}_1, p_1, s_1)$ (satisfying $\boldsymbol{u}_1 \cdot \boldsymbol{n} = 0$) onto the subspace of geostrophic normal modes just described. To do this we first find under what conditions a given flow $\boldsymbol{u}$, p, s is *orthogonal* to *all* geostrophic normal modes. Consider an arbitrary geostrophic normal mode $\hat{\boldsymbol{u}}$, $\hat{p}$, $\hat{s}$ determined in terms of its pressure field $\hat{p}$ by the formulas given above. Using these formulas and judiciously applying the divergence theorem, one finds (using the real form of (12) since geostrophic normal modes may be taken to be real):

$$\text{(17)} \quad \begin{aligned} \langle \hat{\boldsymbol{u}}, \hat{p}, \hat{s} | \boldsymbol{u}, p, s \rangle = {} & -\frac{1}{2}\int_R \hat{p}\Pi \, dV + \int_{B_h} (\varphi_z s_0')^{-1} \hat{p} s \boldsymbol{n} \cdot \boldsymbol{k} \, dS \\ & + \frac{1}{2}\int_{B_n} \varphi_z^{-1} \hat{p} \boldsymbol{n} \cdot \left[\boldsymbol{u} \times \nabla\varphi + \frac{2s}{s_0'}\boldsymbol{k} \right] dS, \end{aligned}$$

where

$$\text{(18)} \qquad \Pi = \rho_0^{-1} \nabla \cdot \left[\rho_0 \varphi_z^{-1} \left(\boldsymbol{u} \times \nabla\varphi + \frac{2s}{s_0'}\boldsymbol{k} \right) \right] - 2\rho_0^{-1}\rho.$$

This quantity Π, which we call the "potential vorticity" of the flow $(\boldsymbol{u}, p, s)$, plays a basic role. Suppose the inner product (17) is zero for an arbitrary geostrophic normal mode $\hat{p}$. We can in particular choose $\hat{p}$ to be zero on the boundary and in R_G and R_B, but otherwise arbitrary in R_F. (17) then shows that Π must be zero throughout R_F. Again, we may choose $\hat{p}$ to be zero in R_G and R_B, so that the volume integral vanishes (Π being zero in R_F), and suppose that $\hat{p}$ is zero on B_n, B_{hG}, and B_{hB} but is arbitrary on B_{hF}. (17) then shows that s must be zero on B_{hF}. Since every geostrophic pressure field is constant along the horizontal curves $\Gamma_n(\varphi)$ lying on B_n, $\hat{p}$ cannot be chosen arbitrarily on B_n. However, since on B_n $\hat{p} = \hat{p}(\varphi)$, the second surface integral in (17) can be evaluated by cutting B_n into infinitesimal strips between adjacent curves Γ_n : in this case one finds $dS = |\nabla\varphi \times \boldsymbol{n}|^{-1}\, d\varphi\, d\lambda$, where λ is arc length along Γ_n, and so

$$\int_{B_n} \varphi_z^{-1}\hat{p}\boldsymbol{n}\cdot\left[\boldsymbol{u}\times\nabla\varphi + \frac{2s}{s_0'}\boldsymbol{k}\right] dS \tag{19}$$
$$= \int_{B_n} \hat{p}(\varphi)\, d\varphi \int_{\Gamma_n(\varphi)} \varphi_z^{-1}|\nabla\varphi \times \boldsymbol{n}|^{-1}\boldsymbol{n}\cdot\left(\boldsymbol{u}\times\nabla\varphi + \frac{2s}{s_0'}\boldsymbol{k}\right) d\lambda.$$

Since we can choose a $\hat{p}$ which is zero in R_G and R_B, zero on B_G and B_B, and an arbitrary function of φ on B_{nF}, we see now (since φ_z is constant along horizontal curves on B_{nF}) that the quantity

$$C(\Gamma_n) = \int_{\Gamma_n} |\nabla\varphi \times \boldsymbol{n}|^{-1}\boldsymbol{n}\cdot\left(\boldsymbol{u}\times\nabla\varphi + \frac{2s}{s_0'}\boldsymbol{k}\right) d\lambda \tag{20}$$

must be zero for each horizontal curve Γ_n on B_{nF}.

Similar arguments, using the available arbitrariness of $\hat{p}$ in the various R's and B's, lead finally to the following:

In order that a flow $(\boldsymbol{u}, p, s)$ should be orthogonal to all geostrophic normal modes, it is necessary and sufficient that each of the following should vanish:

(a) Π (equation (18)) at each point of R_F ;

(21) (b) $$\bar{\Pi}(\Gamma) = \int_{\Gamma} \varphi_z^2|\nabla\varphi \times \nabla\varphi_z|^{-1}\Pi\, d\lambda$$

for each geostrophic curve Γ in R_G ;

(22) (c) $$Q(\Sigma) = \int_{\Sigma} \Pi|\nabla\varphi|^{-1}\, dS - \int_{\Gamma} \varphi_z^{-1}|\nabla\varphi \times \boldsymbol{n}|^{-1}\left(\boldsymbol{u}\times\nabla\varphi + \frac{2s}{s_0'}\boldsymbol{k}\right)\cdot\boldsymbol{n}\, d\lambda$$

for each (connected) horizontal surface Σ in R_B, having boundary Γ on B;

(d) $C(\Gamma_n)$ (equation (20)) for each horizontal curve on B_{nF} or B_{nG} ;

(e) s on B_{hF};

(23) (f) $$D(\Gamma_h) = \int_{\Gamma_h} |\nabla\varphi \times \nabla\varphi_z|^{-1}s\, d\lambda$$

for each geostrophic curve Γ_h on B_{hG} ;

(24) (g) $$F = \int_{B^*_{hB}} |\nabla\varphi|^{-1}s\, dS$$

for each connected component B^*_{hB} of B_{hB}.

The geostrophic part of the solution to the initial value problem is a geostrophic normal mode $\boldsymbol{u}_g$, etc., such that the difference $\boldsymbol{u}_1 - \boldsymbol{u}_g$, etc. between the initial data and $\boldsymbol{u}_g$ is orthogonal to all geostrophic normal modes; consequently $\boldsymbol{u}_g$, etc. must be *that geostrophic normal mode which has the same values of the quantities* (a), . . . , (g) *as the initial flow* $\boldsymbol{u}_1$ *etc. has.* A geostrophic normal mode is uniquely determined by these conditions, for the difference of two such would have zero values of the quantities (a), . . . , (g) and so be self-orthogonal.

To compute the geostrophic part in the free case, for example, we must then solve the elliptic equation $\Pi(p_g) = \Pi_0$ for the geostrophic pressure field p_g, subject to the conditions that p_g is constant along horizontal curves on B_{nF} and gives $C(\Gamma_n) = C_0(\Gamma_n)$ on B_{nF} and $s_g = s_0$ on B_{hF}.

Most of these results come from the paper: "On the initial value problem for rotating stratified flow," by L. N. Howard and W. L. Siegmann, Studies in Applied Mathematics **48** (1969), 153.

Lecture II

1. We turn now to the question of how the conclusions of the previous lecture are affected by the introduction of the dissipative mechanisms. In a sense we are concerned with the perturbation of the normal modes of the nondissipative system by the small dissipation; this involves a change in the eigenfunctions and a change in the frequency. The functions are presumably perturbed only slightly, except in the boundary layers that reflect the fact that it is really a singular perturbation; here the change is, of course, likely to be $O(1)$. The frequencies are presumably also perturbed only a little, yet again the nonuniformities of singular perturbations appear in that though the frequency perturbation formally appears to be $O(E)$, in some cases it is larger: $O(E^{1/2})$. As before, we concentrate on the geostrophic normal modes and ask how they are affected by viscosity—how their "zero frequency" is perturbed into a damping. As in the question of the initial value problem, here also the great degeneracy of the geostrophic normal modes is expected to have interesting consequences. We expect the perturbation to "split" the degeneracy in some sense; however, while this is a helpful way to think about the problem in general terms, we shall not insist on finding a "viscous normal mode" (pure exponential time dependence), but simply take a particular geostrophic normal mode specified by a certain pressure p and investigate the time dependence of this produced by the dissipative terms. The spin-up problem for axisymmetric homogeneous flow indicates that this is likely to give a clearer picture (different spin-up times for different depths—thus a true exponential "mode" has its motion restricted to one depth contour, in general).

The various quantities (a), . . . , (g), such as the potential vorticity, which characterize the geostrophic part of the solution to the initial value problem, are, of course, constant in time for any solution of the non-dissipative equations (with $\boldsymbol{u} \cdot \boldsymbol{n} = 0$). This is because the geostrophic part itself is, but it can naturally also be shown directly from the equations of the time dependent nondissipative model. In the dissipative case, the "interior flow" (away from boundary layers) is still to first order like a geostrophic normal mode but now varies slowly with time. We shall study this evolution by investigating how the dissipative effects produce slow changes in the potential vorticity and other quantities which determine the geostrophic part.

2. Using our basic equations and the definition of the potential vorticity Π, one can derive the following formula:

$$\frac{\partial \Pi}{\partial t} = \frac{1}{\rho_0 \varphi_z^2} \nabla p \cdot \nabla \varphi \times \nabla \varphi_z + \frac{E}{\rho_0} \left\{ \nabla \cdot \left[\frac{1}{\varphi_z} (\nabla \cdot \tau) \times \nabla \varphi + \frac{2\boldsymbol{k}}{\varphi_z \sigma T_0 s_0'} \nabla \cdot \boldsymbol{K} \right] \right\}. \tag{25}$$

In the nondissipative ($E = 0$) free ($\nabla \varphi \times \nabla \varphi_z = 0$) case this exhibits the temporal invariance of Π. In the dissipative free case it shows us that the potential vorticity of the interior flow (away from boundary layers) can change only on the dissipative time-scale, $t = O(E^{-1})$. Thus any evolution of a geostrophic normal mode on a shorter time scale, in particular on the homogeneous spin-up time scale $t = O(E^{-1/2})$ (which we shall see in a moment does in general occur), must take place *without* alteration of the potential vorticity in the interior. In the guided case the potential vorticity itself is not necessarily constant in time because of the $\varphi_z^{-2} \nabla p \cdot \nabla \varphi \times \nabla \varphi_z$ term, but it is easily shown from (25) with $E = 0$ that the "mean potential vorticity" $\overline{\Pi}(\Gamma)$ (equation (21)) is constant in time for every closed geostrophic curve in R_G. The same formula with $E \neq 0$ also then shows that $\overline{\Pi}(\Gamma)$ can vary only on the dissipative time scale for any closed interior geostrophic curve. In the blocked case it is the quantity Q (equation (22)) which is constant for nondissipative flows; since it involves integrations which extend to the boundary, we must consider this a little more closely. What we are doing is this: we are considering a solution of the dissipative equations which can reasonably be called a modified geostrophic normal mode. That is one whose temporal variation is *slow* (on a time scale $O(E^{-1})$ or $O(E^{-1/2})$ or something of the sort). We think of this solution as being the sum of an "interior part" $\boldsymbol{u}_I$ and a boundary-layer part $\boldsymbol{u}_B$, the latter being transcendentally small (in E) off the boundary. $\boldsymbol{u}_I$ itself has the form $u_I^0 + E^{1/2}u_I^1 + Eu_I^2 + \cdots$, or something similar, and at any given time u_I^0 is like a geostrophic normal mode in that its velocity components and entropy are related to its pressure field by the same formulas and its normal component on B is zero. However u_I is changing (slowly) with time. To compute this change we use the dissipative (interior) equations to find the rate of change of the quantities (a), ..., (g) in terms of the current u_I, in particular for the $O(E)$ part in terms of u_I^0. This is straightforward for Π and $\bar{\Pi}$; as just stated we merely use equation (25). But for the other quantities we shall see that we need also some information about boundary values of the flow u_I; insofar as this involves $u_I^0 \cdot n$, there is no problem, for this is zero, but we shall need more. For example, we shall see in a moment that we need $u_I^1 \cdot n$, and this is not necessarily zero; however, its value can be found in terms of u_I^0 by studying the boundary layers. To return to the quantity Q needed in the blocked case, one can show from the equations that we have

$$\frac{dQ}{dt} = 2\int_\Gamma |\nabla\varphi \times n|^{-1} u \cdot n \, d\lambda + E\left\{\int_\Sigma \nabla \cdot \left(\frac{2k\nabla \cdot K}{\sigma\varphi_z T_0 s_0'}\right)\frac{dS}{\rho_0|\nabla\varphi|} - \int_\Gamma \frac{2k \cdot n\nabla \cdot K}{\sigma\varphi_z T_0 s_0'\rho_0}\frac{d\lambda}{|\nabla\varphi \times n|}\right\}. \tag{26}$$

In the nondissipative case this shows $dQ/dt = 0$, for here $u \cdot n = 0$ on the boundary, hence on Γ. In the dissipative case, remembering that we are computing this for the *interior* part u_I, we do not know that $u_I \cdot n = 0$, only that $u_I^0 \cdot n = 0$, on Γ. If there should be, say, a $u_I^1 \cdot n$ not zero on Γ, then we should find Q changing on the $E^{-1/2}$ time scale; this can be ascertained only by examining the boundary layer. In the present case, however, we know that $u_I^0 \equiv 0$ since we have a blocked region. Thus we might expect that with, say, a rigid wall the tangential velocity in the boundary layer would be $O(E^{1/2})$ and (anticipating a boundary layer of thickness $E^{1/2}$) so $u_B \cdot n$ of order E, and consequently also $u_I \cdot n$ would be $O(E)$. This turns out to be true in the case of the *insulated* thermal boundary condition but may not be with a temperature condition. Thus in the insulated case Q, like Π and $\bar{\Pi}$, can vary only on the dissipative time scale, but otherwise it may change on the homogeneous spin-up scale. If we use similarly the dissipative equations (for the interior part of the flow) we find for the quantity (d):

$$\frac{d}{dt}C(\Gamma_n) = -2\int_{\Gamma_n} u \cdot n \frac{\varphi_z \, d\lambda}{|\nabla\varphi \times n|} + E\int_{\Gamma_n}\left[(\nabla \cdot \tau)\cdot(\nabla\varphi \times n) + \frac{2k \cdot n}{\sigma s_0' T_0}\nabla \cdot K\right]\frac{d\lambda}{\rho_0|\nabla\varphi \times n|}. \tag{27}$$

Here again the possibility for evolution on the $E^{-1/2}$ time scale occurs provided $\boldsymbol{u}_I^1 \cdot \boldsymbol{n}$ is not zero. The situation for (e), namely s on B_{hF}, is particularly simple:

$$s_t = -\boldsymbol{u} \cdot \nabla \varphi s_0' + \frac{E}{\rho_0 T_0 \sigma} \nabla \cdot \boldsymbol{K} \tag{28}$$

on B_{hF} (for the interior part). Since here $\nabla\varphi$ is parallel to $\boldsymbol{n}$, again the issue of $E^{-1/2}$ evolution depends on the presence of $\boldsymbol{u}_I^1 \cdot \boldsymbol{n}$ on the boundary; and the situation is the same for the other two quantities (f) and (g), which are obtained from averages of (28). Summarizing, the quantities Π and $\overline{\Pi}$ cannot be changed on the homogeneous spin-up time scale, but the others, (c), . . . , (g), may do so if the boundary layers required to match the interior flow to the true wall boundary conditions lead to interior flow normal velocities of order $E^{1/2}$. All of these quantities will normally change on the diffusive time scale.

3. We must now investigate the boundary layers. It turns out that there is a boundary layer of thickness $E^{1/2}$ which is relevant both on the $\tau = E^{1/2}\, t$ and on the $T = Et$ time scales, and another of thickness $E^{1/4}$ relevant only on the τ scale. The reason for the existence of this $E^{1/4}$ layer can be seen from the potential vorticity equation (25); it is just on this spatial scale that the diffusive term becomes comparable with the time derivative term on the τ scale. Another way to look at it is that in, say, the free case Π is constant on the τ scale and so the interior equation is just $\Pi = \Pi_0$, a second order equation for the interior pressure; whereas on the T scale we have an equation for the interior pressure which is fourth order. In the second case the interior pressure can more completely satisfy the full boundary conditions, and less is required in the way of boundary layer matching; one may say that on the T scale the $E^{1/4}$ layer has diffused into the interior. (On any *long* time scale the interior velocity and entropy are to first order related to the pressure like a geostrophic normal mode, so the interior flow is determined by the pressure. Thus even on the long time scale the interior flow cannot in general satisfy all the boundary conditions, and the $E^{1/2}$ layer is required.)

For the $E^{1/2}$ boundary layer, let the distance from the boundary into the fluid be $E^{1/2}\zeta$ so that ζ is the normal boundary layer coordinate. The boundary layer part of the flow will be denoted by $\hat{\boldsymbol{u}}^0 + E^{1/2}\hat{\boldsymbol{u}}^1 + \cdots$, $\hat{p}^0 + E^{1/2}\hat{p}^1 + \cdots$, etc., assuming that at least one of the flow variables has its largest part $O(E^0)$ and none is larger than this. Of course in matching with the interior and $E^{1/4}$ layers, one might possibly find that the largest part of the $E^{1/2}$ boundary layer flow was not $O(E^0)$, but in that case the whole thing could be appropriately scaled. Then the boundary layer momentum equation gives first

$$\partial \hat{p}^0/\partial \zeta = 0$$

so $\hat{p}^0 = 0$, and to the next order we have

$$2\rho_0 \boldsymbol{k} \times \hat{\boldsymbol{u}}^0 - \boldsymbol{n}(\partial \hat{p}^1/\partial \zeta) + \hat{\rho}^0 \nabla \varphi = \mu_0(\partial^2 \hat{\boldsymbol{u}}^0/\partial \zeta^2). \tag{29}$$

The continuity equation gives first $\partial/\partial \zeta(\boldsymbol{n} \cdot \hat{\boldsymbol{u}}^0) = 0$, so $\boldsymbol{n} \cdot \hat{\boldsymbol{u}}^0 = 0$, and then:

$$-\rho_0(\partial/\partial \zeta)(\boldsymbol{n} \cdot \hat{\boldsymbol{u}}^1) + (\boldsymbol{n} \times \nabla) \cdot \rho_0(\boldsymbol{n} \times \hat{\boldsymbol{u}}^0) = 0. \tag{30}$$

Finally the heat equation is

$$\hat{\boldsymbol{u}} \cdot \nabla \varphi s_0' = (k_0/\sigma \rho_0 T_0)/(\partial^2 \hat{T}^0/\partial \zeta^2). \tag{31}$$

(We here neglect the extra term in $\boldsymbol{K}$ that arises if the conductivity is not constant; this term makes no essential difference but merely complicates the formulas.) We now see an important difference between horizontal and nonhorizontal boundaries. On B_h, $\nabla \varphi \times \boldsymbol{n} = 0$, so the vanishing of $\boldsymbol{n} \cdot \hat{\mathbf{u}}^0$ implies that of $\hat{\boldsymbol{u}}^0 \cdot \nabla \varphi$; and so from (31) also $\hat{T}^0 = 0$. Since $\hat{p}^0 = 0$, the equations of state show that $\hat{\rho}^0$ and $\hat{s}^0$ are zero too and the buoyancy term disappears from (29). On B_h the $E^{1/2}$ layer is essentially the same as the Ekman layer in a homogeneous fluid, to first order. The transverse component of (29) then becomes

$$2\rho_0 \boldsymbol{n} \cdot \boldsymbol{k} \boldsymbol{n} \times \hat{\boldsymbol{u}}^0 = \mu_0(\partial^2 \hat{\boldsymbol{u}}^0/\partial \zeta^2). \tag{32}$$

Taking the cross product of (32) with $\boldsymbol{n}$ and combining we get

$$(\partial^2/\partial \zeta^2)(\hat{\boldsymbol{u}}^0 + i\boldsymbol{n} \times \hat{\boldsymbol{u}}^0) = -2i(\rho_0/\mu_0)\boldsymbol{n} \cdot \boldsymbol{k}(\hat{\boldsymbol{u}}^0 + i\boldsymbol{n} \times \hat{\boldsymbol{u}}^0). \tag{33}$$

Thus when $\boldsymbol{n} \cdot \boldsymbol{k} < 0$,

$$\hat{\boldsymbol{u}}^0 + i\boldsymbol{n} \times \hat{\boldsymbol{u}}^0 = \boldsymbol{C} \exp[-(1 + i)(|\boldsymbol{n} \cdot \boldsymbol{k}|\rho_0/\mu_0)^{1/2}\zeta]; \tag{34}$$

and when $\boldsymbol{n} \cdot \boldsymbol{k} > 0$,

$$\hat{\boldsymbol{u}}^0 + i\boldsymbol{n} \times \hat{\boldsymbol{u}}^0 = \boldsymbol{C} \exp[-(1 - i)(\boldsymbol{n} \cdot \boldsymbol{k}\rho_0/\mu_0)^{1/2}\zeta]. \tag{35}$$

($\boldsymbol{k} \cdot \boldsymbol{n} = 0$, i.e., $\varphi_z = 0$, requires special consideration—also with respect to the interior flow—as noticed in Lecture I.) The complex vector $\boldsymbol{C}$ is to be determined later by the matching process so as to satisfy the viscous boundary conditions on the transverse velocity components: vanishing (total) velocity or tangential stress. But once it is fixed we can calculate $\boldsymbol{n} \cdot \hat{\boldsymbol{u}}^1$ from (30):

$$\begin{aligned} \rho_0 \boldsymbol{n} \cdot \hat{\boldsymbol{u}}^1 &= -\operatorname{Im} \boldsymbol{n} \times \nabla \cdot \rho_0 \left(\frac{1-i}{2}\right)\left(\frac{\mu_0}{|\boldsymbol{n} \cdot \boldsymbol{k}|\rho_0}\right)^{1/2} (\hat{\boldsymbol{u}}^0 + i\boldsymbol{n} \times \hat{\boldsymbol{u}}^0) \\ &= \boldsymbol{n} \times \nabla \cdot \frac{\rho_0}{2}\left(\frac{\mu_0}{|\boldsymbol{n} \cdot \boldsymbol{k}|\rho_0}\right)^{1/2} (\hat{\boldsymbol{u}}^0 - \boldsymbol{n} \times \hat{\boldsymbol{u}}^0) \end{aligned} \tag{36}$$

if $\boldsymbol{n} \cdot \boldsymbol{k} < 0$, and

$$\rho_0 \boldsymbol{n} \cdot \hat{\boldsymbol{u}}^1 = \boldsymbol{n} \times \nabla \cdot \frac{\rho_0}{2} \left(\frac{\mu_0}{\boldsymbol{n} \cdot \boldsymbol{k} \rho_0} \right)^{1/2} (-\hat{\boldsymbol{u}}^0 - \boldsymbol{n} \times \hat{\boldsymbol{u}}^0) \tag{37}$$

if $\boldsymbol{n} \cdot \boldsymbol{k} > 0$. Similarly, $\hat{p}^1$ can be obtained from the normal component of (29), and $\hat{T}^1$ from the $O(E^{1/2})$ heat equation which is just like (31) and expresses $\hat{T}^1_{\zeta\zeta}$ in terms of $\hat{\boldsymbol{u}}^1 \cdot \nabla\varphi$, i.e. in terms of $\hat{\boldsymbol{u}}^1 \cdot \boldsymbol{n}$ in the present case of a horizontal boundary. Finally knowing $\hat{T}^1$ and $\hat{p}^1$ we have $\hat{s}^1$ and $\hat{\rho}^1$ from the equation of state. This completes the description of the $E^{1/2}$ layer on B_h.

On B_n we can no longer conclude from (31) that $\hat{T}^0 = 0$, and the structure is rather different. Taking $\boldsymbol{n} \times$ (29) we get

$$-2\rho_0 \boldsymbol{n} \cdot \boldsymbol{k} \hat{\boldsymbol{u}}^0 + \hat{p}^0 \boldsymbol{n} \times \nabla\varphi = \mu_0 (\partial^2/\partial\zeta^2) \boldsymbol{n} \times \hat{\boldsymbol{u}}^0$$

from which we have:

$$-2\rho_0 \boldsymbol{n} \cdot \boldsymbol{k} \hat{\boldsymbol{u}}^0 \cdot \nabla\varphi = \mu_0 (\partial^2/\partial\zeta^2)(\nabla\varphi \times \boldsymbol{n}) \cdot \hat{\boldsymbol{u}}^0 \tag{38}$$

and

$$2\rho_0 \boldsymbol{n} \cdot \boldsymbol{k} \nabla\varphi \times \boldsymbol{n} \cdot \hat{\boldsymbol{u}}^0 + \hat{p}^0 |\boldsymbol{n} \times \nabla\varphi|^2 = \mu_0 (\partial^2/\partial\zeta^2)(\hat{\boldsymbol{u}}^0 \cdot \nabla\varphi). \tag{39}$$

Eliminating $\hat{\boldsymbol{u}}^0 \cdot \nabla\varphi$ between (31) and (38) we have:

$$\mu_0 \frac{\partial^2}{\partial\zeta^2} (\nabla\varphi \times \boldsymbol{n} \cdot \hat{\boldsymbol{u}}^0) = -\frac{2\boldsymbol{n} \cdot \boldsymbol{k} k_0}{s_0' \sigma T_0} \frac{\partial^2 \hat{T}^0}{\partial\zeta^2},$$

so

$$\boldsymbol{n} \times \nabla\varphi \cdot \hat{\boldsymbol{u}}^0 = \frac{2\boldsymbol{n} \cdot \boldsymbol{k} k_0}{s_0' \sigma T_0 \mu_0} \hat{T}^0. \tag{40}$$

If $\boldsymbol{n} \cdot \boldsymbol{k} = 0$, this velocity component is thus zero, but in any case it is determined by $\hat{T}^0$. We note also that since $\hat{p}^0 = 0$, $\hat{\rho}^0 = (f_s/g_s)\hat{T}_0 = -\alpha \tilde{T} \rho_0 \hat{T}^0$, so using this and (40) in (39) we get

$$\mu_0 \frac{\partial^2}{\partial\zeta^2} (\hat{\boldsymbol{u}} \cdot \nabla\varphi) = -\left[\frac{4(\boldsymbol{n} \cdot \boldsymbol{k})^2 \rho_0 k_0}{s_0' \sigma T_0 \mu_0} + \alpha \tilde{T} \rho_0 |\boldsymbol{n} \times \nabla\varphi|^2 \right] \hat{T}^0. \tag{41}$$

(31) and (41) form a pair of equations for $\hat{T}^0$ and $\hat{\boldsymbol{u}}^0 \cdot \nabla\varphi$ with the same structure as the Ekman layer equations. With some obvious abbreviation they are (assuming for definiteness that α and s_0' are positive):

$$(\partial^2/\partial\zeta^2)(\hat{\boldsymbol{u}}^0 \cdot \nabla\varphi) = -2M^4 N^2 \hat{T}^0, \qquad (\partial^2/\partial\zeta^2)(\hat{T}^0) = 2N^2 \hat{\boldsymbol{u}}^0 \cdot \nabla\varphi$$

or

$$(\partial^2/\partial\zeta^2)(\hat{\boldsymbol{u}}^0 \cdot \nabla\varphi + iM^2 \hat{T}^0) = 2iM^2 N^2 (\hat{\boldsymbol{u}}^0 \cdot \nabla\varphi + iM^2 \hat{T}^0); \tag{42}$$

$$(43) \qquad \therefore \quad \hat{\boldsymbol{u}}^0 \cdot \nabla\varphi + iM^2\hat{T}^0 = Ke^{-(1+i)MN\zeta}.$$

The complex number K is to be determined from matching conditions on the "vertical" velocity $\boldsymbol{u}^0 \cdot \nabla\varphi$ and the temperature. The other tangential velocity component is then given by (40) so that $\boldsymbol{n} \times \hat{\boldsymbol{u}}^0$ is determined, in fact by

$$(44) \quad \boldsymbol{n} \times \hat{\boldsymbol{u}}^0 = |\boldsymbol{n} \times \nabla\varphi|^{-2}\left[\hat{\boldsymbol{u}}^0 \cdot \nabla\varphi \boldsymbol{n} \times \nabla\varphi + \frac{2\boldsymbol{n} \cdot \boldsymbol{k}k_0}{s_0'\sigma T_0\mu_0}\hat{T}^0\boldsymbol{n} \times (\boldsymbol{n} \times \nabla\varphi)\right].$$

Thus $\boldsymbol{n} \cdot \hat{\boldsymbol{u}}^1$ is determined from (30), while $\hat{\rho}^0$ follows from the equation of state and finally $\hat{p}^1$ from the normal component of (29).

4. For the $E^{1/4}$ layer we use $E^{1/4}\xi$ for the distance into the fluid from the boundary and write for *this* boundary layer part $\tilde{\boldsymbol{u}}^0 + E^{1/4}\tilde{\boldsymbol{u}}^{1/2} + \cdots$, etc. Then we find as before that $\tilde{p}^0 = 0$ and $\tilde{\boldsymbol{u}}^0 \cdot \boldsymbol{n} = 0$, and to the next order

$$(45) \qquad 2\rho_0\boldsymbol{k} \times \tilde{\boldsymbol{u}}^0 - \boldsymbol{n}\frac{\partial\tilde{p}^{1/2}}{\partial\xi} + \tilde{\rho}^0\nabla\varphi = 0,$$

$$(46) \qquad -\rho_0\frac{\partial}{\partial\xi}(\boldsymbol{n} \cdot \tilde{\boldsymbol{u}}^{1/2}) + \boldsymbol{n} \times \nabla \cdot \rho_0(\boldsymbol{n} \times \tilde{\boldsymbol{u}}^0) = 0,$$

$$(47) \qquad \tilde{\boldsymbol{u}}^0 \cdot \nabla\varphi s_0' = 0.$$

On B_h, (47) is the same as $\tilde{\boldsymbol{u}}^0 \cdot \boldsymbol{n} = 0$, while taking $\boldsymbol{n} \times (45)$ we get $-2\rho_0\boldsymbol{n} \cdot \boldsymbol{k}\tilde{\boldsymbol{u}}^0 = 0$.
Thus

$$\tilde{\boldsymbol{u}}^0 = 0, \boldsymbol{n} \cdot \tilde{\boldsymbol{u}}^{1/2} = 0,$$

and

$$(48) \qquad \partial\tilde{p}^{1/2}/\partial\xi = \tilde{\rho}^0\boldsymbol{n} \cdot \nabla\varphi.$$

$\tilde{T}^0$ is related to $\tilde{\rho}^0$ by the equation of state.

To the next order we have $\tilde{\boldsymbol{n}} \cdot \tilde{\boldsymbol{u}}^{1/2} = 0$ and

$$(49) \qquad 2\rho_0\boldsymbol{k} \times \tilde{\boldsymbol{u}}^{1/2} - \tilde{\boldsymbol{n}}\frac{\partial\tilde{p}^1}{\partial\zeta} - \boldsymbol{n} \times (\boldsymbol{n} \times \nabla\tilde{p}^{1/2}) + \tilde{\rho}^{1/2}\nabla\varphi = 0,$$

$$(50) \qquad -\rho_0\frac{\partial}{\partial\xi}(\boldsymbol{n} \cdot \tilde{\boldsymbol{u}}^1) + \boldsymbol{n} \times \nabla \cdot (\rho_0\boldsymbol{n} \times \tilde{\boldsymbol{u}}^{1/2}) = 0.$$

$\boldsymbol{n} \times (49)$ gives $-2\rho_0\boldsymbol{n} \cdot \boldsymbol{k}\tilde{\boldsymbol{u}}^{1/2} + \boldsymbol{n} \times \nabla\tilde{p}^{1/2} = 0$, which determines $\tilde{\boldsymbol{u}}^{1/2}$ in terms of $\tilde{p}^{1/2}$; $\boldsymbol{n} \cdot (49)$ then gives $\partial\tilde{p}^1/\partial\xi$ in terms of $\tilde{p}^{1/2}$ and we can continue. The formulas show the relative orders ($\tilde{\rho}^0$, $\tilde{T}^0$, $\tilde{s}^0$, $\tilde{p}^{1/2}$, $\boldsymbol{n} \times \tilde{\boldsymbol{u}}^{1/2}$, $\boldsymbol{n} \cdot \tilde{\boldsymbol{u}}^1$) and indicate that the largest parts are determined from $\tilde{p}^{1/2}$ essentially as in the interior, but to get the boundary layer equation we have to go to still higher order. This can be circumvented by using the potential vorticity equation (25). In the $E^{1/4}$ layer on B_h we find

$$\tilde{\Pi} = -E^{-1/4}\frac{2\boldsymbol{n}\cdot\boldsymbol{k}}{\varphi_z s_0'}\frac{\partial\tilde{s}^0}{\partial\xi} = -E^{-1/4}\frac{2\boldsymbol{n}\cdot\boldsymbol{k}}{\boldsymbol{n}\cdot\nabla\varphi\varphi_z f_s s_0'}\frac{\partial^2\tilde{p}^{1/2}}{\partial\xi^2}. \tag{51}$$

Similarly, the major part of the dissipative term in (25) is

$$-\frac{E^{1/4}}{\rho_0}\frac{2\boldsymbol{n}\cdot\boldsymbol{k}k_0}{\varphi_z\sigma T_0 s_0'}\frac{\partial^3\tilde{T}^0}{\partial\xi^3} = -\frac{E^{1/4}}{\rho_0}\frac{2\boldsymbol{n}\cdot\boldsymbol{k}k_0 g_s}{\varphi_z\sigma T_0 s_0'\boldsymbol{n}\cdot\nabla\varphi f_s}\frac{\partial^4\tilde{p}^{1/2}}{\partial\xi^4}.$$

Thus (25) tells us for this case:

$$\begin{aligned}\frac{\partial}{\partial\tau}\frac{\partial^2\tilde{p}^{1/2}}{\partial\xi^2} = &-\frac{\boldsymbol{n}\cdot\nabla\varphi}{2\boldsymbol{n}\cdot\boldsymbol{k}}\frac{f_s s_0'}{\rho_0\varphi_z}(\nabla\varphi\times\nabla\varphi_z)\cdot\nabla\tilde{p}^{1/2}\\ &+\frac{k_0 g_s}{\sigma\rho_0 T_0}\frac{\partial^4\tilde{p}^{1/2}}{\partial\xi^4}.\end{aligned} \tag{52}$$

In the free case, this can be integrated and we have simply the ordinary heat conduction equation. Otherwise it is somewhat more complicated and can no longer be treated entirely locally.

On B_n, (47) is an independent condition from $\tilde{\boldsymbol{u}}^0\cdot\boldsymbol{n} = 0$; taking $\boldsymbol{n}\times(45)$ we here get

$$-2\rho_0\boldsymbol{n}\cdot\boldsymbol{k}\tilde{\boldsymbol{u}}^0 + \tilde{\rho}^0\boldsymbol{n}\times\nabla\varphi = 0. \tag{53}$$

Then $\boldsymbol{n}\cdot(45)$ gives

$$\partial\tilde{p}^{1/2}/\partial\xi = (2\rho_0\varphi_z/|\boldsymbol{n}\times\nabla\varphi|^2)(\boldsymbol{n}\times\nabla\varphi)\cdot\tilde{\boldsymbol{u}}^0. \tag{54}$$

Thus here $\tilde{\boldsymbol{u}}^0$ can have an $O(1)$ component in the direction $\boldsymbol{n}\times\nabla\varphi$, which then determines $\tilde{p}^{1/2}$ from (54) and $\tilde{\rho}^0$ from (53). Note that $\tilde{\rho}^0 = 0$ if $\boldsymbol{n}\cdot\boldsymbol{k} = 0$ but otherwise may be $O(1)$. The normal component is then given by (46) as $\boldsymbol{n}\cdot\tilde{\boldsymbol{u}}^{1/2}$, and the component along $\nabla\varphi$ from the heat equation is $O(E^{1/2})$ when $\boldsymbol{n}\cdot\boldsymbol{k}\neq 0$ and otherwise is even smaller. Here again we may obtain the boundary layer equation from the potential vorticity equation (25) applied to this $E^{1/4}$ layer on B_n.

5. We conclude by indicating the nature of the matching and its consequences for evolution on the τ scale in the case of rigid insulating boundaries. On B_{nB} we have $\boldsymbol{u}_I^0 = 0$ but $T_I^0 \neq 0$ and in general an $O(1)$ heat flux. This is matched to the wall in the $E^{1/2}$ layer with consequently $\hat{T} = O(E^{1/2})$ and tangential velocity also $O(E^{1/2})$, while $\hat{\boldsymbol{u}}\cdot\boldsymbol{n} = O(E)$. Then the $E^{1/4}$ layer has $\tilde{\boldsymbol{u}}\cdot\boldsymbol{n}\times\nabla\varphi = O(E^{1/2})$; hence this has only $O(E^{1/4})$ heat flux and $\tilde{\boldsymbol{u}}\cdot\boldsymbol{n} = O(E^{3/4})$ or less. We thus see from (26) that Q does not change to first order on the τ time scale. On B_{hF}, say, we have $O(1)\hat{\boldsymbol{u}}$ to

match the interior to the wall, and the Ekman layer formula (36) to show that $\boldsymbol{n} \cdot \hat{\boldsymbol{u}} = O(E^{1/2})$. This is accompanied by an $O(1)$ heat flux in the $E^{1/2}$ layer which, however, probably does not match the interior heat flux to the wall value. We thus have the $E^{1/4}$ layer with heat flux $O(1)$, tangential velocity $O(E^{1/2})$, and normal velocity $O(E^{3/4})$. Thus the interior normal velocity must match that of the Ekman layer alone, and be $O(E^{1/2})$; consequently s on B_h does change on the τ scale, in accordance with equation (28). The other quantities can be investigated similarly.

We see then that although Π and $\overline{\Pi}$ are unchanged to first order on the τ scale, some, at least, of the other quantities which characterize the geostrophic interior flow can change this rapidly; and in general there will be a significant evolution on this scale to a new interior geostrophic flow. This "intermediate" flow appears steady on the τ scale, but becomes the initial condition for subsequent evolution on the T scale. This final phase can be followed in principle in essentially the same way: knowing the interior geostrophic flow at time T, the right-hand side of (25) (now $O(E)$) can be computed, giving $\partial\Pi/\partial T\,(O(1))$ and so $\Pi(T + \Delta T)$, say. Likewise $\overline{\Pi}(T + \Delta T)$ is determined, and after a study of the $E^{1/2}$ boundary layers, which are now the only relevant ones, the values of the other quantities (c), . . . , (g) are also determined at $T + \Delta T$, and finally these determine the new geostrophic interior flow at $T + \Delta T$ as in Lecture I. The process is then repeated for the next "time step," and continued to produce the evolution on the long time scale. Of course, the possibility of the solution, in terms of formulas, of the differential—functional equation which lies behind this seminumerical description is to be expected only in very simple cases.

MASSACHUSETTS INSTITUTE OF TECHNOLOGY

Stability Problems in Fluids

J. T. Stuart

1. **Introduction.** Consideration is given in these lectures to the question of linearization and the development of perturbation procedures for calculating the effects of nonlinear terms. Especially the work of Benney and Bergeron is examined afresh in a way which indicates that their nonlinear critical layer can be considered to be quasi-linear. Problems involved in extending this work are considered.

2. **Basic solutions.** Flows of a homogeneous inviscid fluid in two dimensions are governed by the vorticity equation

$$\partial\zeta/\partial t + \partial(\zeta, \psi)/\partial(x, y) = 0 \tag{2.1}$$

where

$$\partial(\zeta, \psi)/\partial(x, y) = (\partial\psi/\partial y)(\partial\zeta/\partial x) - (\partial\psi/\partial x)(\partial\zeta/\partial y) \tag{2.2}$$

is the usual Jacobian, and

$$\zeta = -\nabla^2\psi \tag{2.3}$$

is the vorticity. The velocity components (u, v) are given in terms of the stream function by

$$u = \partial\psi/\partial y, \qquad v = -\partial\psi/\partial x. \tag{2.4}$$

If the flow is time-independent (in some appropriate reference frame), then (2.1) and (2.2) yield the result

$$\zeta = F(\psi) \tag{2.5}$$

where F denotes any suitable function. Here we wish to consider

$$\nabla^2\psi = e^{-2\psi} \tag{2.6}$$

which follows from (2.3) and (2.5) by assumption of an exponential form for F.

A special solution of (2.6) is given by

$$\psi = \log\cosh y, \qquad u = \partial\psi/\partial y = \tanh y \tag{2.7}$$

AMS 1970 *subject classifications.* Primary 76E30, 76D30, 76D10.

which represents a shear layer between parallel streams flowing in the $\pm x$ directions. However, (2.7) is merely one solution for the case in which (2.6) becomes an ordinary differential equation. But it has been known since the time of Liouville [1853] that the partial differential equation (2.6) also can be solved, and comparatively recently [Stuart, 1967] some solutions have been discovered which are especially relevant for our studies of the stability or instability of fluid flows. In fact (2.6) and, therefore, (2.1) in the steady case are satisfied by

$$\psi = \log[C \cosh y + (C^2 - 1)^{1/2} \cos x]. \tag{2.8}$$

The class of solutions is obtained by variation of C.

Let us consider some possibilities:

(i) If $C = 1$ we retrieve (2.7), while if C is slightly greater than 1 we have

$$\psi - \log C = \log \cosh y + \varepsilon \operatorname{sech} y \cos x \tag{2.9}$$

where $\varepsilon = (C^2 - 1)^{1/2}/C$; this formula is the *mean flow* (2.7) with a *small* perturbation superimposed, the perturbation itself being periodic in a spatial coordinate, namely x. We refer to the term proportional to $\cos x$ as the *fundamental.*

(ii) If we expand in powers of ε, then we obtain

$$\begin{aligned} \psi - \log C = {} & \log \cosh y + \varepsilon \operatorname{sech} y \cos x - \tfrac{1}{4}\varepsilon^2 \operatorname{sech}^2 y(1 + \cos 2x) \\ & + \tfrac{1}{12}\varepsilon^3 \operatorname{sech}^3 y(3 \cos x + \cos 3x) \\ & - \tfrac{1}{32}\varepsilon^4 \operatorname{sech}^4 y(3 + 4 \cos 2x + \cos 4x) + O(\varepsilon^5). \end{aligned} \tag{2.10}$$

Here the fundamental ($\cos x$) has order ε. At order ε^2 there is a correction to the *mean flow* together with the first harmonic ($\cos 2x$), while at order ε^3 there is a correction to the fundamental together with the *second harmonic* ($\cos 3x$); the terms of order $O(\varepsilon^4)$ illustrate both further corrections and higher harmonics generated at higher order.

(iii) If C is large, (2.8) takes on the approximate form

$$\psi - \log C = \log(\cosh y + \cos x) \tag{2.11}$$

which [Lamb, 1932, p. 224] represents the flow due to a set of point vortices of strength $K = -4\pi$ spaced at a distance 2π apart on the x axis. In fact (2.11) satisfies Laplace's equation. Thus a classical solution of the latter equation is seen as a member of a class of *rotational* flows.

The properties of (2.8), as C varies, are most easily discussed by reference to the vorticity, which is given by

$$\zeta = -e^{-2\psi} = -[C \cosh y + (C^2 - 1)^{1/2} \cos x]^{-2}. \tag{2.12}$$

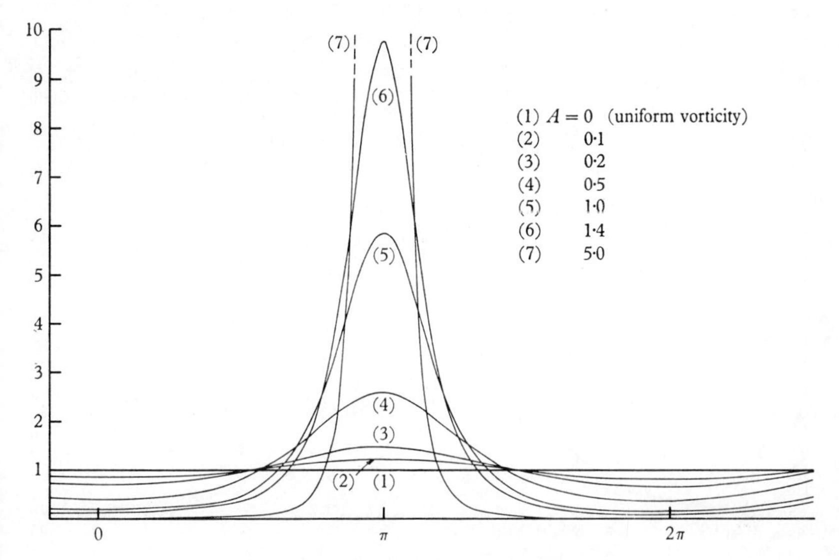

FIGURE 1. Vorticity on $y = 0$ (periodic in x).

The area integral Γ of the vorticity over the rectangle bounded by the contour $\mathscr{C}$, composed of the straight lines $x = 0$ and 2π, $y = \pm\infty$, is $\Gamma = -4\pi$; by Stokes' Theorem this is the circulation around $\mathscr{C}$. On $y = 0$ the vorticity, which is periodic in x, is shown by Figure 1, where $A = (C^2 - 1)^{1/2}$ takes on several values. On $x = \pi$ the vorticity is shown by Figure 2. The strong peaking of the vorticity around $x = \pi$ (and, indeed, by inference around $x = (2n + 1)\pi$) is shown clearly for A large, although the circulation Γ remains unchanged. On the other hand, for A small ($C \to 1$) the vorticity is nearly sinusoidal in x. The streamlines of the flow for a given C are of the celebrated Kelvin's cat's-eye form (see Figure 3).

The above solutions may be generalized in several ways. By setting $x_1 = x + ct$, $\psi_1 = \psi + cy$, we obtain a wave which travels at a speed c, rather than remaining stationary as in (2.8). Moreover, if the wavelength is a and the circulation K, then the solution corresponding to (2.8) is

$$(2.13) \quad \psi = (-K/4\pi)\log[C\cosh(2\pi y/a) + (C^2 - 1)^{1/2}\cos(2\pi x/a)],$$

which satisfies

$$(2.14) \qquad \nabla^2\psi = (-\pi K/a^2)\exp(8\pi\psi/K).$$

An interesting property of (2.13) is that, if $a \to 0$ with K proportional to a, we obtain $\psi \sim -K|y|/2a$, which represents two uniform streams of speeds $-K/2a$ for $y > 0$ and $K/2a$ for $y < 0$, separated by a vortex sheet.

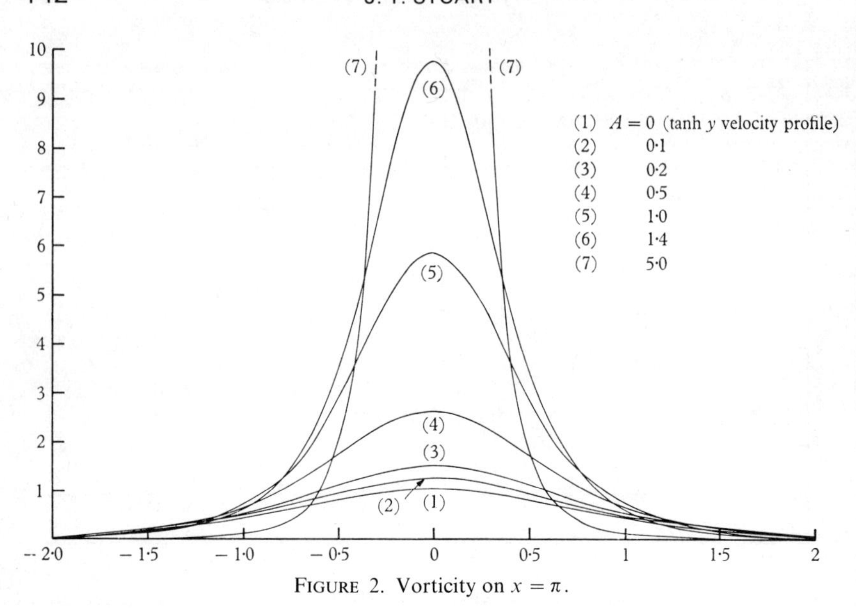

FIGURE 2. Vorticity on $x = \pi$.

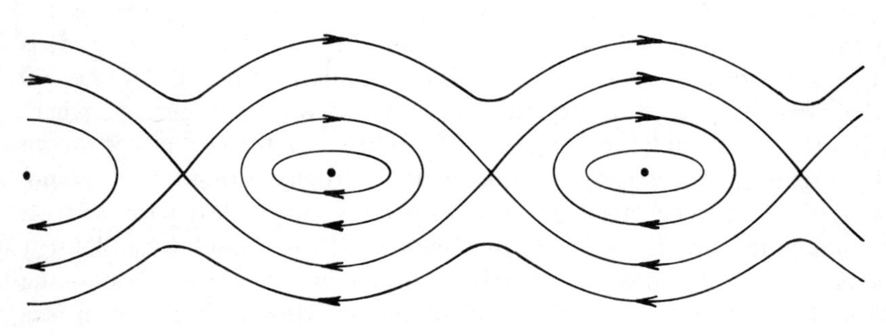

FIGURE 3

Several features of the above class of flows are typical, especially the development of a fundamental oscillation ($\varepsilon \cos x$) on a given mean shear flow ($\psi = \log \cosh y$), followed by the stimulation of harmonics and mean-flow corrections by the nonlinearity of the system (which here is represented by $e^{-2\psi}$). Moreover the streamline pattern, as shown in Figure 3, is typical of that formed by a shear (vorticity) wave propagating on shear flow. However one feature is at variance with generally accepted ideas in fluid mechanics. Prandtl [1927] and Batchelor [1956] have shown that the inclusion of viscosity, even though it may be small, produces a different distribution of vorticity from that shown in Figures 1, 2. Especially they have argued that if a two-dimensional flow has closed streamlines, diffusion will render the vorticity uniform within the region of closed streamlines; this is seen to be contrary to the result indicated by Figures 1, 2, 3 where the vorticity, derived from truly inviscid considerations only, varies across the region of closed streamlines. We return to this matter in §4.

3. **The critical layer in parallel flows.** Consider a flow of boundary-layer type, or one which is exactly parallel, and let the velocity distribution be $\bar{u}(y)$ in the x direction. In two dimensions a *small* perturbation stream function of the form $\varphi(y)\exp(i\alpha(x - ct))$ satisfies the linear differential equation

$$(3.1)\quad (\bar{u} - c)(\varphi'' - \alpha^2\varphi) - \bar{u}''\varphi + (i/\alpha R)(\varphi^{\text{iv}} - 2\alpha^2\varphi'' + \alpha^4\varphi) = 0$$

where R is the Reynolds number based on a characteristic speed, say U_0, a characteristic length, say δ, and the kinematic viscosity ν. For example, if $\bar{u} = \tanh y$, $c = 0$, $\alpha = 1$, then the perturbation part of (2.9) satisfies the inviscid form of (3.1), with $\alpha R \to \infty$. A known necessary requirement for such *neutral* (α and c real) solutions to exist is that the vorticity magnitude of the given mean flow shall have a maximum (at the inflexion point of $\bar{u}$), and this is seen to be true for $\bar{u} = \tanh y$ at the point $y = 0$.

If there is no inflexion point of $\bar{u}$, then a neutral solution of the inviscid equation does not exist. Then the inflexion-point mechanism cannot be invoked to explain the occurrence of instabilities. However, waves, both two- and three-dimensional, are known to occur in such flows, as has been demonstrated by many elegant experiments over the years at the National Bureau of Standards. The essential mechanism involved in the generation of these waves was first described by Heisenberg [1924] and Tollmien [1929] and later clarified by Lin [1945]; the idea is that the momentum required to maintain the growth of the wave is produced by a transport of mean-flow momentum, proportional to $\bar{u}(y)$, by a velocity fluctuation v in the y coordinate, thus yielding an acceleration (in the x direction) proportional to $v\,d\bar{u}/dy$. This effect is present provided that the mean flow $\bar{u}(y)$ has a shear in the y direction, so that $d\bar{u}/dy \neq 0$. Clearly, however, this mechanism can produce growth of perturbations only if the y component of velocity fluctuations (v) has an appropriate phase shift relative to the x component u. (If u and v differed in phase by $\pi/2$, the term $v\,d\bar{u}/dy$ would be proportional to $\partial u/\partial t$.) Such phase shifts, first explained by L. Prandtl, require the action of viscosity, as we shall now see.

As mentioned earlier, the equation governing small-amplitude wavy perturbations from a given mean flow $\bar{u}(y)$ is the Orr-Sommerfeld Equation (3.1). The essential properties that we need are summarized in Figure 4, which shows a boundary-layer flow above a flat wall. The flow $\bar{u}(y)$ is from

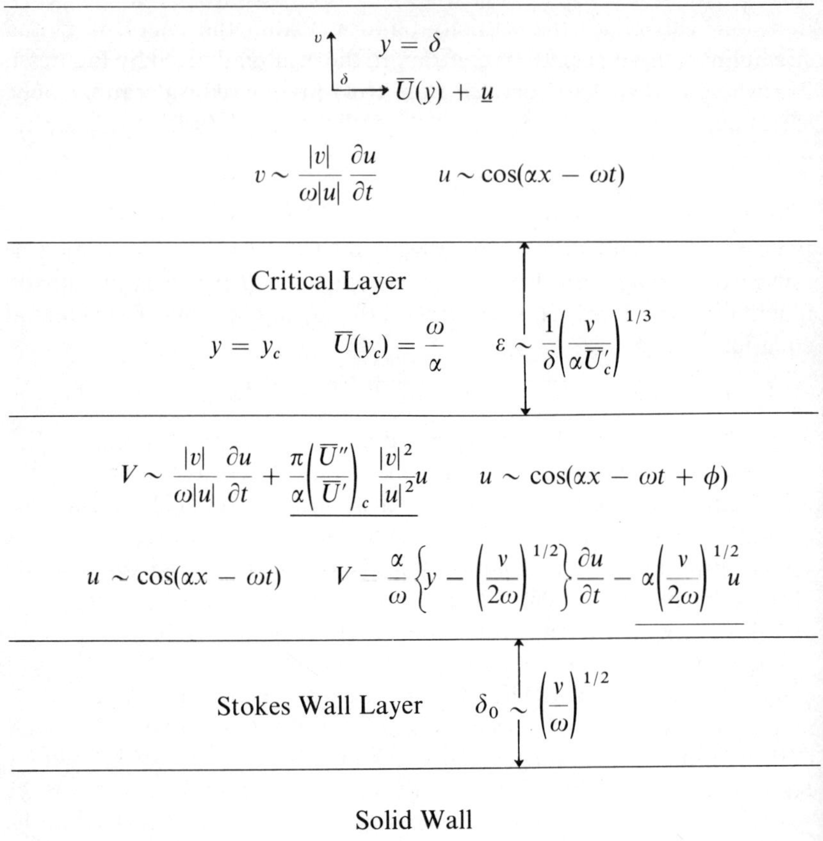

$$y = \delta \qquad \xrightarrow{\delta} \overline{U}(y) + \underline{u}$$

$$v \sim \frac{|v|}{\omega|u|}\frac{\partial u}{\partial t} \qquad u \sim \cos(\alpha x - \omega t)$$

Critical Layer

$$y = y_c \qquad \overline{U}(y_c) = \frac{\omega}{\alpha} \qquad \varepsilon \sim \frac{1}{\delta}\left(\frac{\nu}{\alpha \overline{U}'_c}\right)^{1/3}$$

$$V \sim \frac{|v|}{\omega|u|}\frac{\partial u}{\partial t} + \underline{\frac{\pi}{\alpha}\left(\frac{\overline{U}''}{\overline{U}'}\right)_c \frac{|v|^2}{|u|^2}u} \qquad u \sim \cos(\alpha x - \omega t + \phi)$$

$$u \sim \cos(\alpha x - \omega t) \qquad V - \frac{\alpha}{\omega}\left\{y - \left(\frac{\nu}{2\omega}\right)^{1/2}\right\}\frac{\partial u}{\partial t} - \underline{\alpha\left(\frac{\nu}{2\omega}\right)^{1/2} u}$$

Stokes Wall Layer $\quad \delta_0 \sim \left(\dfrac{\nu}{\omega}\right)^{1/2}$

Solid Wall

FIGURE 4

left to right and is subject to perturbations u, v in the x and y coordinates; $y = 0$ is the wall and $y = \delta$ is the edge of the boundary layer in some sense.

If αR is a large parameter, it might be supposed that (3.1) could be approximated by the inviscid (or Rayleigh) equation obtained by letting $\alpha R \to \infty$. Such an approximation, however, lowers the order of the equation from four to two, so that all four boundary conditions on (3.1) can no longer be imposed. Moreover, it appears that the Rayleigh equation is singular when $\bar{u} = c$ (the critical layer), since the coefficient of the second (and highest) derivative has a zero there. (Such a singularity may be in the complex plane since c may be complex.) For sufficiently large (positive or negative) values of c, this singularity would not interest us since it would be well outside the range of y of interest. For the mechanism under discussion, however, it appears to be crucial that the critical layer should occur in the range $0 < y < \delta$. The upshot of the above facts is that there are two regions where viscosity cannot be ignored, even if αR is large, namely (i) in a layer attached to the wall and (ii) in the critical layer neighboring the point $\bar{u}(y) = c$. We shall now summarize the effects of these two viscous layers.

Suppose that the velocity component u, parallel to the wall and close to the wall, is known from an inviscid calculation (the Rayleigh equation) and is proportional to $\cos(\alpha x - \omega t)$; here ω represents αc of (3.1). Then, provided the frequency is high enough, the effect of viscosity on u is independent of $\bar{u}$ (as Lin [1954] showed) and is given by the simple Stokes layer, which implies a balance between $-c\psi''$ and $(i/\alpha R)\psi^{\text{iv}}$ of (3.1). This layer has thickness $(\nu/\omega)^{1/2}$ and exists to ensure that the velocity is zero at the wall, while the u component tends to the imposed form outside the layer. However the v component of velocity tends to the form (with *dimensional* quantities in use)

(3.2)
$$v = \frac{\alpha}{\omega}\left\{y - \left(\frac{\nu}{2\omega}\right)^{1/2}\right\}\frac{\partial u}{\partial t} - \alpha\left(\frac{\nu}{2\omega}\right)^{1/2} u,$$

where the term independent of viscosity (ν) arises from continuity of flow and the terms proportional to $(\nu/2\omega)^{1/2}$ represent the displacement effect. More important here is to note that in this unsteady boundary layer the displacement effect gives rise to a part of v which is proportional to u itself; this is additional to the (mainly inviscid) part of v proportional to $\partial u/\partial t$. Thus a phase shift between v and u near the wall is achieved to modify the inviscid property that because of flow continuity, u and v differ in phase by $\pi/2$.

Now we turn our attention to the viscous region surrounding the critical point where $\bar{u}(y) = c = \omega/\alpha$. For conceptual simplicity we may suppose that c is real, so that the perturbation neither amplifies nor decays, but there is no essential difficulty in discussing the case of c complex. Let us suppose that, for $y > y_c$, u is proportional to $\cos(\alpha x - \omega t)$, though the phase may be different from that close to the wall for which a similar assumption was made. By continuity of flow v is proportional to $(|v|/\omega|u|)\, \partial u/\partial t$. Our object is to find the corresponding relationship between v and u for $y < y_c$. In the critical layer the important balance in Equation (3.1) is between $(\bar{u} - c)\psi''$ and $(i/\alpha R)\psi^{\text{iv}}$, which yields an Airy equation. The characteristic thickness of this layer is $(\nu/\alpha\bar{u}'_c)^{1/3}$, where the suffix c denotes conditions at $y = y_c$. The solution of this equation shows that, for $y < y_c$, u is proportional to $\cos(\alpha x - \omega t + \phi)$, where ϕ is a phase angle. On the other hand

(3.3)
$$v = \frac{|v|}{\omega|u|}\frac{\partial u}{\partial t} + \frac{\pi}{\alpha}\frac{\bar{u}''_c}{\bar{u}'_c}\frac{|v|^2}{|u|^2}u,$$

for y just less than y_c. Again, due to the presence of viscosity, we see that a phase shift between u and v has been brought about so that their phase difference is no longer $\pi/2$. Since $\bar{u}$ is negative, the coefficient of u in (3.3) is negative, just as in (3.2).

The theory of instability based on (4.4) may be summarized roughly as stating that if the equation and its boundary conditions permit a choice of ω, α, and R so that in the region between the Stokes layer and the critical layer a matching is possible between the two forms of both u and v, neutral perturbations are possible. Perhaps the most important feature of this matching is that involving the phase difference between u and v; this is a very delicate matter, and yields a complicated relation between α, ω, and R, as we shall see. By modifying the choice of α and R, complex values of ω result and can give instability by growth in time.

The ideas described above may be given alternatively as follows. The x-momentum equation for the perturbation may be written

$$\frac{\partial u}{\partial t} + \bar{u}\frac{\partial u}{\partial x} + v\frac{d\bar{u}}{dy} = -\frac{1}{\rho}\frac{\partial p}{\partial x} + \nu\nabla^2 u. \tag{3.4}$$

If

$$v = A\partial u/\partial t - Bu, \qquad A > 0, \quad B > 0, \tag{3.5}$$

as we have seen from the arguments of the previous paragraph, (3.4) becomes

$$\left(1 + A\frac{d\bar{u}}{dy}\right)\frac{\partial u}{\partial t} + \bar{u}\frac{\partial u}{\partial x} = Bu\frac{d\bar{u}}{dy} - \frac{1}{\rho}\frac{\partial p}{\partial x} + \nu\nabla^2 u. \tag{3.6}$$

We see that the term $A\,d\bar{u}/dy$ represents a virtual inertia, whereas $Bu\,d\bar{u}/dy$ contributes to growth or decay of the perturbation; this, of course, is the essential difference between the two parts of (4.12), with their phase difference. The phase shift produces growth of u (since $A\,d\bar{u}/dy > 0$) if

$$-\overline{uv} = B\overline{u^2} > 0 \tag{3.7}$$

over a substantial region of the flow. (The overbar denotes an average in x.) Since we have seen that both phase shifting processes ensure that $B > 0$, conditions are favorable for the satisfaction of (3.7). The expression $(-\overline{uv})$ is known more familiarly as the Reynolds stress if it is multiplied by ρ. It appears in the energy equation

$$\frac{\partial}{\partial t}\iint \frac{1}{2}\rho(u^2 + v^2)\,dx\,dy = \iint (-\rho\overline{uv})\frac{d\bar{u}}{dy}\,dx\,dy - \mu\iint \zeta^2\,dx\,dy, \tag{3.8}$$

which can be obtained from (3.4) and its companion, ζ being the fluctuating vorticity. For instability the first integral on the right-hand side of (3.8),

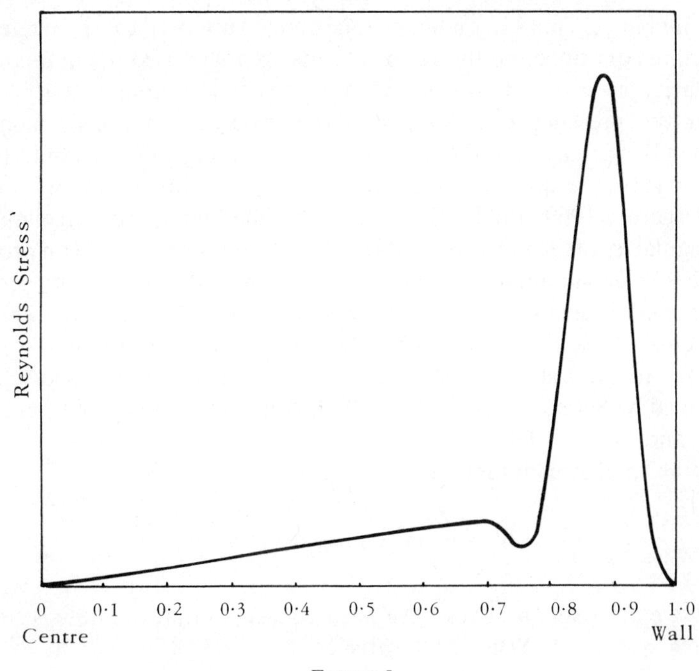

FIGURE 5

representing energy input to the perturbation through transfer of energy by the Reynolds stress from the mean flow $\bar{u}$, has to overcome energy dissipation, which is given by the second integral on the right-hand side of (3.8).

It is clear that the mechanism described above is one of great subtlety, involving, as it does, the presence of viscosity to produce the correct phase shifts between the velocity components. Figure 5 shows the kinematic Reynolds stress ($-\overline{uv}$) for plane Poiseuille flow in a channel; the Reynolds number, based on maximum speed and half-width of the channel, is 10^4, while the corresponding nondimensional wave number is 1. The sharp peak in the Reynolds stress occurs near the critical point $y = y_c$ and illustrates the importance of viscosity. In this case, as in many others of practical importance, the wall layer and the critical layer merge together.

Such is one description of the (now classical) theory of the viscous critical layer in the instability of boundary-layer flows. There is, however, another possible form of critical layer, which we discuss in the next section.

4. **Nonlinear critical layer.** The discussion of the previous section centers on linearized theory and pays little or no regard to the nonlinear phenomena

illustrated in §2. Thus (3.1), a linear equation, is supposed to be the relevant governing equation, and the viscous terms are there used to eliminate the singularity at $\bar{u} = c$. However, §2 indicates that nonlinear effects can permit the presence of solutions of the inviscid equations which are completely regular, even though inviscid amplifying solutions of (3.1) are singular (§3). The question thus raised, and partly answered, by Benney and Bergeron [1969] and by Davis [1969] is whether, in order to eliminate the singularity of the inviscid solution of (3.1), nonlinear effects can provide a critical-layer balance without viscosity playing the dominant role ascribed to it in §3. It transpires that the answer pivots on the Prandtl–Batchelor Theorem on the uniformity of vorticity within the regions of closed streamlines. We now pursue this matter further by means of a modified scheme developed by Schneyer and Stuart [1970], following Benney and Bergeron [1969] and Davis [1969].

Let the total stream function be

$$\Psi = \int_0^y (\bar{u} - c)\, dy + \varepsilon\psi(x, y) \tag{4.1}$$

where α, c are real, and axes have been chosen so that the flow is steady with $\bar{u} = c$ at $y = 0$. Whereas Ψ satisfies

$$\Psi_y \nabla^2 \Psi_x - \Psi_x \nabla^2 \Psi_y = R^{-1}\nabla^4\Psi, \tag{4.2}$$

the perturbation stream function satisfies

$$(\bar{u} - c)\nabla^2\psi_x - \bar{u}_{yy}\psi_x + \varepsilon(\psi_y \nabla^2\psi_x - \psi_x \nabla^2\psi_y) = R^{-1}\nabla^4\psi, \tag{4.3}$$

suffixes denoting derivatives. If ε, the perturbation amplitude, and R^{-1} are so small as to be negligible, then (4.3) yields the inviscid form of (3.1) for disturbances periodic in x. An attempt can be made, as follows, to use the ε terms of (4.3) to eliminate the singularity in the inviscid solution of (3.1). Let us set $y = \varepsilon^\alpha Y$, where α is to be determined as a positive number. The four main terms of (4.3) have dominant parts as follows:

$$\bar{u}_c' \varepsilon^{-\alpha} Y \psi_{xYY},$$

$$\bar{u}_c'' \psi_x,$$

$$\varepsilon^{1-3\alpha}(\psi_Y \psi_{xYY} - \psi_x \psi_{YYY}),$$

$$R^{-1}\varepsilon^{-4\alpha}\psi_{YYYY}.$$

The singularity, which arises from the factor Y multiplying the second Y derivative of the first term, can be dealt with, it is suggested, by balance of that term against the nonlinear term. Thus $-\alpha = 1 - 3\alpha$, which yields $\alpha = \frac{1}{2}$. The viscous term is negligible provided $R^{-1}\varepsilon^{-4\alpha} \ll \varepsilon^{-\alpha}$, which yields

$$\lambda = (\varepsilon^{3/2}R)^{-1} \ll 1. \tag{4.4}$$

The parameter λ^{-1} is a *local* Reynolds number for the flow near $y = 0$ in the presence of a critical layer of thickness $\varepsilon^{1/2}$.

Another approach to the meaning of the parameter λ shows that $\lambda^{-2/3}$ also is another relevant local Reynolds number. From (4.1) the velocity in the neighborhood of $y = 0$ is εU_0 in dimensional terms. On the other hand, §3 has shown that a typical thickness of the *viscous* critical layer (if one exists) in that neighborhood is $\delta R^{-1/3}$. Therefore a characteristic Reynolds number for the *flow in the neighborhood of the viscous critical layer* is $\varepsilon U_0 \delta R^{-1/3}/\nu$, which equals $\varepsilon R^{2/3} = \lambda^{-2/3}$. For viscous to dominate over nonlinear effects, this Reynolds number $\lambda^{-2/3}$ has to be small, and this is the rationale of §3.

If λ is small, we can expect to be able to ignore the viscous terms of (4.2) and (4.3). With the transformation

$$y = \varepsilon^{1/2} Y \tag{4.5}$$

applied to (4.2), together with the use of (4.4), those equations become

$$\begin{aligned} &\Psi_Y \Psi_{YYx} - \Psi_x \Psi_{YYY} + \varepsilon(\Psi_Y \Psi_{xxx} - \Psi_x \Psi_{Yxx}) \\ &\quad = \lambda\varepsilon(\Psi_{YYYY} + 2\varepsilon \Psi_{YYxx} + \varepsilon^2 \Psi_{xxxx}), \end{aligned} \tag{4.6}$$

$$(\bar{u} - c)\nabla^2 \psi_x - \bar{u}_{yy}\psi_x + \varepsilon(\psi_y \nabla^2 \psi_x - \psi_x \nabla^2 \psi_y) = \lambda \varepsilon^{3/2} \nabla^4 \psi. \tag{4.7}$$

Equations (4.6) and (4.7) are "inner" and "outer" equations in the usual sense. The point is that we can solve (4.7) in a power series in ε with or without recourse to (4.4), for both y positive and y negative, but the solutions are singular at $y = 0$. Can we then utilize the inner equation (4.6), in which the length scale is now $\varepsilon^{1/2}\delta$, to match the outer solutions and eliminate the singularity? It is this question which was answered, up to a certain point, by Benney and Bergeron and by Davis.

Outer region. We expand ψ in the outer (o) region as

$$\psi = \psi_{os}^{(1o)} + \varepsilon \psi_{os}^{(2o)} + \varepsilon^2 \psi_{os}^{(3o)} + \cdots + \lambda \varepsilon^{3/2} \psi_{os}^{(o1)} + \cdots \tag{4.8}$$

in which $s = \pm$ for $y \gtrless 0$. The function $\psi_{os}^{(1o)}$ satisfies the Rayleigh equation, the inviscid form of (3.1), and is given by

$$\psi_{os}^{(1o)} = (A_s^{(1o)}\phi_a + B_s^{(1o)}\phi_b)e^{i\xi} + (\tilde{A}_s^{(1o)}\phi_a + \tilde{B}_s^{(1o)}\phi_b)e^{-i\xi}, \tag{4.9}$$

where A_s, B_s are constants and a tilde ($\sim$) denotes a complex conjugate. Moreover $\xi = \alpha x$ and

$$\begin{aligned} \phi_a &= y + \tfrac{1}{2}Ky^2 + (\tfrac{1}{6}\alpha^2 + \tfrac{1}{6}K_1)y^3 + \cdots, \\ \phi_b &= 1 + (\tfrac{1}{2}\alpha^2 + \tfrac{1}{2}K_1 - K^2)y^2 + \cdots + K\phi_a \log|y|, \end{aligned} \tag{4.10}$$

where $K = \bar{u}_c''/\bar{u}_c'$ and $K_1 = \bar{u}_c'''/\bar{u}_c'$. In order to determine the amplitude ε as a function of α, it is necessary also to expand α in terms of ε

$$\alpha = \alpha_{1o} + \varepsilon\alpha_{2o} + \varepsilon^2\alpha_{3o} + \cdots + \lambda\varepsilon^{3/2}\alpha_{o1} + \cdots. \tag{4.11}$$

The equations for the inviscid and viscous functions in (4.8) are

$$(\bar{u} - c)\left(\frac{\partial^2}{\partial y^2} + \alpha_{1o}^2\frac{\partial^2}{\partial \xi^2}\right)\psi_{os\xi}^{(1o)} - \bar{u}_{yy}\psi_{os\xi}^{(1o)} = 0, \tag{4.12}$$

$$\begin{aligned}(\bar{u} - c)\left(\frac{\partial^2}{\partial y^2} + \alpha_{1o}^2\frac{\partial^2}{\partial \xi^2}\right)\psi_{os\xi}^{(2o)} - \bar{u}_{yy}\psi_{os\xi}^{(2o)} &\\ = -\psi_{osy}^{(1o)}\left(\frac{\partial^2}{\partial y^2} + \alpha_{1o}\frac{\partial^2}{\partial \xi^2}\right)\psi_{os\xi}^{(1o)} + \psi_{os\xi}^{(1o)}\left(\frac{\partial^2}{\partial y^2} + \alpha_{1o}^2\frac{\partial^2}{\partial \xi^2}\right)\psi_{osy}^{(1o)} &\\ - 2(\bar{u} - c)\alpha_{1o}\alpha_{2o}\frac{\partial^2}{\partial \xi^2}\psi_{os\xi}^{(1o)},&\end{aligned} \tag{4.13}$$

$$\begin{aligned}(\bar{u} - c)\left(\frac{\partial^2}{\partial y^2} + \alpha_{1o}^2\frac{\partial^2}{\partial \xi^2}\right)\psi_{os\xi}^{(o1)} - \bar{u}_{yy}\psi_{os\xi}^{(o1)} &\\ = \frac{1}{\alpha_{1o}}\left(\frac{\partial^4}{\partial y^4} + 2\alpha_{1o}^2\frac{\partial^4}{\partial y^2\partial \xi^2} + \alpha_{1o}^4\frac{\partial^4}{\partial \xi^4}\right)\psi_{os}^{(1o)}.&\end{aligned} \tag{4.14}$$

Once $\psi_{os}^{(1o)}$ is known from (4.9) and (4.10), then higher order inviscid and viscous functions can be determined successively from (4.13), (4.14), and related equations. Matching of these functions is then required to the solution in the inner region, and especially α_{2o} and the arbitrary constants in (4.9) have to be determined.

Inner region. Here we write

$$\xi = \alpha x, \qquad \eta = \Psi/\varepsilon, \qquad \Psi_{YY} = \Phi, \qquad \Psi_{\xi\xi} = \chi, \tag{4.15}$$

in which the quantities Φ and χ are to be regarded as functions of ξ and η while η (or Ψ) is a function of ξ and Y. The reason for this transformation is the knowledge, gained from §2, that in two-dimensional steady flows the vorticity, which is proportional to $\varepsilon^{-1}\Phi + \alpha^2\chi$, is a function of Ψ only. Then (4.6) can be written

$$\begin{aligned}\Phi_\xi + \varepsilon\alpha^2\chi_\xi = \frac{\lambda}{\alpha}\left\{\frac{\partial}{\partial \eta}(\eta_Y\Phi_\eta) + 2\varepsilon\alpha^2\frac{\partial}{\partial \eta}(\eta_Y\chi_\eta)\right.&\\ \left. + (\varepsilon^2\alpha^4/\eta_Y)\left(\frac{\partial}{\partial \xi} + \eta_\xi\frac{\partial}{\partial \eta}\right)^2\chi\right\}.&\end{aligned} \tag{4.16}$$

If ε and λ are small enough to be ignored, then (4.16) reduces to the simple *linear* equation

$$\Phi_\xi = 0. \tag{4.17}$$

It is the elegance of this result which is responsible for the transformations (4.15) which were used to achieve it. To the order of approximation (4.17), Φ, which is proportional to the vorticity, is a function of η only. In the expansion of the solution for Ψ, Φ, and χ, the form (4.17) will play an important role in which its linearity is crucial. Our problem, in truth, becomes quasi-linear.

In order to see the form which the inner expansion takes, we must revert to (4.10), for this yields a form of solution when $y \to 0$ which is required to match that of the inner solution when $|Y| \to \infty$.

From (4.1), (4.8), (4.9), (4.10), together with (4.5), we find that to $O(\varepsilon^{3/2})$ the *outer* solution for $y \to 0$ written in terms of the inner variables becomes

$$\begin{aligned}\Psi_{os} &= \varepsilon(\tfrac{1}{2}\bar{u}_c' Y^2 + B^{(1o)} e^{i\xi} + \tilde{B}^{(1o)} e^{-i\xi}) \\ &\quad + \varepsilon^{3/2} \log \varepsilon(\tfrac{1}{2}KY)(B^{(1o)} e^{i\xi} + \tilde{B}^{(1o)} e^{-i\xi}) \\ &\quad + \varepsilon^{3/2}\Big[\tfrac{1}{6}\bar{u}_c'' Y^3 + Y(A^{(1o)} e^{i\xi} + \tilde{A}^{(1o)} e^{-i\xi}) \\ &\quad + KY\log|Y|(B^{(1o)} e^{i\xi} + \tilde{B}^{(1o)} e^{-i\xi}) \\ &\quad - \tfrac{1}{4}K\frac{B^{(1o)2}}{\bar{u}_c'} Y^{-1} e^{2i\xi} - \tfrac{1}{4}K\frac{\tilde{B}^{(1o)2}}{\bar{u}_c'} Y^{-1} e^{-2i\xi}\Big] + O(\varepsilon^2(\log\varepsilon)^2).\end{aligned} \tag{4.18}$$

The first terms of order ε and $\varepsilon^{3/2}$ arise from the basic flow, while most of the other terms of orders ε, $\varepsilon^{3/2}\log\varepsilon$, and $\varepsilon^{3/2}$ arise from the fundamental $\psi_{os}^{(1o)}$ except that the harmonic $\psi_{os}^{(2o)}$ gives rise to the last two terms of order $\varepsilon^{3/2}$. Terms of order $\varepsilon^2(\log\varepsilon)^2$, as well as higher-order terms, require the harmonics also.

The form of (4.18) together with higher-order terms makes it clear that for λ fixed Φ and χ should be expanded in the inner (i) region as

$$\begin{aligned}\Phi &= \varepsilon\Phi_i^{(0\lambda)} + \varepsilon^{3/2}\log\varepsilon\Phi_i^{(1\lambda)} + \varepsilon^{3/2}\Phi_i^{(2\lambda)} \\ &\quad + \varepsilon^2(\log\varepsilon)^2\Phi_i^{(3\lambda)} + \varepsilon^2\log\varepsilon\Phi_i^{(4\lambda)} \\ &\quad + \varepsilon^2\Phi_i^{5\lambda} + O[\varepsilon^{5/2}(\log\varepsilon)^3],\end{aligned} \tag{4.19}$$

$$\chi = \varepsilon\chi_i^{(0\lambda)} + \varepsilon^{3/2}\log\varepsilon\chi_i^{(1\lambda)} + \varepsilon^{3/2}\chi_i^{(2\lambda)} + O[\varepsilon^2(\log\varepsilon)^2] \tag{4.20}$$

$$\eta = \eta_0(Y,\xi) + \varepsilon^{1/2}\log\varepsilon\eta_1(Y,\xi) + \varepsilon^{1/2}\eta_2(Y,\xi) + \cdots \tag{4.20a}$$

where χ is related to Φ by (4.15) and η is Ψ/ε. In addition α has to be expanded by (4.11).

The equations for the functions occurring in (4.19) are

$$O(\varepsilon) : \Phi_{i\xi}^{(0\lambda)} = \frac{\lambda}{\alpha_{io}} \frac{\partial}{\partial \eta}(\eta_{oY}\Phi_{i\eta}^{(0\lambda)}),$$

$$O(\varepsilon^{3/2} \log \varepsilon) : \Phi_{i\xi}^{(1\lambda)} = \frac{\lambda}{\alpha_{io}} \frac{\partial}{\partial \eta}(\eta_{oY}\Phi_{i\eta}^{(1\lambda)} + \eta_{1Y}\Phi_{i\eta}^{(0\lambda)}), \tag{4.21}$$

$$O(\varepsilon^{3/2}) : \Phi_{i\xi}^{(2\lambda)} = \frac{\lambda}{\alpha_{io}} \frac{\partial}{\partial \eta}(\eta_{oY}\Phi_{i\eta}^{(2\lambda)} + \eta_{2Y}\Phi_{i\eta}^{(0\lambda)})$$

and so forth. The work of Benney and Bergeron goes to the order shown by (4.21), but by other methods.

Solution in the inner region. We need to solve (4.21) and match to (4.18) when $Y \to \infty$. Some laborious but relatively straightforward calculations show that the functions shown in (4.21) can be obtained. If λ is sufficiently small, the viscous terms may be neglected, and then (4.21) take on the especially simple form epitomized by (4.17). For example if we write

$$\Phi_i^{(o\lambda)} = \Phi_i^{(o)} + (\lambda/\alpha_{io})X_{o\eta}, \tag{4.22}$$

then $\Phi_i^{(o)}$ satisfies (4.17). Thus

$$\Phi_i^{(o)} = G'(\eta), \tag{4.23}$$

where G is some function to be determined, while (4.18) indicates that

$$G'(\eta) \sim \bar{u}_c' \quad \text{when} \quad |Y| \to \infty. \tag{4.24}$$

It is necessary to invoke viscosity in order to determine $G(\eta)$, and at this point the analysis runs in parallel with that of Prandtl and Batchelor.

Thus X_o satisfies

$$X_{o\eta\xi} = (\partial/\partial\eta)[\eta_{oY}\Phi_{i\eta}^{(o)}], \tag{4.25}$$

which integrates to yield

$$X_{o\xi} = \eta_{oY}G''(\eta) + \mathscr{H}_o(\xi), \tag{4.26}$$

where $\mathscr{H}_o(\xi)$ is arbitrary. Now we integrate over ξ and insist that X_0 shall be periodic with wavelength 2π for a fixed value of η such that the streamline goes to infinity (but see later). Then it follows that

$$\int_0^{2\pi} \eta_{oY}\, d\xi \cdot G''(\eta) + \mathscr{H}_{oI}(2\pi) = 0. \tag{4.27}$$

But when $|Y| \to \infty$, $G'' \to 0$, so that $\mathscr{H}_{oI}(2\pi) = 0$. On the assumption, which can be justified later, that the integral in (4.27) is not zero, it follows that

$$G''(\eta) = 0. \tag{4.28}$$

This implies that

(4.29) $$\partial^2\Psi_i^{(0)}/\partial Y^2 = \Phi_i^{(0)} = \bar{u}'_c,$$

so that by matching with (4.18),

(4.30) $$\Psi_i^{(0)} = \tfrac{1}{2}\bar{u}'_c Y^2 + B^{(1o)}e^{1\xi} + \tilde{B}^{(1o)}e^{-i\xi}.$$

For simplicity we may define ε and the phase so that

(4.31) $$\Psi_i^{(0)} = \tfrac{1}{2}\bar{u}'_c Y^2 + \cos\xi.$$

Since, in general, viscous terms can be ignored, (4.31) represents η_0 (but not the whole of η, because of higher-order terms in ε).

Further work indicates that in the inviscid case

(4.32) $$\Psi_i^{(1)} = \tfrac{1}{2}(\bar{u}''_c/\bar{u}'_c)Y\cos\xi,$$

so that to order $\varepsilon^{1/2}\log\varepsilon$, η is given by

(4.33) $$\eta = \tfrac{1}{2}\bar{u}'_c Y^2 + \cos\xi + \varepsilon^{1/2}\log\varepsilon\cdot\tfrac{1}{2}(\bar{u}''_c/\bar{u}'_c)Y\cos\xi + O(\varepsilon^{1/2}).$$

The integral involved in (4.27) is given by

(4.34) $$\int_0^{2\pi}\eta_{oY}\,d\xi = \int_0^{2\pi}\bar{u}'_c Y\,d\xi$$

where Y is obtained by inversion of (4.33) as

(4.35) $$Y = (2/\bar{u}'_c)^{1/2}(\eta-\cos\xi)^{1/2} - \varepsilon^{1/2}\log\varepsilon(\bar{u}''_c/2\bar{u}'^2_c)\cos\xi + O(\varepsilon^{1/2}).$$

It follows that (4.34) gives

(4.36) $$(2\bar{u}'_c)^{1/2}\int_0^{2\pi}(\eta-\cos\xi)^{1/2}\,d\xi + O(\varepsilon^{1/2}),$$

which can be evaluated by means of an elliptic integral.

The next stage in the expansion shows that in the inviscid case

(4.37) $$\Phi_i^{(2)} = K(\eta)$$

where K is a function of η, which has to be matched to (4.18) at order $\varepsilon^{3/2}$. Moreover, it is necessary to ensure that on streamlines which go to infinity (as in Figure 3), $\Phi_i^{(2)}$ has period 2π in ξ. On the other hand, for streamlines which are closed, it is necessary to perform an integration around the circuit of a streamline (η fixed) in order to determine K. The results are that

(4.38) $$K = M_1\int_{\eta_{ce}}^{\eta}\frac{d\eta}{\int_0^{2\pi}Y(\xi,\eta)\,d\xi} + N_1,\quad \eta > \eta_{ce},$$
$$K = N_0,\quad \eta < \eta_{ce},$$

where Y is given by (4.35) plus higher-order terms, N_0, N_1, M_1 are constants, and η_{ce} is the value which determines the edge of the cat's eye.

Benney and Bergeron have taken their analysis to the point of showing that the logarithmic term in (4.10) is $\log |y|$, whether y be positive or negative. This is in direct contrast to the result of the *viscous* critical-layer analysis, in which $\log y$ for $y > 0$ becomes $[\log |y| - \pi i]$ for $y < 0$. This result enabled Benney and Bergeron to calculate eigenvalues for α as a function of c, the latter being the wave speed.

5. **Discussion.** As is indicated by the transformation (4.15), in association with (4.21), the perturbation problem is essentially linear. It is this feature which enables a simple exploitation to be made of the concept of a nonlinear critical layer. The formulae derived in the last section enables one to show that the flow within the cat's eyes matches velocity and pressure with the flow outside; but a viscous vorticity layer is necessary in order to match the vorticity, which is discontinuous according to the (essentially) inviscid theory.

There are, however, two serious defects in theories of this type, and these are currently undergoing investigation. One is that the amplitude ε is not determined at the stage to which Benney and Bergeron take their calculations. Higher order ε terms are needed in order to do this, as indicated by the expansion (4.19). The second defect is that the solution given by Benney and Bergeron is *neutral* and neither amplifies nor decays. How, if at all, is this solution connected with known results for amplified disturbances of linearized theory? An associated problem arises from the fact that the πi logarithmic phase shift of linearized theory would have to disappear if an equilibrium value of ε exists. Another possibility is that a different small-amplitude solution, quasi-linear in the sense of (4.15) and (4.21), exists, which never has any connection with normal linearized theory. These possibilities are undergoing investigation.

REFERENCES

G. K. Batchelor 1956, *On steady laminar flow with closed streamlines at large Reynolds number*, J. Fluid Mech. **1** (1956), 177–190. MR **18**, 840.

D. J. Benney and R. F. Bergeron 1969, *A new class of nonlinear waves in parallel flows*, Studies Appl. Math. **48** (1969), 181–204.

R. E. Davis 1969, *On the high Reynolds number flow over a wavy boundary*, J. Fluid Mech. **36** (1969), 337–346.

W. Heisenberg 1924, *Über Stabilität und Turbulenz von Flüssigkeitsströmen*, Ann. Phys. Lpz (4) **74** (1924), 577–627 (in English as N.A.C.A. T.M. 1291).

H. Lamb 1932, *Hydrodynamics*, 6th ed., Dover, New York, 1945.

C. C. Lin 1945, *On the stability of two-dimensional parallel flows*. I, II, III, Quart. Appl. Math. **3** (1945), 117–301. MR **7**, 225; 226; 346.

——— 1954, *Some physical aspects of the stability of parallel flows*, Proc. Nat. Acad. Sci. U.S.A. **40** (1954), 741–747. MR **16**, 759.

J. Liouville 1853, *Sur l'équation aux différences partielles* $\partial^2 \log \lambda/\partial u\, \partial v \pm \lambda/2a^2 = 0$, J. Math. (1) **18** (1853), 71–72.

L. Prandtl 1927, *On the motion of fluids of small viscosity*, In English as N.A.C.A. T.M. 452 (1928).

G. P. Schneyer and J. T. Stuart 1970, *On fluid oscillations associated with nonlinear critical layers* (to appear).

J. T. Stuart 1967, *On finite amplitude oscillations in laminar mixing layers*, J. Fluid Mech. **29** (1967), 417–440.

W. Tollmien 1929, *Über die Enstehung der Turbulenz*, Nachr. Ges. Wiss. Gött. **1929**, 21–44 (in English as N.A.C.A. T.M. 609).

IMPERIAL COLLEGE, LONDON

The Dynamics of Tsunamis

G. F. Carrier

1. **Introduction.** When an undersea earthquake displaces a significant sample of the ocean floor, a gravity wave is generated and propagates across the ocean. Rarely, even after the event is over, does one know the spacial and temporal distribution of this initiating displacement. However, we do know that in the Alaskan earthquake of 1964, the displacement had roughly the form shown in Figure 1b. The area on which the ground motion occurred was about 100 miles by 800 miles, and the average displacement normal to the bottom was very small compared to the largest displacement. One can infer from energy estimates that areas of this general size must accompany those earthquakes which generate the more devastating tsunamis.

Thus, the lateral scale of the topographical changes which initiate intense tsunamis may be smaller than 30 miles. Accordingly, we must include an estimate of the extent to which the main features of the generated waves are sensitive to the lateral scales of the generating ground motion.

The propagation across the ocean is dispersive and, since the propagation path of interest can be several thousand miles in length, this dispersion can play an important role. On the other hand, the spectral region of interest is typified by wave lengths of many miles. The depth of the water is approximately 3 miles, and the wave heights are of the order of one foot, except during the run-up phase of the phenomenon; accordingly, the propagation over the deep water can be treated within the framework of a linear, inviscid theory. In [**16**] this linear dispersive theory was carried out without resorting to a "shallow water" or "long wave" approximation; no significant loss in accuracy is incurred, however, when one does use the Boussinesque form of shallow water theory, even though this formalism does exaggerate the dispersion of the shorter waves. We will start out here with the classical theory and revert to the Boussinesque long wave theory

AMS 1970 *subject classifications*. Primary 76B15, 86A05.

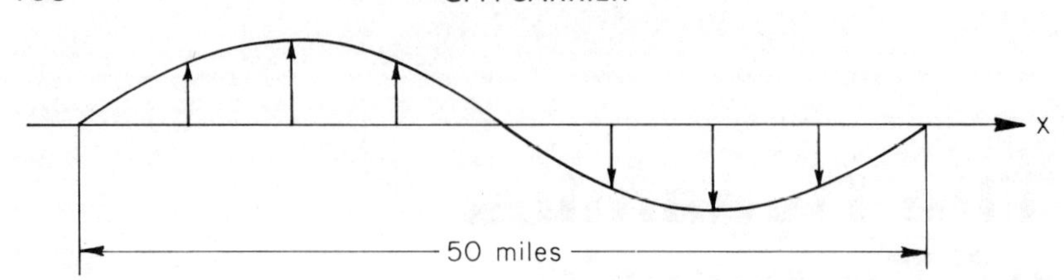

FIGURE 1a. Schematic view underwater ground displacement for 1964 Alaskan earthquake.

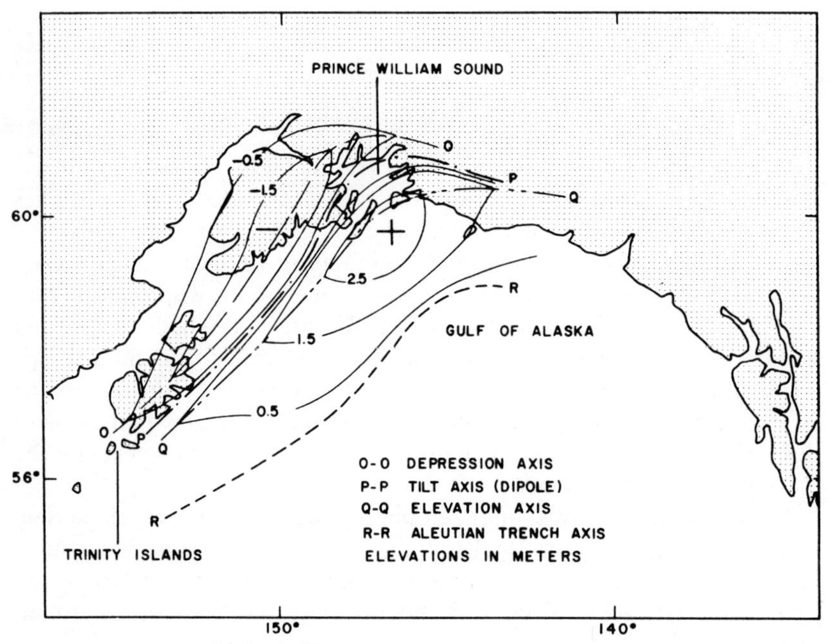

FIGURE 1b. Hypothetical contours of vertical land dislocation in Gulf of Alaska during earthquake of 3/28/64, as deduced from shoreline changes. Pattern is dipolar, with positive pole on shallow shelf bordering Gulf.

at a point where it is natural to do so, and where the nature of the approximation is particularly evident. In this way we avoid the more cumbersome manipulations of the classical theory; the reader who wishes to see the continuation of the classical theory can find it in [**16**].

In this dispersive analysis we will ignore, tentatively, the irregularities of the bottom topography. We then assess the importance of this topographical irregularity by estimating the cumulative reflection of individual monochromatic waves by a stochastically described bottom topography. The results (§3) indicate that, unless there are significant systematic topographical features which are not revealed by maps of the Pacific, reflections associated with the nonuniform depth are of no real importance.

When the tsunami encounters continental or island boundaries, the waves propagate into more and more shallow water, and, despite the fact that some of the energy is reflected back toward the deep water during this process, the energy per unit depth increases toward the shoreline and the waves become so intense that a nonlinear theory is required. Fortunately, the nonlinear contributions become important only at distances from the shorelines which are short compared to the radius of curvature of the major features of that shoreline. This permits a run-up inference based on a linear theory for the refractive effects of the general shoreline geometry and a one-dimensional, nonlinear run-up theory.

The run-up can differ considerably at different, closely spaced places on a given shoreline. This is almost surely associated with complications due to the smaller scale local topography. The effects of resonance of well-defined features such as narrow mouthed harbors can be assessed in a reasonable way for low intensity incident waves, but the overall local response problem, especially for intense inputs, is prohibitively complex.

The energy in a tsunami is dissipated very slowly (the e-folding time is several days) and the energy distribution over the globe after one day is more dependent on the geometry of the entire basin and on the dissipative mechanisms than on the dynamics of the propagation from the source to any particular "target." In this paper we deal only with the early phases of the phenomenon and are concerned only with waves which arrive early and directly from the source.

2. **Generation and constant depth propagation.** The linear, inviscid theory which governs the propagation of gravity waves requires that one define a velocity potential, ϕ, such that the particle velocity $\mathbf{v}$, is given by

$$\mathbf{v} = \operatorname{grad} \phi.$$

The conservation of mass then implies that

$$\Delta\phi \equiv \phi_{xx} + \phi_{yy} + \phi_{zz} = 0 \tag{2.1}$$

in the region occupied by the fluid. When the coordinate system is such that gravity acts in the negative y direction, the (linearized) boundary conditions at the free surface require that[1]

$$\phi_y(x, 0, z, t) + \phi_{tt}(x, 0, z, t) = 0. \tag{2.2}$$

Here, we have adopted the convention that the rest position of the free surface is at $y = 0$ and we have adopted a unit of time such that $g/L = 1$

[1] All of this is documented in, e.g. Stoker [**19**].

where L is the unit of length; it will be convenient in our problem to take L to be the (dimensional) depth of the basin so that, in our coordinate system, the bottom of the basin is at $y = -1$ after the ground motion has ceased.

The free surface displacement, η, (i.e. the wave height) is given by

$$\eta(x, z, t) = \phi_y(x, 0, z, t). \tag{2.3}$$

On the bottom, we require that

$$\phi_y(x, -1, z, t) = F(x, z, t) \tag{2.4}$$

where F is the prescribed vertical component of velocity of the ground motion. We shall always take this motion to occur in some time interval prior to $t = 0$. It is clear that different functions F will lead to an enormous variety of wave motions and, unless we confine our attention to a reasonable variety of such functions, F, we won't be able to characterize the results in any very simple way. Accordingly, we will adopt a very simple "fundamental source," F, and try to infer from its implications the motion which can be built up from simple combinations of such fundamental sources. We choose

$$F_0(x, z, t) = \tfrac{1}{2}(\pi\alpha)^{-1/2} \exp(-x^2/4\alpha)g(t) \tag{2.5}$$

where $g(t)$ is any function whose area $\int_{-\infty}^{0} g(t)$ is unity, whose value is zero in $t \geqq 0$ and in $t < -T_0$, where T_0 is short compared to the time it takes a gravity wave to cross the area where the generation occurs.

Suppose now that the ground motion in which we are interested has the distribution in x which is depicted in Figure 1, but that it occurs in the interval $-z_0 < z < z_0$ as a torsional wave proceeding from $-z_0$ to z_0 (a distance of, say, 800 miles) at such a speed that the displacement at z_0 is finished five minutes after it began at $-z_0$. This is roughly the process which is believed to have occurred in the Alaskan earthquake of 1964. Figure 2 is a top view of the region of width (in the x direction) $4\alpha^{1/2}$ and of length $2z_0 = 800$ miles[2] on which this motion occurs. When the ground motion wave has just reached z_0, the wave emanating from $-z_0$ has had five minutes to travel and the leftward moving crest will have moved to some point C, whereas the rightward moving crest will have moved to C'; C and C' differ in position by $(gL)^{1/2}$ times 10 minutes. Furthermore, the left moving crest locus will lie along the line connecting C and D and the leftward wave motion will proceed as though the ground motion

[2] When it serves a descriptive purpose, we mix dimensional quantities and nondimensional quantities without apology.

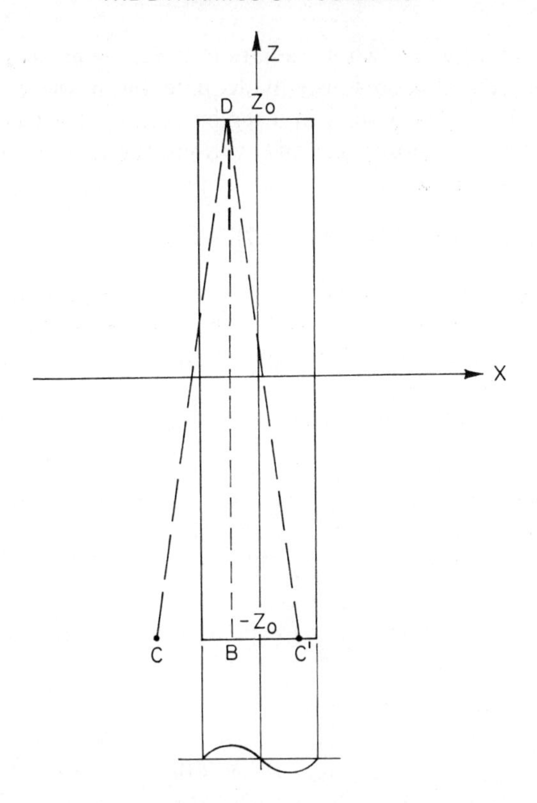

FIGURE 2. Early wave motion as described in text. BD is the line on which the crest of the ground displacement forms, starting at Z_0. CD is the low of the gravity wave crest at the time the ground motion wave reaches Z_0.

had been simultaneous at all z in $(-z_0, z_0)$ but had occurred along a strip parallel to the line from D to C.

Furthermore, the attenuation of the wave which is implied by the fact that the ground motion didn't occur along the whole line $-\infty < z < \infty$ will be well approximated by an amplitude proportional to $(A^{1/2}/x)^{1/2}$ for x much larger than $A^{1/2}$, (where A is the area over which most of the generation occurs), whereas the change in waveform and the further attenuation due to dispersion will be *very* nearly the same as though the ground motion *had* occurred all along $-\infty < z < \infty$. That is, the solution, η_{3D}, of the three-dimensional problem that we really are faced with can be nicely approximated by writing

$$\eta_{3D}(x, y, z, t) = (A^{1/2}/x)^{1/2}\eta_{2D}(x, y, t)$$

for values of z in the wedge-shaped region $z < x/3$, $x > 2A^{1/2}$.

If one really wants the whole radiation pattern from such a source, the foregoing procedure is obviously inadequate, but in the tsunami context we want the radiation pattern in a region comparable to that described above. Thus, with the foregoing observations, the solution of the problem in which F is given by

$$F(x, z, t) = F_0(x, z, t) - F_0(x + b, z, t)$$

with judicious choices of α and b will permit us to make the appropriate inferences. Accordingly, we proceed to the problem defined by Equations (2.1), (2.2), (2.3), and (2.4) with F replaced by F_0, and the observation that ϕ and η do not depend on z.

Defining the composite Fourier transform with regard to t and x as

$$\bar{\phi}(\xi, y, \omega) = \int_{-\infty}^{\infty} \int \exp(-i\omega t - i\xi x)\phi(x, y, t)\, dx\, dt$$

so that

$$\phi(x, y, t) = (1/4\pi^2) \int_{-\infty}^{\infty} \int \exp(i\omega t + i\xi x)\bar{\phi}(\xi, y, \omega)\, d\omega\, d\xi,$$

Equations (2.1) to (2.4) are replaced by

$$\bar{\phi}_{yy} - \xi^2\bar{\phi} = 0, \tag{2.6}$$

$$\bar{\phi}_y(\xi, 0, \omega) - \omega^2\bar{\phi}(\xi, 0, \omega) = 0, \tag{2.7}$$

$$\bar{\eta}(\xi, \omega) = \bar{\phi}_y(\xi, 0, \omega), \tag{2.8}$$

and

$$\bar{\phi}_y(\xi, -1, \omega) = \bar{F}_0(\xi, \omega) = \exp(-\alpha\xi^2)\bar{g}(\omega). \tag{2.9}$$

Because $g(t)$ is narrow in t and is zero after $t = 0$, the wave behavior in $t > 0$ will be virtually unaffected by the extreme choice $g(t) = \delta(t)$; i.e. by $\bar{g}(\omega) \equiv 1$. With this choice (and $s = i\omega$),

$$\bar{\phi}(\xi, y, \omega) = -(\xi \cosh \xi y + s^2 \sinh \xi y) \exp(-\alpha\xi^2)/\xi(s^2 \cosh \xi + \xi \sinh \xi) \tag{2.10}$$

and, in particular,

$$\bar{\eta}(\xi, \omega) = -s \exp(-\alpha\xi^2)/(s^2 \cosh \xi + \xi \sinh \xi); \tag{2.11}$$

the inversion of $\bar{\eta}$ over ω involves only the evaluation of the residues of the integrand $I = -s\exp(-\alpha\xi^2 + i\omega t)/(s^2\cosh\xi + \xi\sinh\xi)$ at $\omega = \pm(\xi\tanh\xi)^{1/2}$ and produces the reduction

$$\eta(x, t) = \frac{1}{4\pi^2}\int\int_{-\infty}^{\infty} I\, d\xi\, d\omega$$

(2.12)

$$= \frac{1}{2\pi}\int_{-\infty}^{\infty} \frac{\exp(-\alpha\xi^2 + i\xi x)(\exp(itk(\xi)) + \exp(-itk(\xi)))}{2\cosh\xi}\, d\xi.$$

In the dispersive long wave theory known as the Boussinesque theory [**19**], one pretends that the horizontal particle velocity is independent of the depth coordinate, but that the vertical velocity, w, is given by $w = (y + 1)\eta_t(x, z, t)$. An appropriate average of the momentum equations and the conservation of mass then imply a partial differential equation in x, z, and t for the wave height η. Its solution under initial conditions equivalent to the boundary condition of Equations (2.4) provides a solution which differs from Equation (2.12) in that $\cosh\xi$ is replaced by unity and $(\xi\tanh\xi)^{1/2}$ is replaced by $(\xi - \xi^3/6)$. The large wave-length portion of the spectrum, which corresponds to small ξ, is very well reproduced by such a theory, but it is clear that the dispersion of short waves will be exaggerated by the Boussinesque formalism and that these same short waves will appear to be more intense under that formalism than under the more correct one. However, we shall see that only the longer wave-length end of the spectrum is of interest, and it is well worth the sacrifice in accuracy in the description of the short wave content to adopt the Boussinesque theory and its inherent manipulative simplicity. Accordingly, we replace Equations (2.12) as indicated above. While we are at it, we also eliminate the term in $\exp(+it(\xi\tanh\xi)^{1/2})$. That term contributes only to the wave in $x < 0$ and we are interested only in $x \gg 1$. Thus, we have

$$\eta = \frac{1}{4\pi}\int_{-\infty}^{\infty} \exp[-\alpha\xi^2 + i\xi x - it(\xi - \xi^3/6)]\, d\xi. \tag{2.13}$$

According to [**20**], this is equivalent to

(2.14)

$$\eta = \tfrac{1}{2}(2/t)^{1/3}A_i\{(2/t)^{1/3}[x - t + (2\alpha^2/t)]\}\exp[(8\alpha^3/3t^2) - (2\alpha(t - x)/t)]$$

where A_i denotes the Airy function.

The effect of the lateral scale of the generating ground motion can now be assessed. In Figures 3a to 3e are plotted $\eta(1000, t)$ for $\alpha = 90, 50, 30, 10, 0$. For a basin of 3-mile depth these values correspond to "widths" of the generating region which are 114, 85, 65, 38, and 0 miles respectively. The

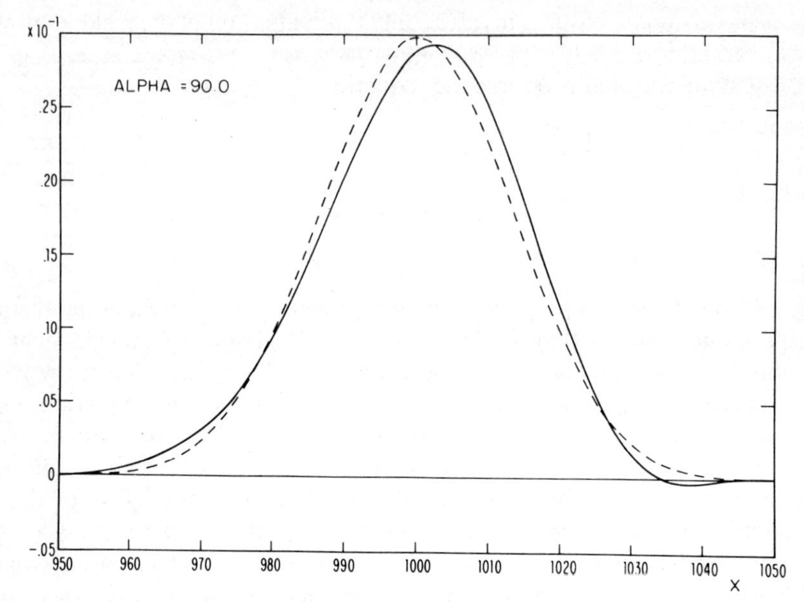

FIGURE 3a. Wave form at $t = 1000$ and at $t = 0$.

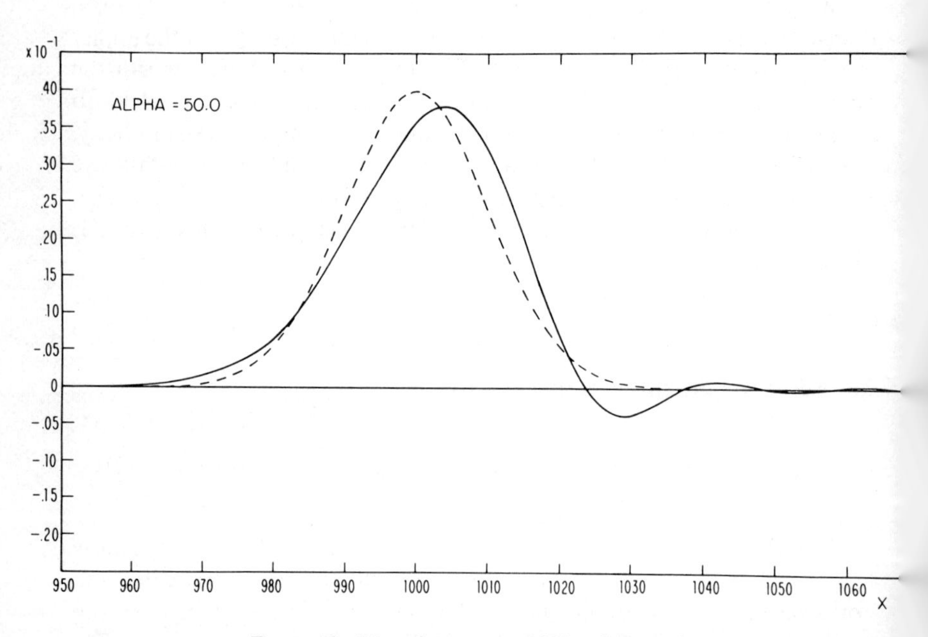

FIGURE 3b. Wave form at $t = 1000$ and at $t = 0$.

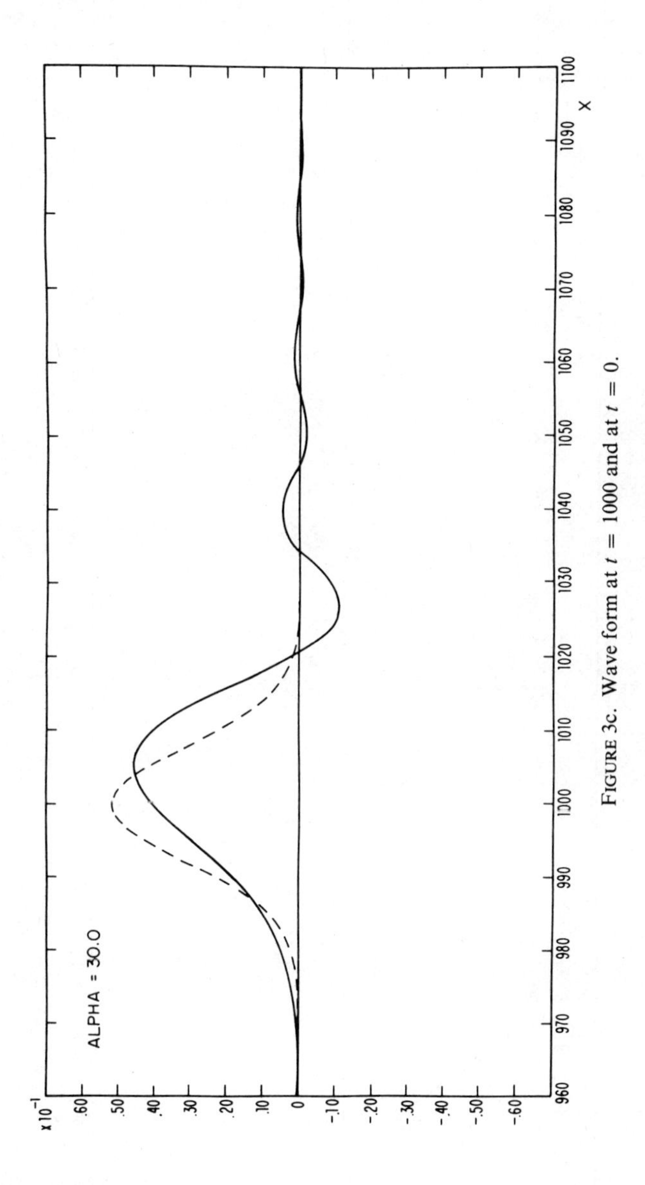

FIGURE 3c. Wave form at $t = 1000$ and at $t = 0$.

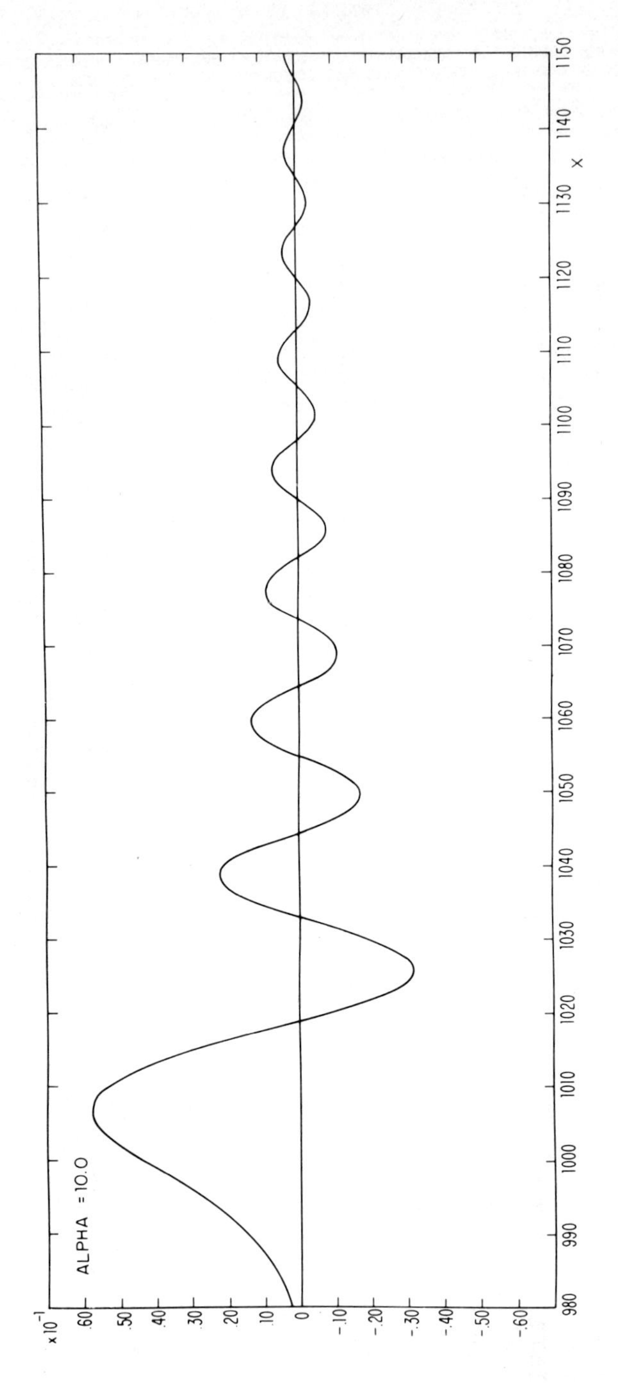

FIGURE 3d. Wave form at $t = 1000$.

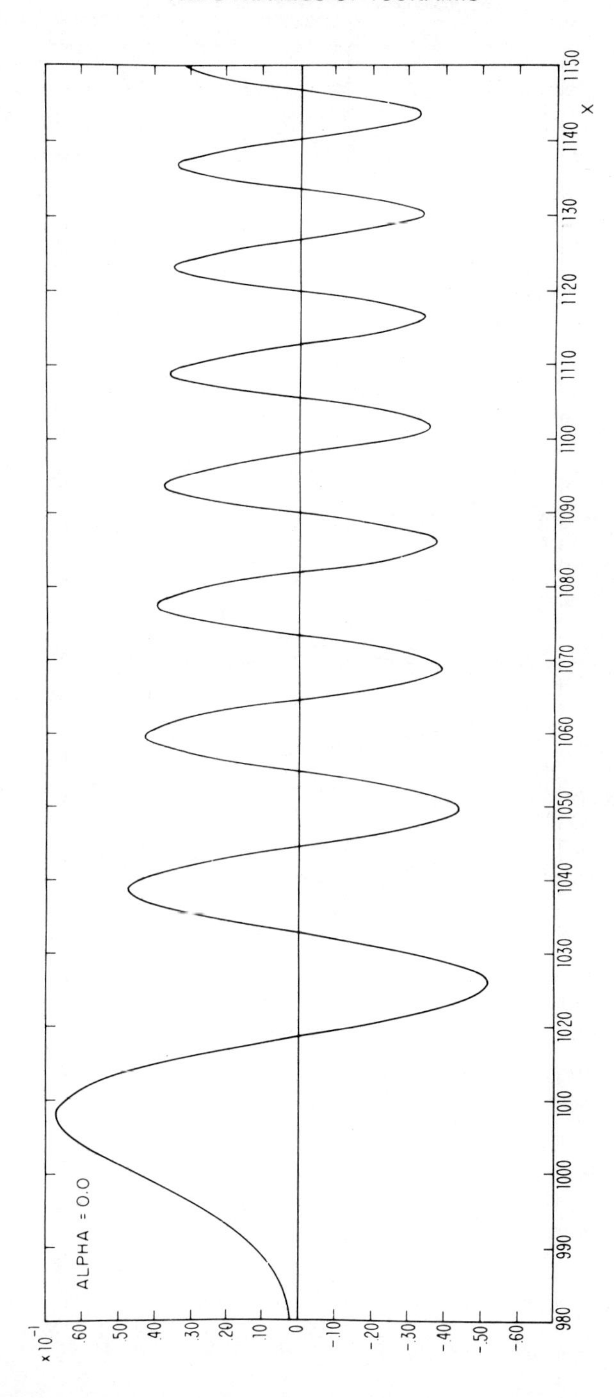

FIGURE 3e. Wave form at $t = 1000$.

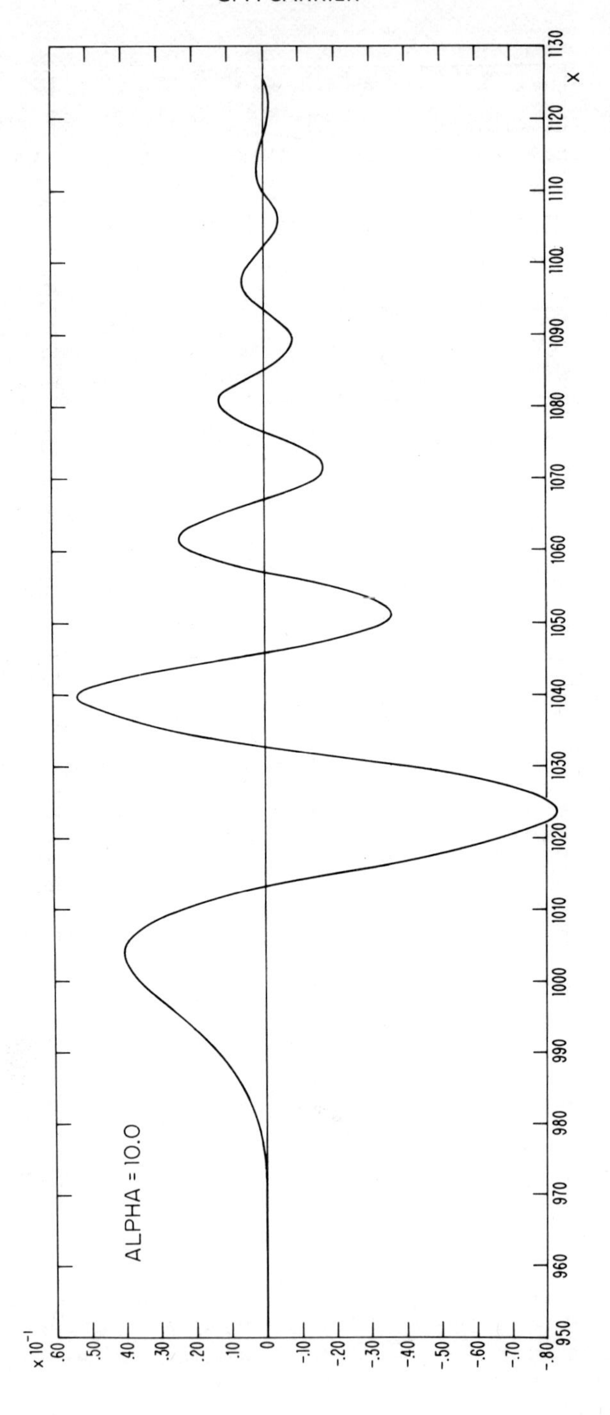

FIGURE 3f. Wave form at $t = 1000$ for source of Equation (2.15).

ground displacement is dotted in three of these graphs; it does not fit on the others. Note that dispersion plays no role at all for broad initial motions but is overwhelmingly important for narrow ground displacements (the implication in [**16**] that all ground motions of interest could be treated as though they were narrow was a bit naive). In particular, in Figure 3f we plot the values of $\eta(1000, t)$ for the ground displacement described by

$$F = F_0(x, z, t) - F_0(x + 14, z, t) \tag{2.15}$$

with $\alpha = 10$. This initiating motion corresponds closely to that shown schematically in Figure 1a. Note that the second crest is higher than the first crest. Note also that $x = 1000$ corresponds to the 3000-mile distance between the seismically active Aleutian region and the Hawaiian Islands.

One frequently hears the statement that the second and/or third crests are higher than the first in the intense tsunamis which reach the Hawaiian shores, but whether this is *really* a characterizing feature of such phenomena, I do not know. If it is consistently true, it suggests that ground motions of the type shown in Figure 1a might be more typical than those of the curves in Figures 3a–3e. At this stage such a claim concerning the nature of undersea earthquakes would be unfounded speculation, but the possibility that tsunami wave trains may permit some inference concerning such earthquakes should not be entirely overlooked.

3. **Variable depth effects.** The wave which arrives at $x = 1000$ according to the results of §2 is not that which would arrive if the bottom topography were nonuniform. Some distributions of bottom topography would imply negligible differences from those described, but others could have a profound influence on the transmission properties of the basin. Since, for large x, the behavior of η is dominated by the behavior of the integrand of Equation (2.13) at the saddle points of its exponent, the effect of a given topography on $\eta(x, t)$ can be inferred from a knowledge of the effects of that topography on a monochromatic wave whose (average depth) wave number, ξ, has the value $\xi_0(x, t)$ where ξ_0 is the saddle point of the exponent of the integral of Equation (2.13). Thus, for our purposes, it suffices to study the propagation of a monochromatic wave over various topographies, and since we have already assessed the effects of dispersion, it also suffices to use a nondispersive theory. Accordingly, we adopt the conventional, shallow water, nondispersive water wave theory [**19**] which requires that

$$[h(x)\eta'(\alpha)]' + g^{-1}\omega^2\eta(\alpha) = 0 \tag{3.1}$$

where $\eta(\alpha)\exp(i\omega t)$ is again the description of the surface displacement.

It is advantageous to define a new horizontal coordinate x according to the rule[3]

$$x = \left(\frac{h_0\omega^2}{g}\right)^{1/2} \int_0^{\alpha} [h(\alpha')]^{-1}\, d\alpha'$$

and to write

$$h(\alpha) = H_0(1 + \varepsilon f(x)) \quad \text{and} \quad \eta(\alpha) = u(x).$$

Equation (3.1) becomes

$$u''(x) + [1 + \varepsilon f(x)]u(x) = 0. \tag{3.2}$$

The topographies which are useful in this study are those depicted in Figure 4. The depth is constant outside of $0 < x < L$, but varies in $0 < x < L$.

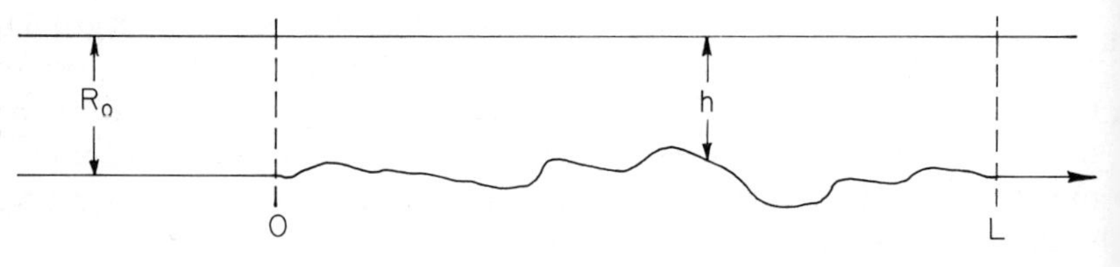

FIGURE 4

We allow an incident wave of intensity I to approach from the right; a reflected wave of intensity R is reflected to the right and a transmitted wave of unit intensity propagates to the left in $x < 0$. That is,

$$\begin{aligned} u(x) &= e^{ix} && \text{in } x < 0, \\ &= Ie^{ix} + Re^{-ix} && \text{in } x > L. \end{aligned} \tag{3.3}$$

Since $u(x)$ must be continuous at $x = 0$ (as well as at $x = L$), the first of Equations (3.3) implies that the solution of Equation (3.2) in $0 < x < L$ must obey the boundary conditions $u(0) = 1$ and $u'(0) = i$. Thus, we need only to solve an initial value problem rather than a two-point boundary value problem; this will be very advantageous before this section is completed and is the reason we choose the wave directions which led to Equation (3.3). The situation in which the lateral scale in x of $f(x)$ is large compared to all wave lengths in the energetic part of the tsunami spectrum is easily treated. The solutions of Equation (3.2) can be approximated very

[3] h_0 is defined in Figure 4.

accurately by the WKB method [**22**], and within the framework of that method, the predicted reflection is zero.

In [**16**], a more elaborate treatment of this "large scale topography" problem can be found; in that treatment the full dispersive and transient problem is solved but the zero reflection result also emerges there. However, it should have been noted there (but was not) that the analysis is informative only for that part of the spectrum in which the wavelength is small compared to the horizontal topographic scale.

Alternatively, if the topography were suitably approximated by $f(x) = \sin ax$, Equation (3.2) is Mathieu's equation, and for small ε, solutions of that equation which are uniformly valid in x can be constructed. The reflection is strongest when $a = 2$. In fact, with $a = 2$ and with $u(0) = 1$, $u'(0) = i$,

$$u \simeq \cosh(\varepsilon x/4)e^{ix} + \sinh(\varepsilon x/4)e^{-ix} \tag{3.4}$$

in $0 < x < L$, so that

$$u = \cosh(\varepsilon L/4)e^{ix} + \sinh(\varepsilon L/4)e^{-ix} \quad \text{in } x > L. \tag{3.5}$$

Thus, the reflection coefficient, R/I, is $\tanh(\varepsilon L/4)$. Since ε is the ratio of the topographical depth fluctuation to the average depth and L is 2π times the number of wavelengths in the distance from source to target, it is sensible to use $\varepsilon = 1/50$ and $L = 300$ which leads to $R > .9$. If this number had been small, then one could reason that the real and highly irregular topography would reflect even less and that no further analysis was necessary. However, the large reflection associated with the resonant situation of Equation (3.2) with $f(x) = \sin 2x$ suggests that even an irregular topography *may* imply significant reflections.

When the lateral scale of $f(x)$ in Equation (3.2) is of order unity, the problem must be solved individually for each $f(x)$ of interest. In the physical problem under study, the topography is not really one-dimensional and, in fact, the analysis of each source-target combination implies that a different topography characterizes the path. Thus, the fact that we are concerned with an ensemble of relatively uncharted topographies suggests that we might adopt a probabilistic description of $f(x)$ and attempt to predict the statistics of $u(x)$. In doing this, one should recognize that the observational data would probably allow us to estimate $\langle f(x)f(y)\rangle = R(x - y)$, but that more refined statistical properties would not be accessible. Here, R is the auto-correlation of f which, *by hypothesis*, is stationary and $\langle\ \ \rangle$ denotes "the ensemble average of." An analysis which allows the prediction of $\langle u(x)u(y)\rangle$ has been given in [**23**] and will be repeated here.

As we noted before, the normalization implied by the choice $u(x) = e^{ix}$ in $x < 0$ is such that we need not solve a 2-point boundary value problem. The fact that $u(x)$ must be continuous at $x = 0$ implies that we can study the problem posed by

$$u''(x) + [1 + \varepsilon f(x)]u(x) = 0 \quad \text{in } x > 0 \tag{3.6}$$

with

$$u(0) = 1, \qquad u'(0) = i. \tag{3.7}$$

When this has been done, the behavior of $u(x)$ at L will imply the information we want concerning the reflection coefficient just as it did in the situation where $f(x) = \sin ax$.

The problem we must solve, then, is: How can one describe statistical properties of the ensemble of solutions $u(x)$ of the initial value problem given by Equations (3.6) and (3.7) when we know nothing about the ensemble of functions $f(x)$ except that

$$\langle f(x)f(y)\rangle = R(x - y)? \tag{3.8}$$

We initiate such a description by noting that Equations (3.6) and (3.7) imply

$$u(x) = e^{ix} - \varepsilon \int_0^x \sin(x - x')f(x')u(x')\,dx' \tag{3.9}$$

and, therefore, that (when an asterisk denotes the complex conjugate)

$$\begin{aligned} u(x)u^*(y) &= \exp[i(x - y)] - \varepsilon e^{ix} \int \sin(y - y')f(y')u^*(y')\,dy' \\ &\quad - \varepsilon e^{-iy} \int_0^x \sin(x - x')f(x')u(x')\,dx' \\ &\quad + \varepsilon^2 \int_0^x \int_0^y \sin(x - x')\sin(y'y')f(x')f(y')\,u(x')u^*(y')\,dx'\,dy \end{aligned} \tag{3.10}$$

The ensemble average of Equation (3.10) contains three different unknown statistical quantities, i.e. $\langle u(x)u(y)\rangle$, $\langle f(x')u(x')\rangle$ and $\langle f(x')f(y')u(x')u(y')\rangle$. It is useful to inquire whether, in the spirit of the "truncated hierarchy techniques" the last two of these quantities can be suitably approximated in terms of $\langle f(x')f(y')\rangle$ and of $\langle u(x')u(y')\rangle$.

When $\varepsilon \ll 1$ (as it is in the tsunami context) Equation (3.2) implies that $u(x)$ can differ but little from a linear combination of e^{ix} and e^{-ix} over an interval, Δ, for which $\Delta \ll \varepsilon^{-1}$. Thus, for a given small $x = x_0$, each member of the ensemble of functions $u(x_0)$ has nearly the same value whereas the ensemble average of $f(x_0)$ is zero. It follows that one could hypothesize that $\langle f(x)u(x)\rangle = 0$ for small x. For large x_0, on the other hand, the behavior of $u(x_0)$ depends on the cumulative effects of $f(x)$ in $0 < x < x_0$; since $R(x - y)$ is narrow, this "history" of $f(x)$ is not correlated with $f(x_0)$, and again one could guess that $\langle f(x)u(x)\rangle \simeq 0$.

One can also write

$$f(x)f(y) = \langle f(x)f(y)\rangle - \rho(x, y),$$
$$u(x)u(y) = \langle u(x)u(y)\rangle + p(x, y),$$

and, on basically the same arguments as those given above, one can hypothesize that

$$\langle \rho(x, y)p(x, y)\rangle \simeq 0,$$

so that

$$\langle (f(x)f(y)u(x)u(y)\rangle \simeq \langle f(x)f(y)\rangle\langle u(x)u(y)\rangle.$$

It is shown in [**23**] that both of these hypotheses are very accurate indeed, and we can replace the ensemble average of Equation (3.10) by

$$\langle u(x)u^*(y)\rangle = e^{i(x-y)} + \varepsilon^2 \int_0^x \int_0^y \sin(x - x')\sin(y - y') \times \langle u(x')u^*(y')\rangle R(x' - y')\,dx'\,dy' \tag{3.11}$$

with confidence.

Since the manipulational details associated with the solution of this problem are documented in [**23**], we will confine ourselves here to a verbal argument as to what one can expect. Let δ be the "width" of $R(z)$; that is to say, let almost all of the area under $R(z)$ lie over the interval $|z| < \delta$. Then if $\delta \ll 2\pi$ and $x, y > \delta$, the integral over y' of Equation (3.11) proceeds as though R were a delta function. Thus,

$$|u(x)|^2 = 1 + \varepsilon^2 \int_0^x \sin^2(x - x')\langle |u(x')|^2\rangle\,dx'. \tag{3.12}$$

Except near $x = 0$, the oscillatory part of $\sin^2(x - x') = \frac{1}{2}(1 - \cos[2(x - x')])$ is relatively unimportant and

$$\langle |u(x)|^2\rangle \simeq 1 + \frac{\varepsilon^2}{2}\int_0^x \langle |u(x')|^2\rangle\,dx'$$

so that

$$\langle |u(x)|^2\rangle \simeq \exp(\varepsilon^2 x/2). \tag{3.13}$$

Alternatively, if δ is of order unity or larger (but not of order ε^{-1}), the integral over y' will reflect the fact that R is multiplied by various trigonometric quantities.

It requires the extensive manipulations documented in [**23**], of course, to show that these trigonometric contributions lead to the result

$$\langle |u(x)|^2 \rangle \simeq \exp([\bar{R}(0) + \bar{R}(2)]\varepsilon^2 x/4) \tag{3.14}$$

(where $\bar{R}(\xi)$ denotes the Fourier transform of $R(x)$) instead of the result in Equation (3.13).

Once we know $\langle |u(x)|^2 \rangle$ we can also find $\langle |u'(x)|^2 \rangle$ as follows. Equation (3.9) implies that

$$u'(x) = ie^{ix} + \varepsilon \int_0^x \cos(x - x')f(x')u(x')\,dx' \tag{3.15}$$

and that

$$\langle u'(x)u^{*\prime}(y) \rangle = \exp[i(x - y)]$$

$$+ \varepsilon^2 \int\int \cos(x - x')\cos(y - y')R(x' - y')\langle u(x')u^*(y') \rangle\,dy'\,dx', \tag{3.16}$$

and this provides a quadrature for $\langle u'(x)u^{*\prime}(y) \rangle$.

The boundary conditions at $x = L$ imply that

$$|R^2| + |I^2| = \tfrac{1}{2}\{\langle |u(L)|^2 \rangle + \langle |u'(L)|^2 \rangle\} = \exp(\tfrac{1}{4}\varepsilon^2 L[\bar{R}(2) + \bar{R}(0)]),$$

and the conservation of energy implies that $|I^2| - |R|^2 = 1$ so that the reflection coefficient is given by

$$|R/I|^2 = O[(e^{\varepsilon^2 L/2} - 1)/(e^{\varepsilon^2 L/2} + 1)].$$

Using the same values of ε and L as in §2, we find

$$|R/I|^2 = O(.03).$$

Thus, despite all the analysis, it seems very doubtful that scattering by bottom irregularities is very important in the propagation of tsunamis.

There is another type of topographical feature which might have some importance. If the source were located near $x = 0$, $y = 0$ and if the depth were given by

$$h = h_0[1 + \varepsilon f(x, y) + \alpha g(y)],$$

where

$$g(y) = 0 \quad \text{in } |y| > a,$$
$$\quad\;\; = 1 \quad \text{in } |y| < a$$

and where f describes irregularities with zero average, the ridge so defined could act as a wave guide and could thereby provide, in some spectral region, larger amplitudes at $x = 0$, $y = y_0$ than, for example, at $x = 3a$, $y = y_0$. Preliminary estimates, with $\alpha \leqq 1/30$, suggest that such ridges would not be very important either, but the transcendental mess which characterizes the analysis of that problem has not been so accurately unraveled yet that this can be said with confidence.

4. **Run-up on an inclined plane.** When the wave discussed in §2 encounters the shelf of a continent or an island, the wave intensifies and reflects as it encroaches into shallower water. The distance which is travelled by the wave on such a shelf is short enough so that dispersion is not important but the intensity (free surface slope) becomes so great that a nonlinear treatment is required. Accordingly, we adopt the conventional, non-dispersive, nonlinear shallow water theory in which (in its one-dimensional form)

$$[(h + \eta)u]_x + \eta_t = 0, \qquad u_t + uu_x + \eta_x = 0. \tag{4.1}$$

η is the wave height and u the horizontal particle velocity.

In this section we measure x and t in the same units as in §2 so that, in particular, (see Figure 5)

$$h = (x_0 - x)\tan\theta = (x_1 - x)/(x_1 - x_0).$$

It happens [**18**] that the following is true.

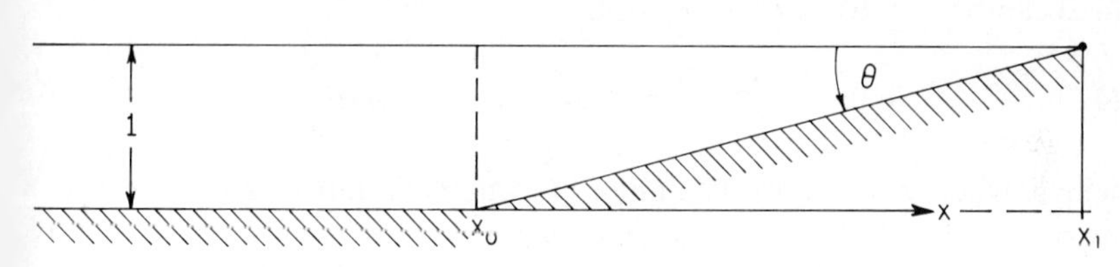

FIGURE 5. The analysis of §4 applies to the region $x_0 < x < x_1$.

Let $\psi(\sigma, \lambda)$ be almost[4] any bounded solution of

$$(\sigma\psi_\sigma)_\sigma - \sigma\psi_{\lambda\lambda} = 0 \tag{4.2}$$

in $0 < \sigma < 4(x_1 - x_0)^{1/2}$, $0 < \lambda$, and let

$$u(\sigma, \lambda) = \psi_\sigma/\sigma\theta^{1/2}, \tag{4.3}$$

$$x - x_1 = \psi_\lambda/4\theta - \sigma^2/16 - u^2/2\theta, \tag{4.4}$$

[4] One constraint will be appended later in this paragraph.

(4.5) $$\eta(\sigma, \lambda) = \tfrac{1}{4}\psi_\lambda - \tfrac{1}{2}u^2,$$

(4.6) $$t = \lambda/2\theta^{1/2} - u/\theta.$$

Then the parametric description of u and η as functions of x and t which is described by Equations (4.2) through (4.6) is a single-valued solution of Equation (4.1) provided only that the Jacobian $J(x, t/\sigma, \lambda)$ is greater than zero in $0 < \sigma < 4(x_1 - x_0)^{1/2} \equiv \sigma_0$.

This representation has some entirely fortuitous but remarkably convenient features; at any point x, t, the local water depth, $h + \eta$, is given by $\theta\sigma^2/16$. That is, $\sigma = 0$ defines the $x(t)$ which describes the time dependent position of the edge of the region occupied by water. Furthermore, when $h = O(1)$, the wave height and slope in our problems will be very small and Equations (4.4) and (4.6) degenerate to

$$x_1 - x \simeq \sigma^2/16, \qquad t = \lambda/2\theta^{1/2}.$$

Finally, when $\eta(0, \lambda)$ is the maximum or a minimum, the velocity, u, is zero. When this is the case, and also when $\sigma = \sigma_0$, η is given by $\tfrac{1}{4}\psi_\lambda$. In other words, if we care only about the maximum excursion of the water's edge, we care only about the λ derivative of the solution of the linear equation (4.2) subject to appropriate linear boundary conditions at $\sigma = \sigma_0$, and $\lambda \simeq 2t(\theta)^{1/2}$.

It is also particularly advantageous to note that solutions of Equation (4.2) can be constructed in the form

(4.7) $$\tfrac{1}{4}\psi_\lambda = \int_{-\infty}^{\infty} \exp(-i\lambda p(\xi))H(\xi)J_0(\sigma p(\xi))\, d\xi$$

which, when $\sigma = \sigma_0$, is very similar in form to Equation (2.13) since λ can be replaced by $t/2\theta^{1/2}$ for that value of σ.

Actually, the wave in $x < x_0$ must consist of that described by Equation (2.13) plus a term which depicts the reflected wave moving to the left. This reflected wave can be written:

$$\eta_{\text{left}} = \int F(\xi)\exp(-i\xi(x - 2x_0) - it(\xi - \xi^3/6))\, d\xi.$$

Both η and η_x must be continuous at $x = x_0$ (i.e. at $\sigma = \sigma_0$) and this must be true for each Fourier component (in time) of the wave. Hence

$$\exp(-a\xi^2)\exp(i\xi x_0 - it(\xi - \xi^3/6)) + F(\xi)\exp(i\xi x_0 - it(\xi - \xi^3/6))$$
$$= H(\xi)J_0(\sigma_0 p(\xi))\exp(-2i(\theta)^{1/2}tp(\xi))$$

Thus,

$$H(\xi) = (2e - \alpha\xi^2 + i\xi x_0)/(J_0[\sigma_0 p(\xi)] + i(1 - \xi^2/6)J_0'[\sigma_0 p(\xi)]).$$

Again, I am reluctant to repeat details which have been published elsewhere [**16**] so we note only that the integral of Equation (4.7) can be estimated by the method of stationary phase. When $\alpha = 0$,

$$\frac{\eta_{\max} \text{ at } \sigma = 0}{\eta_{\max} \text{ at } \sigma = \sigma_0}$$

has been calculated [**16**], and depending on whether the ground motion associated with F_0 is up or down, this ratio takes the value $4.2\theta^{-1/2}x_0^{-1/3}$ or $5.6\theta^{-1/2}x_0^{-1/3}$. For values of α greater than 30 the $x_0^{-1/3}$ will not appear (it arises only because of the dispersion) but I know of no calculations which have treated such cases.

The most fortunate implication of this "run-up" theory is the fact that the maximum run-up is given correctly by a linear theory in which $\eta = \psi_\lambda/4$, Equation (4.2) governs ψ, $x_1 - x = \sigma^2/16$, $t = \lambda/2\theta^{1/2}$ and the run-up is the largest value of $\psi_\lambda/4$ at $\sigma = 0$. Furthermore, it is readily checked that the nonlinearities become important only in the very shallow water. These facts make one optimistic that, when two-dimensional shelf topographies are of interest, one can use, and interpret meaningfully, a linear theory. That we do in the next section.

5. **Refraction by islands.** When a plane incident gravity wave encounters an island whose underwater constant depth curves are of the order of 100 miles across, the one-dimensional run-up treatment is inadequate. Accordingly, following Lautenbacher [**24**], we try to infer the effects of the refractive influence of such islands by treating the following problem.

The geometry is given by

$$h = 1 \qquad \text{in } r \geqq L/2,$$

$$1 - h = (L - 2r)/(L - R) \qquad \text{in } 0 < r \leqq L/2.$$

Note that the water extends only into $r = R/2$ where the depth becomes zero.

The incident wave is given by $\eta_i = \exp(i\omega(x - t))$, and the wave height of the complete wave field, $N = \eta + \eta_i$, must satisfy the nondispersive linear shallow water requirement that

$$(5.1) \qquad (hN_x)_x + (hN_z)_z + k^2 N = 0.$$

It must also satisfy the radiation condition which states that $\eta_i + \eta \equiv N = \eta_i +$ outgoing wave, as well as the requirement, implied by the fact that there is no frictional dissipation, that η be bounded on $r = R/2$.

There seems to be no effective way to treat this problem analytically and numerical methods have been used. The major difficulty in a numerical treatment arises in connection with the infinite domain and the radiation condition. The idea that Lautenbaucher adopted to defeat these difficulties is outlined here. We write

$$\eta_{xx} + \eta_{zz} + k^2\eta = ((1 - h)\eta_x)_x + ((1 - h)\eta_z)_z + ((1 - h)\eta_{i,x})_x \tag{5.2}$$

to replace (5.1) and we note that the right-hand side of Equation (5.2) vanishes outside of $r = L/2$. We can now use Green's theorem in conjunction with the outgoing wave Green's function, $H_0^{(1)}(kr)$, of Equation (5.2) to form an integral equation. A considerable amount of manipulation accompanied by fortunate cancellations (all of which are too messy to be worth repeating here but are fully documented in [**24**]) lead finally to the equation

$$\begin{aligned} h\eta(x, z) = -\frac{i}{4}\int\int\Big\{ & -k^2 H_0^{(1)}(kR_1)h' \\ & + kH_1^{(1)}(kR_1)\left[\frac{x - x'}{R_1}h'_{x'} + \frac{z - z'}{R_1}h'_{z'}\right]\Big\}\eta(x', z')\,dx'\,dz' \\ & + \frac{1}{4}\int\int kH_0^{(1)}(kR_1)e^{ikx'}(h'_{x'} + ikh')\,dx'\,dz'. \end{aligned} \tag{5.3}$$

In this equation, $h' = 1 - h$, the integration takes place over the region $R/2 < r < L/2$ and $R_1^2 = (x - x')^2 + (z - z')^2$. The notation differs only slightly from Lautenbacher's.

The following description of the integration process is taken verbatim from [**24**].

> Equation (5.3) was solved numerically with the aid of an IBM 7094 computer. For circular islands, polar coordinates and circular grids were employed. The unknown was determined by assuming a biquadratic polynomial form over every mesh division of nine points. Quadrature coefficients were calculated by integrating over the block of the dual grid centered on the middle of the nine points under consideration. In this manner, a set of complex, linear, algebraic equations was developed for the values of the unknown at each grid point. This set of equations was solved for varying island size, slope, height and wavelength.

The results of this analysis for various choices of parameter are detailed in [**24**].

We reproduce here, in Figures 6 and 7, illustrative samples of those results. Note, as shown in Figure 6, that the wavelength to radius ratio has a profound effect on the run-up variation with coastal position and note from Figure 7 that the size of the maximum run-up is also sensitive to that ratio.

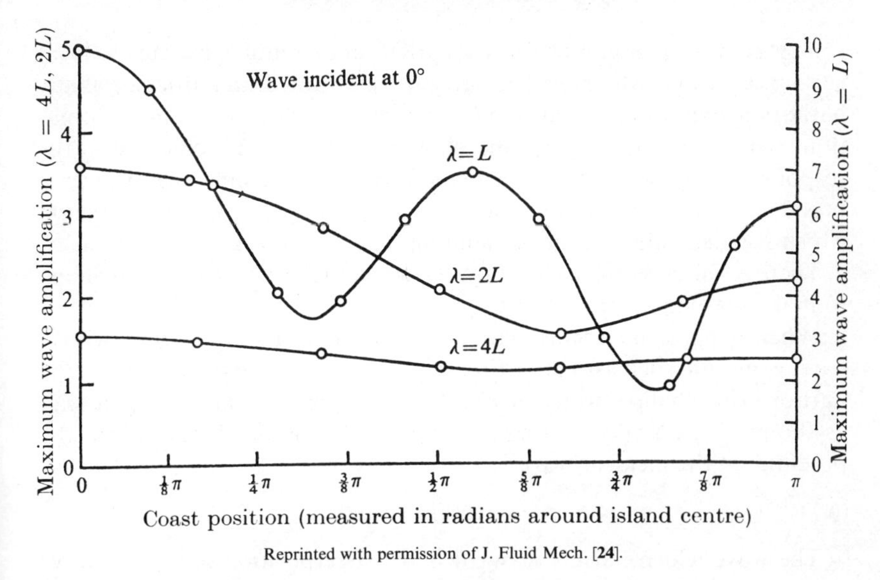

Reprinted with permission of J. Fluid Mech. [**24**].

FIGURE 6. Maximum wave amplification at coast (Oahu).

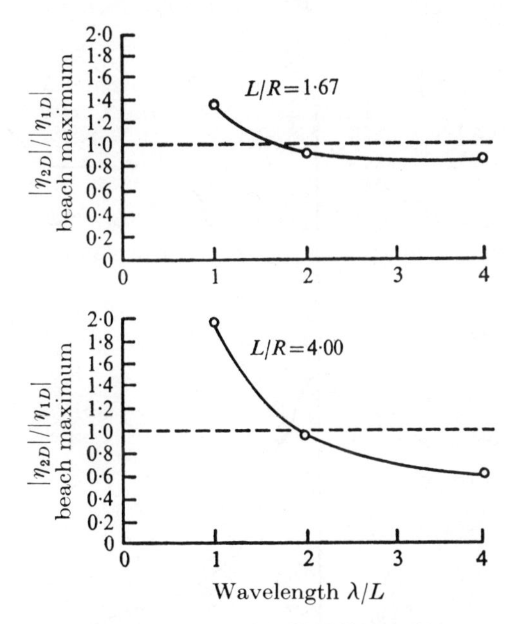

Reprinted with permission of J. Fluid Mech. [**24**].

FIGURE 7. Ratio of two-dimensional to one-dimensional maximum wave amplitude on beach. L, island diameter at ocean floor; R, island diameter at beach.

6. **Harbor response.** As we noted earlier, one cannot calculate the local run-up in regions where the geometry is complicated. In particular, many harbors respond to tsunami inputs in a way which involves bore formation and propagation along the shoreline. In fact, in some areas, the response seems to be that of a black box which accepts energy over a broad spectral range and re-issues that energy in a narrow spectral region which is independent of the tsunami input.

There is one class of local geometries, however, for which one could hope to make reasonable predictions.

When a plane monochromatic wave is incident on a harbor with a narrow mouth, such as that of Figure 8, we can analyze the motion in the harbor using shallow water theory. We treat the wave in the "open sea" as though it were propagating in a region of constant depth so that the potential of the incident wave is

$$\phi_i = \exp(i(kx - \omega t)). \tag{6.1}$$

The wave which reflects seaward is the superposition of a plane wave and one which emanates from the mouth of the harbor. We *approximate* this wave by

$$\phi' = e^{ikx} + AH_0^{(1)}(kr)$$

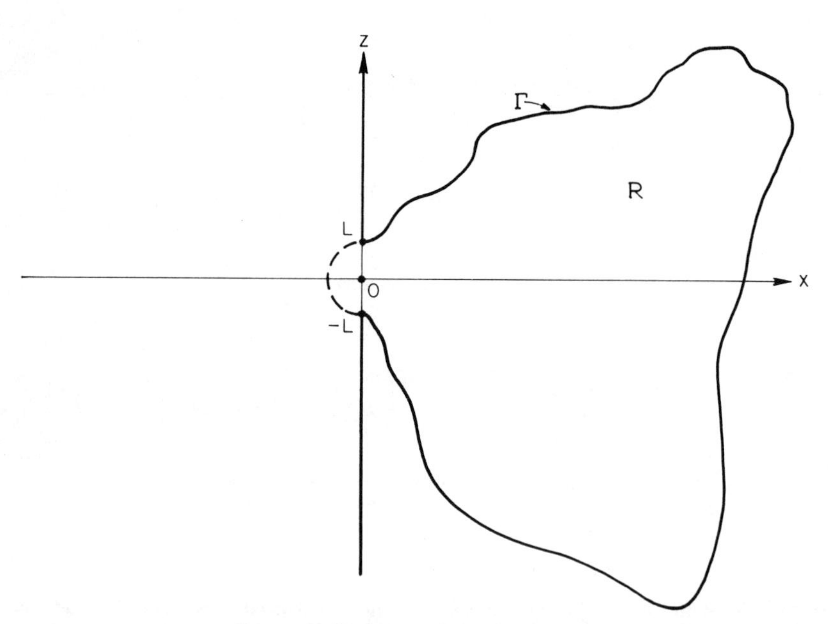

FIGURE 8. Harbor configurations of §6.

so that the potential ϕ, in $x > 0$ is

$$\phi = e^{ikx} + e^{-ikx} + AH_0^{(1)}(kr). \tag{6.2}$$

Clearly, this approximation is no good when $kL \geqq \pi$ but it should be an excellent description in $r > L$ when $kL \ll \pi$, i.e. when the wavelength is small compared to the aperture.

Inside the basin we have

$$\phi_{xx} + \phi_{yy} + k^2\phi = 0 \quad \text{in } R \tag{6.3}$$

with $\phi_n = 0$ on the boundary Γ.

There is a flow through the aperture, of course, which must be consistent with the potential of Equation (6.2). Another approximation which is very useful and which should be comparable in accuracy with that already invoked can be adopted here. We note that the larger scale modes of motion in the real basin should be nearly the same as those which would arise if the basin were closed (i.e. as if there were a wall across the aperture), but the flow came in through a small source located near the aperture but inside the basin. In this approximation, of course, the pressure at the source location must match that in $x > 0$ near the aperture. With this approximation, the potential in R must obey the equation

$$\phi_{xx} + \phi_{yy} + k^2\phi = -F_0 m(\bar{r}) \tag{6.4}$$

where F_0 is the flux of fluid into the basin (volume per unit time) and $m(\bar{r})$ denotes the distribution of the entering fluid. Clearly, to be consistent with the foregoing words, the integral of m over R is unity and $m(\bar{r})$ is nonzero only in a small region near the aperture. The boundary condition appropriate to this model requires that

$$\partial\phi/\partial n = 0 \quad \text{on } \Gamma' \tag{6.5}$$

where Γ' is the closed boundary consisting of Γ plus the aperture.

The seaward flux of fluid across the aperture which is associated with the potential of Equation (6.2) is

$$F = A\pi kL[H_0^{(1)}(kL)] \tag{6.6}$$

and the pressure on the semicircle where $r = L$ and $x < 0$ is

$$p = i\omega\rho[2 + AH_0^{(1)}(kL)]. \tag{6.7}$$

The potential of the motion in R can be described by

$$\phi = \sum b_i\phi_i(\bar{r}) \tag{6.8}$$

where the $\phi_i(\bar{r})$ are the eigenmodes of the homogeneous counterpart of Equations (6.4) and the boundary condition (6.5).

If we denote by a_i the coefficients of the expansion

$$m(\bar{r}) = \sum a_i\phi_i(\bar{r}), \tag{6.9}$$

we have

$$a_i = \int_R m(\bar{r})\phi_i(\bar{r})\,dA \Big/ \int_R \phi_i^2(\bar{r})\,dA$$

and, directly from (6.4), (6.8) and (6.9),

$$B_i = -F_o \int_R m(\bar{r})\phi_i(\bar{r})\,dA \Big/ (k^2 - k_i^2) \int \phi_i^2(\bar{r})\,dA, \tag{6.10}$$

where k_i is the eigenvalue which accompanies the ith eigenmode. However, the fact that m is nonzero on only a small region implies that

$$m(r)\phi_i(\bar{r})\,dA = B_i\phi_i(\bar{r}_0)$$

where $\bar{r}_0$ characterizes the source location (e.g. $\bar{r}_0 = 0$ is an appropriate choice) and B_i is very nearly unity for the lower modes but becomes very small for those modes whose lateral scale is smaller than the aperture size.

Thus

$$b_i \simeq -F_o\phi_i(\bar{r}_0)B_i \Big/ (k^2 - k_i^2) \int_R \phi_i^2(\bar{r})\,dA \tag{6.11}$$

and

$$\phi(\bar{r}) = -F_0 \sum \frac{B_i\phi_i(\bar{r}_0)\phi_i(\bar{r})}{(k^2 - k_i^2)\int \phi_i^2(\bar{r})\,dA} = -R_0\sigma(\bar{r}). \tag{6.12}$$

We can match the flux of fluid across the aperture by equating the quantity F of Equation (6.6) to the F_0 of Equation (6.10) et seq., and we can match pressures by equating $p = i\omega\rho\phi(\bar{r}_0)$ to its counterpart in Equation (6.7).

These lead to

$$\phi(\bar{r}) = 2kLH_0^{(1)\prime}(kL)\sigma(\bar{r})/D \tag{6.13}$$

where

$$D = H_0^{(1)}(kL) + kL\sigma(\bar{r}_0)H_0^{(1)\prime}(kL). \tag{6.14}$$

A careful examination reveals that, for $|kL| \ll 1$, there are zeros, $k_j + i\beta_j$ of D, for which the imaginary point is very small. This implies a large steady response at frequencies close to $k = k_j$. In the neighborhood of such frequencies, $\sigma(\bar{r})$ is dominated by at most two terms of the series which define $\sigma(\bar{r})$, and at such frequencies the calculation of the response is straightforward.

Suppose now that the incident wave is not monochromatic, but that, at $x = 0$, it is given by

$$\phi_i(0, t) = \frac{1}{2\pi} \int_{-\infty}^{\infty} Q_i(\omega) e^{i\omega t} \, d\omega.$$

The response of the harbor to this wave is given by the potential

$$\Phi(\bar{r}) = \frac{1}{2\pi} \int_{-\infty}^{\infty} Q_i(\omega) e^{i\omega t} \phi(\bar{r}) \, d\omega \tag{6.15}$$

where $\phi(\bar{r})$ is given by Equation (6.13) and where we must recall that $\phi(\bar{r})$ depends on k where $k = \omega/(gh)^{1/2}$.

We note, furthermore, that $\phi(\bar{r})$ can be decomposed into a sum

$$\phi(\bar{r}) = \sum_{i=1}^{n} \frac{a_i \phi_i(\bar{r})}{\omega - \omega_i} + S \tag{6.16}$$

where the ω_i are roots of D and where a_i is the residue of $\phi(\bar{r})$ (considered as a function of ω) at ω_i. In general one is interested primarily in a few of the lowest "modes" and $n = 3$ or 4 is a useful choice.

Equation (6.16) implies (to no one's surprise) that the harbor responds approximately as though it were a collection of simple oscillators, and one can examine the response of each such oscillator of interest by replacing $\phi(\bar{r})$ in Equation (6.15) by the form given in Equation (6.16) and evaluating the integral piece by piece.

A palatable way to see the implications of this is to look at the solution of a schematic model of the harbor problem.

Suppose that the "harbor" is a wall of mass m which can translate as indicated in Figure 9. Its motion is resisted by a linear spring with spring constant K and the water to the left of the wall admits gravity waves which imply a varying force on the wall.

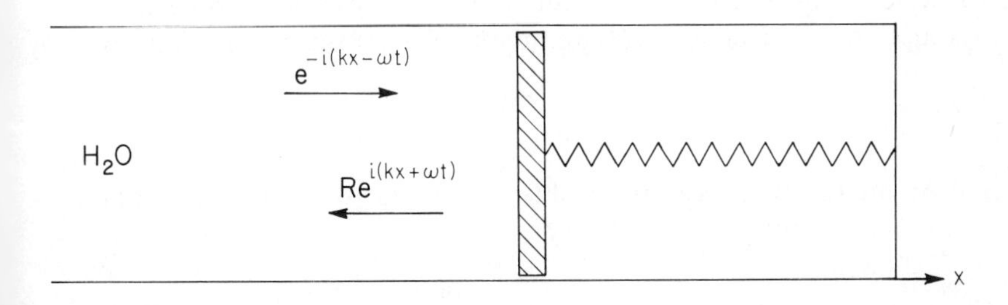

FIGURE 9. Schematic harbor for transient response analysis.

The potential of the incoming wave can be written

$$\phi_i = (iV_0/k)\exp(i(\omega t - kx)) \tag{6.17}$$

where $k = \omega/(gh)^{1/2} = \omega/c$.

With this input and with the one-dimensional geometry, the reflected wave is a multiple of $\exp(i(\omega t + kx))$ so the full potential ϕ is

$$\phi = (iV_0/k)(e^{-ikx} + Re^{ikx})e^{i\omega t}. \tag{6.18}$$

The particle velocity at $x = 0$ is

$$v = V_0(1 - R)e^{i\omega t} \tag{6.19}$$

and the pressure is

$$p = \rho V_0 c(1 + R)e^{i\omega t}. \tag{6.20}$$

For this monochromatic input, the dynamics of the piston are described by

$$-\omega^2 mX + KX = \rho V_0 c(1 + R) \tag{6.21}$$

where m is the (mass/unit area of the piston) and K is the (spring force/unit displacement, unit area of piston). However, v must be equal to dX/dt, so we also have

$$i\omega X = V_0(1 - R). \tag{6.22}$$

From (6.21) and (6.22) we get

$$1 - R = 4i\beta\omega/(\omega_1^2 - \omega^2 + 2i\beta\omega)$$

where $2\beta = \rho c/m$, $\omega_1^2 = K/m$, i.e.

$$1 - R = -4i\beta\omega/(\omega - i\beta - \Omega)(\omega - i\beta + \Omega) \tag{6.23}$$

where $\Omega = (\omega_1^2 - \beta^2)^{1/2}$. Thus, $A = (dX/dt)/V_0 e^{i\omega t}$ is a good measure of the "amplification" and is just the expression in Equation (6.23). Note that when $\omega = \Omega$, the response is at its largest value, which is very nearly 2.

Now suppose the incoming wave is not monochromatic but that its Fourier transform in t is $V_0 F(\omega)$. Then the "response" to each Fourier component is given by (6.23) and the motion of the piston, A, is given by

$$A = \frac{1}{2\pi}\int_{-\infty}^{\infty} \frac{-4i\beta\omega F(\omega)e^{i\omega t}}{(\omega - i\beta - \Omega)(\omega - i\beta + \Omega)}\,d\omega. \tag{6.24}$$

An interesting illustrative problem is that for which the incoming wave at $x = 0$ has the particle velocity

$$v_i = V_0 e^{i\omega_0 t}\exp(-at^2) \tag{6.25}$$

and particular interest goes with $\omega_0 \simeq \omega_1$.

The Fourier transform of v_i is

$$v_i = V_0(\pi/a)^{1/2} \exp(-(\omega - \omega_0)^2/4a)$$

and

$$A = \frac{-i\alpha}{(\pi N)^{1/2}} \int_{-\infty}^{\infty} \left(\frac{1 + i\alpha}{\xi - i\alpha - 1} + \frac{1 - i\alpha}{\xi - i\alpha + 1} \right) \exp(i\xi\tau - (\xi - \sigma)^2/4N)\, d\xi$$

(6.26)

where $\xi = \omega/\Omega$, $\Omega t = \tau$, $\alpha = \beta/\Omega$, $\sigma = \omega_0/\Omega$, $N = a/\Omega^2$. A can be written (using Cauchy's integral theorem)

$$A = \frac{-i\alpha}{(\pi N)^{1/2}} \exp(i\sigma\tau - N\tau^2) \int_{-\infty}^{\infty} \exp(-\zeta^2/4N)$$
$$\times \left(\frac{1 + i\alpha}{\zeta + (\sigma - 1) + i(2N\tau - \alpha)} + \frac{1 - i\alpha}{\zeta + (\sigma + 1) + i(2N\tau - \alpha)} \right) d\zeta$$

where we have used $\zeta = \xi - \sigma - 2iN\tau$ and we have moved the path of integration to the real axis in the z-plane except that, when necessary, the path must be indented to pass below $\zeta = -i(2N\tau - \alpha) - (\sigma \pm 1)$.

Thus, when $2N\tau$ is enough bigger than α, the residue will be important.

The most interesting case is $\sigma = 1$, and for this the second fraction does not contribute enough to bother with.

Thus, we must evaluate

$$\bar{b} = \int_{\infty}^{\infty} \frac{\exp(-\zeta^2/4N)}{\zeta + i}\, d\zeta$$

which gives

$$I = i\pi \exp(\lambda^2/4N)(1 + \operatorname{erf}[\lambda/2N^{1/2}])$$

so that in the original coordinates

$$A = (B/a^{1/2})\pi^{1/2} \exp(i\omega_1 t - \beta t + \beta^2/4a)(1 + \operatorname{erf}[ta^{1/2} - \beta/2a^{1/2}).$$

Thus, the oscillation proceeds with period $2\pi/\omega_1$ as one would expect, and it decays at large time like $e^{-\beta t}$ as one would also expect. The size of the oscillation depends, however, on $\beta/a^{1/2}$; that is, it depends on the ratio of time during which the input is appreciable (i.e. $1/a^{1/2}$) and the time needed to bring the oscillation nearly up to equilibrium amplitude which is equal to the damping time $1/\beta$.

For $\beta/a^{1/2} \geqq 1$, $A_{\max}$ is approximately 2, and for $\beta/a^{1/2} < 1$, $A_{\max}$ is smaller and occurs later; note that for $-t > 2/a^{1/2} - \beta/2a$

$$A \simeq \exp(i\omega_1 t)(2\exp(-at^2)/(2at/\beta - 1)$$

and for $t > 2/a^{1/2} + \beta/2a$

$$A \simeq \exp(i\omega_1 t)\exp(-\beta(t - \beta/4a)2\beta/\pi^{1/2}/a\exp(i\omega_1 t)$$
$$\times\ 2\exp(-\beta(t - \beta/4a)\ln\beta(\pi^{1/2}/a)/\beta).$$

One can verify readily that if the input is

$$v_i = V_0\begin{cases}0 & \text{in} \quad t < -T,\\ \exp(i\omega_1 t) & \text{in} \quad -T < t < T,\\ 0 & \text{in} \quad t > T,\end{cases}$$

then, in $t > T$, A is given by

$$A = \underset{\substack{\text{radiation}\\ \text{damping}}}{2e^{i\omega_1 t - \beta(t-T)}}\underset{\substack{\text{input}\\ \text{effectiveness}}}{(1 - e^{-2\beta T})};$$

and this is an even simpler way to note that an equilibrium level response can be obtained only if the time, T_1, during which the spectrum of the input is strong near the resonant frequency, is comparable to the natural damping time of the oscillator.

The results in this paper which are new were carried out under the sponsorship in part of the office of Naval Research under Grant N0014-67-A-0298-002, and by the Division of Engineering and Applied Physics, Harvard University.

Bibliography

1. W. H. Munk, *Increase in the period of waves traveling over large distances with applications to tsunamis, swell, and seismic surface waves*, Trans. Amer. Geophys. Union **28** (1947), 198–217.

2. C. Eckart, *The ray-particle analogy*, J. Marine Res. **9** (1950), 139–144. (Sears Foundation).

3. R. S. Arthur, W. H. Munk and J. D. Isaacs, *The direct construction of wave rays*, Trans. Amer. Geophys. Union **33** (1952), 855–865.

4. T. Hirono and S. Hisamoto, *A method of drawing the wavefronts of tsunami on a chart*, Geophys. Mag. **23** (1952), 399–406.

5. K. Kajiura, *On the partial reflection of water waves passing over a bottom of variable depth*, Proc. Tenth Tsunami Meetings Pacific Sci. Congr., I.U.G.G. Monographs, no. 24, Inst. Geogr. Natl., Paris, 1961.

6. H. C. Kranzer and J. B. Keller, *Water waves produced by explosions*, J. Appl. Phys. **30** (1959), 398–407. MR **21** #1071.

7. W. G. Van Dorn, *The source motion of the tsunami of March* 9, 1957 *as deduced from wave measurements at Wake Island*, Proc. Tenth Tsunami Meetings Pacific Sci. Congr., I.U.G.G. Monographs, no. 24, Inst. Geogr. Natl., Paris, 1961, pp. 39–48.

8. J. B. Keller, *Tsunamis-water waves produced by earthquakes*, Proc. Tenth Tsunami Meetings Pacific Sci. Congr., I.U.G.G. Monographs, no. 24, Inst. Geogr. Natl., Paris, 1961, pp. 154–166.

9. H. B. Keller, D. A. Levine and G. B. Whitham, *Motion of a bore over a sloping beach*, J. Fluid Mech. **7** (1960), 302–316. MR **22** #1246.

10. W. H. Munk, *Some comments regarding diffusion and absorption of tsunamis*, Proc. Tenth Tsunami Meetings Pacific Sci. Congr., I.U.G.G. Monographs, no. 24, Inst. Geogr. Natl., Paris, 1961, pp. 53–72.

11. G. R. Miller, W. H. Munk and F. E. Snodgrass, *Long-period waves over California's continental borderland*. II: *Tsunamis*, J. Marine Res. **20** (1962), 31–41. (Sears Foundation).

12. K. K. Wong, A. T. Ippen and D. R. F. Harleman, *Interaction of tsunamis with oceanic islands submaring topographies*, Hydrodynamics Lab. Report #62, M.I.T., Cambridge, Mass., 1963.

13. D. V. Ho and R. E. Mayer, *Climb of a bore on a beach*. I. *Uniform beach slope*, J. Fluid Mech. **14** (1962), 305–318; M. C. Shen and R. E. Meyer, *Climb of a bore on a beach*. II. *Non-uniform beach slope*, J. Fluid Mech. **16** (1963), 108–112; M. C. Shen and R. E. Meyer, *Climb of a bore on a beach*, III. *Run-up*, J. Fluid Mech. **16** (1963), 113–125. MR 26#5818; MR **27** #6446; 6447.

14. H. M. Iyer and V. W. Punton, *A computer program for plotting wavefronts and rays from a point source in dispersive mediums*, J. Geophys. Res. **68** (1963), 3473–3482.

15. K. Kajiura, *The leading wave of a tsunami*, Bull. Earthquake Res. Inst. Tokyo Univ. **41** (1963), 535–571.

16. G. F. Carrier, *Gravity waves on water of variable depth*, J. Fluid Mech. **24** (1966), 641–659. MR **33** #8149.

17. W. G. Van Dorn, *Tsunamis*, Advances in Hydroscience, vol. 2, Academic Press, New York, 1965.

18. G. F. Carrier and H. P. Greenspan, *Water waves of finite amplitude on a sloping beach*, J. Fluid Mech. **4** (1958), 97–109. MR **20** #2945.

19. J. J. Stoker, *Water waves: The mathematical theory with applications*, Pure and Appl. Math., vol. 4, Interscience, New York, 1957. MR **21** #2438.

20. M. Abramowitz and I. A. Stegun (Editors), *Handbook of mathematical functions, with formulas, graphs and mathematical tables*, Nat. Bur. Standards Appl. Math. Series, 55, Superintendent of Documents, U.S. Government Printing Office, Washington, D.C., 1964; 3rd printing with corrections, 1965. MR **29** #4914; MR **31** #1400.

21. G. F. Carrier, R. P. Shaw and M. Miyata, *The response of narrow mouthed harbors in a straight coastline to periodic incident waves*, J. Appl. Mech. (to appear).

22. G. F. Carrier and C. E. Pearson, *Ordinary differential equations*, Blaisdell, Waltham, Mass., 1968.

23. G. F. Carrier, *Stochastically driven dynamical systems*, J. Fluid Mech. (to appear).

24. C. C. Lautenbacher, *Gravity wave refraction by islands*, J. Fluid Mech. **41** (1970), 655–672.

HARVARD UNIVERSITY

Resonance of Unbounded Water Bodies

R. E. Meyer

1. **Introduction.** I owe to Professor R. L. Miller communication of the experience of a party of officers and scientists caught by a Nor'easter during inspection of a Texas tower on Brown's Bank off the East Coast. The tower stood on a sand bank on columns of which the feet were protected by ring-shaped iron collars weighing many tons. The waves raised locally by the storm were so strong that they carried the iron collars up 100 feet along the columns and pounded them against the platform, threatening to break up the whole structure.

It now appears virtually certain that those waves were of a type which was not only unknown 20 years ago, but indeed, was very well known to be impossible! It appeared that resonant wave motion—natural in a basin or lake—does not make physical sense in a body of water extending to infinity and thus possessing infinite mass. For such a body, the physically natural wave motions are those resulting from the incidence of waves from infinity and from their reflection and diffraction by shores and other obstacles.

These observed facts were supported by (in truth, they derived from) the classical, mathematical theory of water waves, including the tidal theory. They were strongly confirmed by the great advances of the Courant Group, who solved the classical problem of waves on a plane beach, first for two-dimensional motion, and then even for three-dimensional motion [**1**]. The spectrum of those solutions of the classical water wave equations was indeed purely continuous.

The near-catastrophic waves on Brown's Bank are of a nonclassical type, and there is mounting evidence [**8**], [**17**], [**23**], [**25**] that similar waves may play a critical role in the disastrous effects of tsunamis.

AMS 1970 *subject classifications*. Primary 76B15, 86A05; Secondary 35P20, 86-02, 76-02.

Remarkably enough, this new wavetype was discovered through mathematical rigor. Ursell [**2**] undertook to put the heuristic solution of a basic, classical ship wave problem on a rigorous basis. This required proof that the problem has an entirely continuous spectrum, and he found himself unable to establish such a proof. In fact, he came to the conclusion that the theorem is false [**3**]. Since the analysis was elaborate, he turned to a simpler problem from which more readily convincing results could be hoped for.

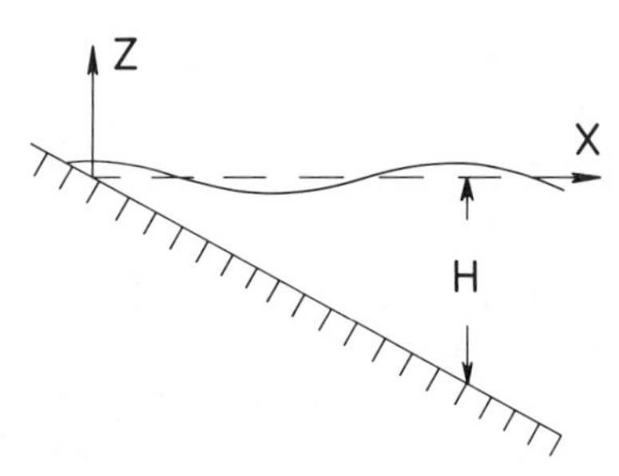

FIGURE 1

2. **Simple edge wave resonance.** He chose surface waves over a plane beach (Figure 1) in a laboratory channel bounded by rigid, vertical side walls. On the classical small amplitude theory [**5**], such surface waves possess a velocity potential $\Phi(X, Y, Z, T)$ governed by Laplace's equation

$$\Phi_{XX} + \Phi_{YY} + \Phi_{ZZ} = 0 \tag{1}$$

in the fluid domain $0 > Z > -H(X, Y)$, $0 < Y < B$, $0 < X$ (Figure 1); by the conditions of zero normal velocity,

$$\Phi_Z + H_X\Phi_X + H_Y\Phi_Y = 0 \quad \text{for } Z = -H(X, Y), \tag{2}$$

$$\Phi_Y = 0 \quad \text{for } Y = 0, \quad Y = B \tag{2a}$$

on the seabed and the vertical side walls; and by the linearized surface condition,

$$\Phi_{TT} + g\Phi_Z = 0 \quad \text{for } Z = 0 \tag{3}$$

applied at the equilibrium surface. Here g denotes the (constant) gravitational acceleration.

For Ursell's idealized beach (Figure 1), $H(X, Y) = X \tan \varepsilon$, and it is convenient, for a start, to choose a length scale L such that the channel width is $B = \pi L$, and to adopt nondimensional variables

$$(x, y, z) = (X, Y, Z)/L, \qquad t = T(g/L)^{1/2}, \qquad \Phi = (L^3 g)\phi.$$

Then [**3**] if $\varepsilon \leqq \pi/2$, Stokes' standing edge wave potential [**6**]

$$\phi = \sin(\omega t)\cos y \exp[z \sin\varepsilon - x\cos\varepsilon], \qquad \omega^2 = \sin\varepsilon \tag{4}$$

satisfies (1) to (3) with *finite kinetic energy*, due to the exponential decay with distance from the shore $X = Z = 0$. It took over 100 years to appreciate that this is a natural mode of oscillation of the water body in the semi-infinite channel, which, if once excited, *can persist* (except for slow viscous dissipation) and hence, that $(g\sin\varepsilon)^{1/2}(4\pi B)^{-1/2}$ is an eigenfrequency of the system (1) to (3). Accordingly also, it is a frequency of potential resonance.

The edge wave mode is quite different from the more familiar wave modes which tend, far from shore, to a superposition of plane waves on deep water, and hence cannot have finite energy. These exist for all $\omega^2 \geqq 1$, and only for those [**1**], and hence correspond to a *continuous spectrum* of frequencies with *cut-off* $\omega_0^2 = 1$. Such a cut-off frequency is also a resonant frequency [**4**]. Physically, these natural modes represent the process of reflection from the beach (Figure 1) of waves incident from the deep ocean. Radiation to and from infinity is thus a characteristic feature of these modes, and they are not resonant. The discrete edge wave mode (4), by contrast, involves no such radiation because, to all intent and purpose, only a finite part of this water mass really moves.

Ursell [**4**] pointed out that Stokes edge wave mode (4) is by no means the only discrete, natural mode, but that, e.g.,

$$\begin{aligned}\phi = \sin(\omega t)\cos y \Big\{&\exp[z\sin\varepsilon - x\cos\varepsilon] \\ &+ \sum_{m=1}^{n} A_{mn}(\exp[-z\sin(2m-1)\varepsilon - x\cos(2m-1)\varepsilon] \\ &+ \exp[z\sin(2m+1)\varepsilon - x\cos(2m+1)\varepsilon])\Big\}\end{aligned} \tag{5}$$

with

$$A_{mn} = (-1)^m \prod_{k=1}^{m} \frac{\tan(n-k+1)\varepsilon}{\tan(n+k)\varepsilon}$$

satisfies (1) to (3) for any integer $n \geqq 0$, provided

$$\omega^2 = \sin(2n+1)\varepsilon. \tag{6}$$

If

$$(2n+1)\varepsilon \leqq \pi/2, \tag{7}$$

then (5) is an edge wave mode decaying exponentially with distance from shore and hence possessing finite energy. Accordingly, the discrete eigenvalues (6) satisfying (7) are also potentially resonant. Stokes's wave (4) is the case $n = 0$ of (5). For $\varepsilon > 30°$, (7) admits only $n = 0$; but for small beach angles ε, (5) and (6) demonstrate the existence of a large number of discrete eigenfrequencies below the cut-off of the continuous spectrum.

Not all the relevant questions, however, can be answered by the classical theory. Instead of water, it concerns idealized, inviscid fluid. Even then, it is only a linearized, small-amplitude approximation, and even as such, not generally valid at the shore (although the direct objections to it at the shore are not applicable to edge waves). Ursell therefore made a further important contribution by proceeding promptly to an experiment testing these issues.

This raised another practical objection to the predictions. Since the exponential decay of edge waves eliminates any radiation from them, these waves cannot, on the classical linear theory, be excited by waves incident from infinity! Instead of using a conventional wave maker, Ursell therefore oscillated the beach itself.

In this way, he was able to confirm the predicted resonances [**4**]. He was led to discover, moreover, that viscosity dominates the motion at frequencies close to the cut-off [**4**]. Far from suppressing the cut-off resonance, however, viscosity raises this resonant response to a level similar to that obtained at the discrete eigenfrequencies. In addition, Ursell tested the *critical beach angles* $\varepsilon = \pi/(4n + 2), n = 1, 2, \ldots$, at which a discrete eigenvalue (6) coincides with the cut-off $\omega_0^2 = 1$, and confirmed his conjecture that these *critical cut-off resonances* are exceptionally dangerous.

Ursell's experiment thus removed any doubt about the validity of the classical theory for edge wave resonance on a beach—except that it left the waves academic for lack of an oceanographical mechanism for exciting them. This last difficulty was removed by Galvin [7], who showed experimentally how the resonance can be excited subharmonically by plane waves incident from infinity.

Further theoretical progress, however, was arrested by the great technical difficulty of the classical theory of surface waves on water of nonuniform depth [**1**]. It became clear that further approximations are necessary for the effective analysis of realistic topographies, and Longuet-Higgins [**8**] and Shen [**17**] undertook such investigations independently and simultaneously. Fortunately they chose quite different approaches and their results are complementary, overlapping only enough to give strong support to each other.

For an introduction to the subject opened up by their work, it is helpful to begin by looking at it from the point of view of Longuet-Higgins' original approach. But an attempt to distinguish precisely the discoveries and contributions of the respective authors is inconsistent with a straightforward presentation and has therefore been abandoned here.

3. **Long-wave trapping.** Longuet-Higgins' investigations were actually sparked by an observation, at Macquarie Island, of long waves of puzzling origin [**8**]. He based his work [**8**], [**9**], [**10**], [**11**] on the classical long-wave equations of tidal theory, which are a first approximation to (1), (2), (3) for wavelength large compared to water depth. Vertical averaging [**12**] then permits a formulation of (1) to (3) in terms of the horizontal velocity components u, v, the pressure p and (constant) density ρ, and surface elevation ζ by

$$\begin{aligned} \partial u/\partial t &= -(g\rho)^{-1}\partial p/\partial x = -\partial\zeta/\partial x, \\ \partial v/\partial t &= -(g\rho)^{-1}\partial p/\partial x = -\partial\zeta/\partial y, \\ \partial\zeta/\partial t &+ \partial(hu)/\partial x + \partial(hv)/\partial y = 0. \end{aligned} \tag{8}$$

Horizontal lengths are here again made nondimensional by reference to a scale L, but vertical lengths, by reference to a water depth, in terms of which the seabed is given by $z = -h(x, y)$. Further differentiation of (8) to eliminate u and v yields

$$(\nabla^2 - h^{-1}\partial^2/\partial t^2)\zeta + h^{-1}(\nabla h)\cdot(\nabla\zeta) = 0, \tag{9}$$

where the Laplacian ∇^2 and gradient operators ∇ are with respect to the horizontal variables x, y.

It is observed immediately that (9) is basically a wave equation with propagation velocity

$$c = (gH)^{1/2} \tag{10}$$

depending on the local water depth $H(X, Y)$. This phase velocity increases with the local depth so that the wave crests and troughs are seen (Figure 2) to have a tendency always to turn towards the shallows during propagation.

This is a basic observation of the theory. Quite generally, surface waves due to gravity have a *propagation velocity dependent on water depth*, provided the depth is not too large compared with the wave length.

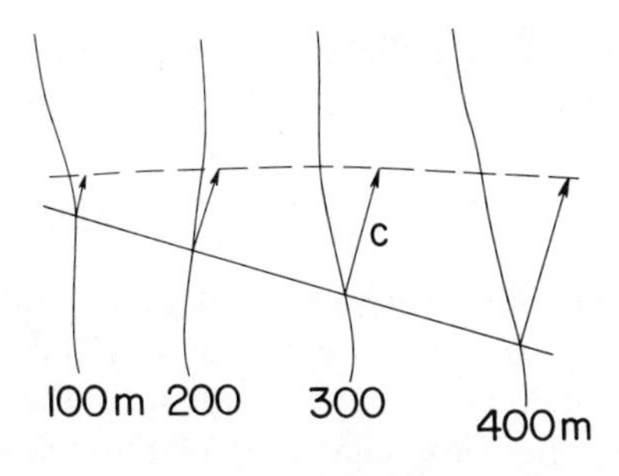

FIGURE 2

Variation of water depth may therefore cause phenomena of which classical oceanography is essentially ignorant. Further, while (10) is only an approximation, it is found quite generally that the propagation velocity of surface waves due to gravity increases monotonically with water depth. It follows (Figure 2) that, all other things being equal, *the wave crests and troughs turn toward the shallows during propagation.*

The possibility therefore arises that waves of a given type and propagating seaward, to begin with, *may be gradually turned around* to propagate landward. It would then look as if there was an invisible barrier beyond which those waves cannot penetrate seaward. After being turned around, such waves would travel landward and then be (partially) reflected by the shore, and the process would start over again. In a channel, it would become possible that a wave crest, after a cycle of reflections from the side walls, the invisible barrier, and the shore, would end up facing in the same direction as at the beginning of the cycle. If the phase relationships of such a recirculating wave pattern were right, resonance would become a very plausible possibility.

These considerations apply only to waves on which the seabed has a significant influence, i.e. to long waves. Near coasts and islands, however, any wave length is large compared to the depth sufficiently close to a natural shore. Edge waves of all lengths may thus be surmised to ply along shores; and in the case of an island, if the perimeter is suitably related to the wave length, edge wave resonance becomes conceivable.

For long waves, a first impression of quantitative features corresponding to these physical considerations may be obtained from plane waves near a straight coast. For if the water depth $h(x)$ depends on only one coordinate, waves of the form

$$\zeta(x, y, t) = A(x) \exp[imy - i\omega t] \tag{11}$$

with constants m, ω are formally possible, and from (9), they must satisfy

$$\frac{d}{dx}\left(\frac{h\, dA}{dx}\right) + (\omega^2 - m^2 h)A = 0. \tag{12}$$

This is a typical Sturm-Liouville equation, generalizing the elementary harmonic oscillator equation $d^2F/dx^2 + qF = 0$, and the nature of the solution of (12) depends critically on the sign of the coefficient

$$\omega^2 - m^2 h = q(x). \tag{13}$$

As for the harmonic oscillator, an oscillatory solution $A(x)$ must be anticipated, if $q \geqq \varepsilon > 0$, but an exponentially damped (or enhanced) solution must be anticipated if $q \leqq -\varepsilon < 0$. If the depth variation is sufficiently gradual, the local character of the solution is determined similarly by the local sign of $q(x)$.

On a continental slope, for instance, where $h(x)$ increases from $h(0) = 0$ to $\lim_{x\to\infty} h(x) = h_\infty > 0$, there exist positive number pairs (ω, m) such that $q(x) > 0$ near the coast, and $q(x) < 0$ far from the coast. In fact, all the pairs (ω, m) such that

$$0 < \omega^2/m^2 < h_\infty \tag{14}$$

have this property. The waves corresponding to any such number pair are periodic in y, but are oscillatory in x only near the shore; far from the shore, they are exponentially damped. (The solutions which increase exponentially with x cannot satisfy the physical conditions at infinity and will be disregarded in the following without further mention.) The transition from oscillatory to exponential behavior occurs at the "caustic" $x = x_c$ defined by

$$q(x_c) = \omega^2 - m^2 h(x_c) = 0,$$

which is thus seen to be the mathematical counterpart of the more indefinite, "invisible barrier." Waves which are thus essentially restricted to a part of the water surface are called "trapped" by Shen [**17**]. Ursell's edge waves (§2) belong to this class. By contrast, pairs (ω, m) such that $(\omega/m)^2 > h_\infty$ correspond over a continental slope to wave forms (11) which are oscillatory with respect to both x and y on the whole water surface. Such solutions describe the reflection from the coast of plane waves incident from infinity, and waves of this type are called "progressive" by Shen [**17**].

It is a small step now to conjecture that "trapping" is a necessary condition for resonance, but it is far from sufficient since it gives no information on the questions of phase relationships crucial to resonance. It should also be noted, however, that the physical existence of trapped waves need not depend on resonance. Shelf-waves, e.g., are trapped but not always resonant [**17**].

4. **Axially symmetric trapping of long waves.** One of Longuet-Higgins' main discoveries [**8**] was that the same considerations lead to still much more unexpected conclusions for round islands or reefs. For axially symmetrical sea-bed topographies, the (nondimensional) water depth is $h(r)$ in terms of polar coordinates r, θ in the equilibrium water surface. Long-wave solutions of surface elevation

$$\zeta(r, \theta) = P(r) \exp[in\theta - i\omega t] \tag{15}$$

are then formally admissible, and by (9), their dependence on r is governed by

$$\frac{d}{dr}\left(\frac{rh\, dP}{dr}\right) + (\omega^2 - (n/r)^2 h)P = 0. \tag{16}$$

This is another Sturm-Liouville equation, and the character of the solution is again determined by the sign of the coefficient

$$q(r) = \omega^2 - n^2 h(r)/r^2. \tag{17}$$

It can be concluded immediately that, sufficiently far from an island or reef, all solutions are oscillatory because

$$\lim_{r \to \infty} h(r) = h_\infty > 0, \qquad \lim_{r \to \infty} q(r) = \omega^2 > 0.$$

This stands in sharp contrast to the pattern recognized in the preceding two sections as characteristic of trapping and necessary for resonance. Ursell's edge waves in a channel turn out, after all, not to be representative of general three-dimensional wave patterns.

For most submerged reefs, e.g., the curve of $h(r)/r^2$ is monotone (Figure 3), asymptoting $h(0)/r^2$ as $r \to 0$ and h_∞/r^2 as $r \to \infty$. It follows that any solution of (16) is oscillatory outside the caustic at the root of

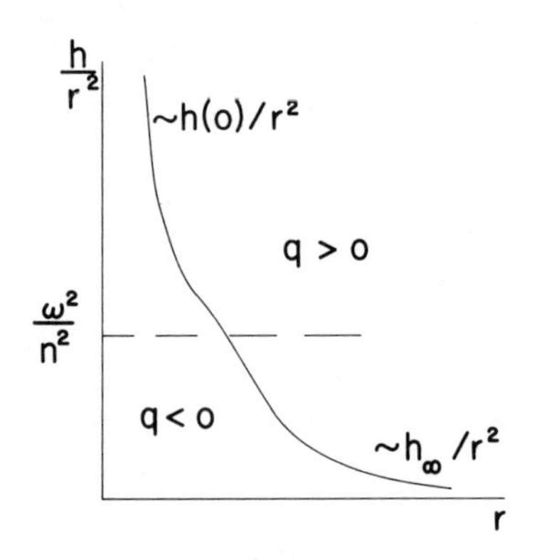

FIGURE 3

$q(r)$, and damped inside that caustic (Figure 3). Little can be conclude directly from this, because the argument concerns only special solution of the form (15), which do not satisfy the physical conditions of mai interest as $r \to \infty$. The more general theory [**17**], however, extends th indication just obtained to general, three-dimensional wave motio (§9). Its long-wave limit admits decomposition into waves of the type (1 and shows—when $r^{-2}h(r)$ is monotone decreasing (Figure 3)—any suc wave to have precisely one caustic circle outside of which the motion oscillatory with respect to r, but inside of which it is damped. For fixe frequency ω, different wave numbers n correspond to different causti radii, by (17), but the total result is that the long-wave equations (

describe the diffraction by such reefs of waves incident on them from infinity. All the solutions, therefore, must be regarded as progressive rather than trapped. This suggests that a necessary condition for resonance cannot be met by such reefs, and Shen's approximation [**17**] confirms it (§9).

Nonetheless, it is bound to be important in marine engineering that $q(r) \to -\infty$ as $r \to 0$ at the center of any submerged round reef, by (17), and hence that the crown of a submerged round reef is always a region of wave damping.

The situation changes drastically, however, if the tide should lower the water level so that the reef becomes an island. Then $r^{-2}h(r)$ must be nonmonotone, because it vanishes at the shore $h(r) = 0$, and still asymptotes h_∞/r^2 as $r \to \infty$. The simplest case is shown in Figure 4, but in any case, from (17) for all positive pairs (ω, n), $q(r) > 0$ at sufficiently small distances from the shore $r = r_s$, and *all solutions* of (16) are therefore *oscillatory in a neighborhood of the shore.* Furthermore, there is always a critical value of ω/n such that $q > 0$ for all $r > r_s$ for larger ω/n, but for smaller ω/n, there is an interval of r on which $q(r) < 0$ (Figure 4). For larger ω/n, therefore, the solutions of (16) are oscillatory for all $r > r_s$, and thus represent progressive waves. For smaller ω/n, on the other hand, the solutions *exhibit a ring of wave trapping near shore*, separated from the outer wave region by a ring of damping; there are now two caustics at which $q(r) = 0$ (Figure 4).

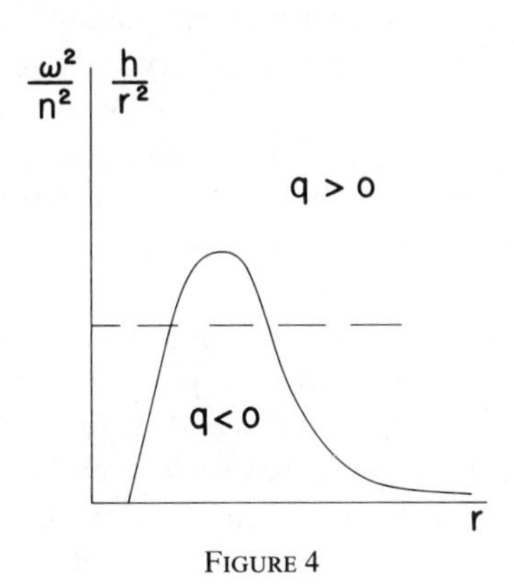

FIGURE 4

The argument refers only to the special solutions (15), but the more general theory (§9) extends the conclusions just drawn to general, three-dimensional long-wave motions. Theoretical indication of trapping near shore is therefore strong for small frequencies ω or large azimuthal wave numbers n. Due to their exponential damping beyond the inner caustic, these waves clearly resemble edge waves, and their existence is physically quite plausible.

While trapping is again far from implying resonance, the trapped edge waves are seen to ply around the island, and for integer n, existence of some resonant frequencies ω thus becomes plausible. It was a very important observation of Longuet-Higgins [**8**] that a new phenomenon now enters oceanography. The annulus of damping (Figure 4) is of finite width, and while the inner and outer wave motions decay exponentially with distance from the respective caustics, such decay cannot be strictly complete within a finite distance. Accordingly, the trapped wave motion inside the inner caustic and the progressive wave motion beyond the outer caustic *cannot be independent.* A trapped motion, therefore, cannot exist without a progressive motion outside the damping annulus, and the latter motion must involve the radiation of wave energy towards infinity. Hence, trapping cannot be complete in the sense that the energy is trapped and finite. In particular, any natural wave mode must be a *leaky mode* in which energy leaks from the trapped wave through the damping annulus to the progressive wave and to infinity, so that the natural frequency must be complex. In fact, the trapping annulus near the shore (Figure 4) is analogous to a potential well, and the damping annulus, to a potential barrier, in quantum mechanics.

The rate of leakage may be safely conjectured to depend on the width of the damping annulus. For values of ω/n near the critical one, this annulus is narrow (Figure 4) and massive leakage must be anticipated, so that even if discrete eigenvalues of this type exist, the rate of decay of the eigenfunctions must be expected to be too fast for resonance to be effectively observable. For very small values of ω/n, by contrast, the damping annulus is very wide (Figure 4), and extremely small leakage only is to be anticipated. At any discrete eigenfrequency of such kind clear resonance must be expected.

While (8) and (15) are too narrow a basis to support all these conclusions properly, it will be seen below that they are essentially valid, and it is worth noting how much they change the ideas just introduced by Ursell's discoveries (§2). There it was seen that resonance of an unbounded water body is possible because the natural modes have finite energy since they involve, effectively, only a finite part of the infinite water mass. Now it emerges that, in natural oceanographical circumstances, any natural modes must be expected to involve, effectively, almost all of the infinite water mass and to have infinite energy. Resonance, moreover, emerges to be a matter of degree, instead of the clear-cut, classical phenomenon.

Conversely, the leakage of natural wave modes removes the difficulty of exciting them (§2), since it provides a mechanism for direct excitation by waves of the same frequency incident from infinity.

Figures 3 and 4 represent only the simplest type of dependence of $r^{-2}h(r)$ on r. It is not necessary, e.g., that this dependence be monotone (Figure 4) for a reef, and if it is not (Figure 5), then the question of trapping must be examined afresh. The condition of non-monotonicity (Figure 5) is that there be an interval of r on which

$$d(r^{-2}h(r))/dr > 0 \tag{18}$$

and Longuet-Higgins calls such an interval a "hedge." The kinematical, necessary condition for trapping is precisely the same, namely that the angular velocity of propagation of the wave crests and troughs increases with r. Since the phase velocity is $c = (gh)^{1/2}$ and the angular phase velocity c/r, that condition is again (18).

For a reef with a single hedge (Figure 5), solutions of (16) for large and for small values of ω/n are progressive waves outside a damping disc, for the same reasons as for a simple reef (Figure 3). But there are now two critical values of ω/n, and for values of ω/n between those (Figure 5), there are now two intervals $q(r) > 0$ and two intervals $q(r) < 0$. The solutions of (16) are oscillatory in the intervals $q > 0$, and the outer of these corresponds to the progressive wave region far from any reef or island, but the inner one (Figure 5) represents an annulus of trapping. The solutions are exponentially damped in the intervals $q < 0$; the inner of these corresponds to the damping disc over the crown of any reef, but the outer one (Figure 5) represents a potential barrier through which the trapped waves leak energy to the progressive waves.

The hedge condition (18) requires a relatively flat-topped reef with relatively steep slopes. Similarly, it can be satisfied by an island topography with a pronounced shelf bounded by steep slopes. It will then add a second annulus to some of the wave patterns of the island. And further

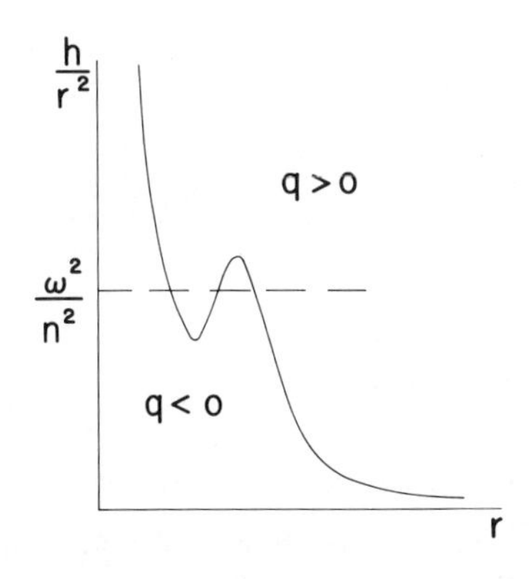

FIGURE 5

hedges for islands or reefs, with corresponding annuli of trapping and damping, are possible in principle. As before, of course, (15) affords too narrow a basis for the desired physical conclusions, but the more general theory [**17**] confirms the conjectures. The more exact hedge condition (51) is more complicated and slightly more stringent, but (18) is its long-wave limit. Each trapping annulus adds the possibility of additional resonant frequencies, but the more detailed analysis [**17**] indicates that a very pronounced shelf structure is necessary to add discrete eigenfrequencies with a degree of resonance likely to be observable (§9).

While all these considerations are of much physical interest on account of the insight they offer into the physical wave patterns near shores and reefs, the primary oceanographical interest is in the resonant frequencies.

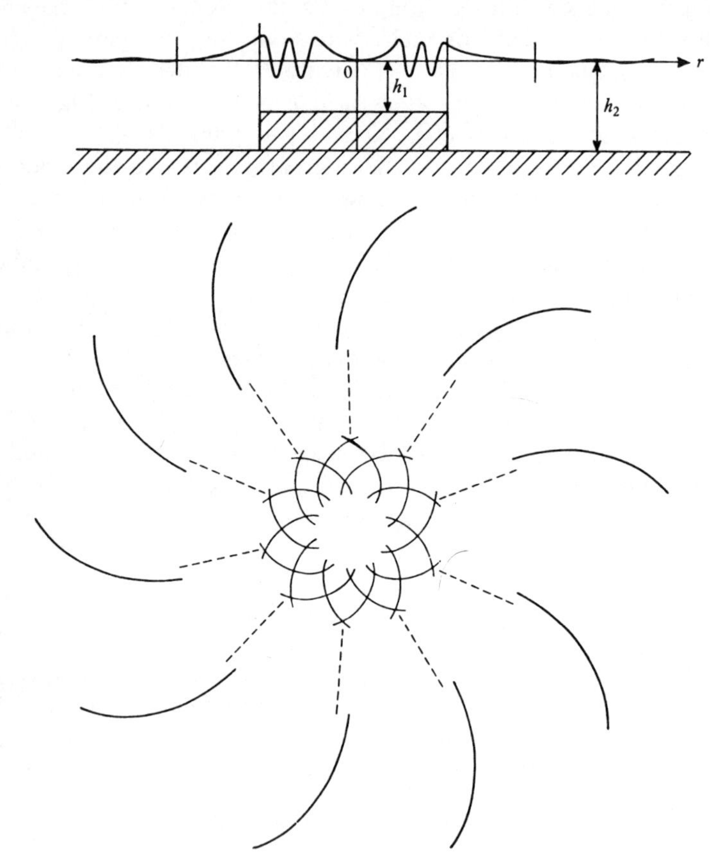

Reprinted with permission of J. Fluid Mech. [**8**, p. 794]

FIGURE 6. The form of free waves trapped by a circular sill. (*a*) A radial cross-section; (*b*) plan view of the wave crests.

At first sight, the Sturm-Liouville equation (16) suggests that both the decision whether eigenvalues exist for a given type of topography, and also their computation, e.g., by the standard variational methods, should be a straightforward matter. This optimism, however, is short-lived. First, (16) governs only eigenfunctions of the form (15); and secondly, the eigenvalue problem involves the secondary parameter n and, for practical usefulness, further topographical parameters (slopes, etc.). But above all, the Sturm-Liouville problem is *singular*, as in quantum mechanics, precisely because an infinite r-interval is an essential feature of the oceanographical problem. And the spectrum will indeed emerge (§§8, 9) to be much more complicated than for a classical Sturm-Liouville problem.

The first reported computation of eigenvalues was made for a circular sill (Figure 6) on a flat ocean bottom. The vertical side of the sill represents an extreme case of a hedge, and the wave pattern is therefore that of a round reef with one hedge (Figure 6). For sufficiently long waves, this is a good approximation to a flat reef with very steep sides, and the long-wave limit can be obtained by matching solutions for constant depth across the sill boundary with continuous surface elevation and normal mass-flow rate; the eigenfunctions are thus expressible in terms of Bessel functions. The computation [**8**] stands out even now, as the only one in which the whole relevant, discrete *complex* spectrum has been determined. That is, both the resonant frequencies and the decay rates due to leakage are reported, carrying the theory to a rather triumphant conclusion, at least for this limiting problem. Many eigenvalues are found to be strongly damped, many others to be moderately strongly damped, but a few are found to be virtually free of damping.

In addition to the long-wave spectrum, Longuet-Higgins also computed [**8**] the response coefficient of the circular sill (Figure 6) to a plane wave train incident from infinity. Figure 7 (from [**8**]) shows the result; a very strong, even if narrow, resonant response is found at the eigenfrequencies for which the leakage is small.

An analogous calculation of response factors for conical islands of round or elliptic shore line, based on more direct numerical analysis of (9), has been reported by Lautenbacher [**23**].[1] It confirms the excitation of trapped modes with $n > 1$ by incident plane waves, and that this may be a process of major importance in tsunami amplification. But the results

[1] In contrast to Longuet-Higgins' circular sill, this computation involves a shore line; it is represented (as in much other theoretical work) by the boundary condition of perfect reflection of waves at the shore. This is, of course, at variance with the fact that the beach has proved, over many decades, to be the most efficient *wave absorber* in the marine laboratory. There are now indications (§7), however, that a proper shore boundary condition may need to distinguish between progressive waves and trapped edge waves.

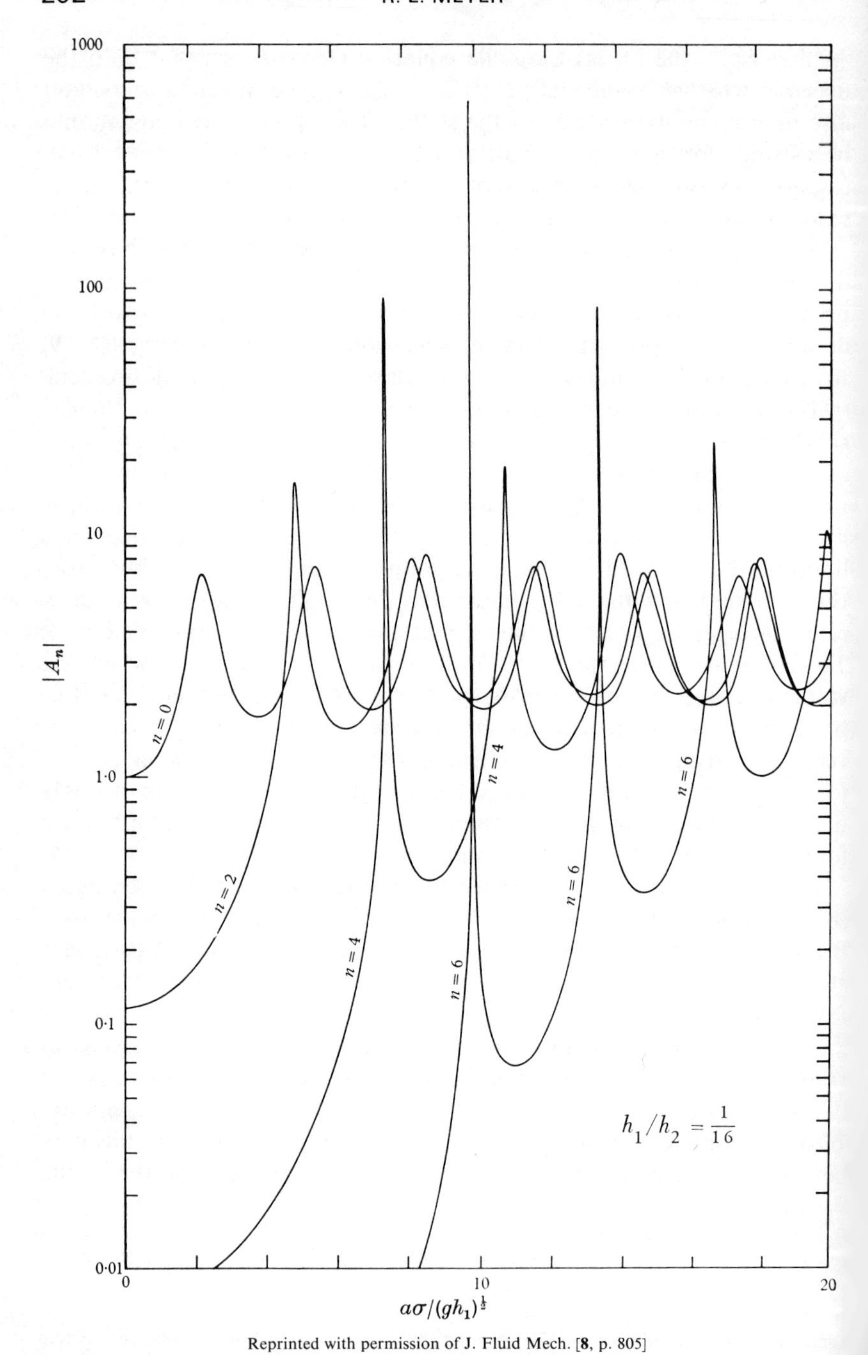

Reprinted with permission of J. Fluid Mech. [**8**, p. 805]

FIGURE 7. Graphs of the amplitude of the response as a function of the frequency of the incident wave.

also illustrate the weakness of primarily numerical approaches. The number of variables and parameters made it impractical [**23**] to test for more than three frequencies, and such a sample cannot give a guide to a response structure of the selectivity observed in Figure 7. Even then, the practicable mesh size [**23**] leaves some doubt whether components with higher n have been properly detected.

The practical importance of results on resonant response is further emphasized by a recent long-wave computation by Duff [**13**] showing that the Gulf of Maine has a natural mode of frequency close to that of the semidiurnal lunar tide, and that a substantial part of the famous tidal amplitudes of the Bay of Fundy are generated by resonance at this frequency.

5. **Long-wave trapping in a rotating ocean.** Longuet-Higgins proceeded to estimate the effect of various other factors on the results just discussed [**8**] but, above all, undertook a major investigation of the influence of the earth's rotation on long-wave trapping [**9**], [**10**], [**11**]. The first two equations of the long-wave approximation (8) then become

$$\partial u/\partial t - fv = -\partial\zeta/\partial x, \qquad \partial v/\partial t + fu = -\partial\zeta/\partial x,$$

where $g^{1/2}f$ denotes the Coriolis parameter, dependent on latitude (and assumed constant in what follows); the last equation of (8) is unchanged. Attention will be restricted to simple periodic waves, and elimination of u and v from the long-wave equation leads [**8**] to

$$\left(\nabla^2 + \frac{\omega^2 - f^2}{h}\right)\zeta^* + \frac{1}{h}(\nabla h)\cdot(\nabla\zeta^*) = \frac{f}{i\omega h}[\mathbf{k}, \nabla h, \nabla\zeta^*] \tag{19}$$

for $\zeta = \zeta^*(x, y)\exp[-i\omega t]$ in the place of (9); the operators are again horizontal, $\mathbf{k}$ is the vertical unit vector, and the bracket denotes the mixed triple product.

Now, most waves have a period much smaller than a day, and the Coriolis effect is then a small perturbation, for a first approximation to which $(f/\omega)^2$ may be neglected in (19), so that the effect depends decisively on the seabed slope. For Stokes's edge wave over a plane beach, Reid showed that the speed of propagation is increased for waves in the northern hemisphere travelling with the shallows to their left, but is decreased for waves travelling in the opposite direction. For edge waves plying around an island, this implies a conjecture that the time taken to travel once around depends on the sense of travel, and hence, that the Coriolis perturbation splits any natural frequencies.

For application to a round sill (Figure 6), where the water depth is piecewise constant, the first-order Coriolis term is eliminated from (19);

but such a term appears in the continuity condition for the normal mass-flow at discontinuities of the depth, and Reid's rule is recovered. Longuet-Higgins [**8**] thus estimated the frequency split for the sill in terms of the depth ratio $\gamma\,(\geqq 1)$, the propagation speed c_1 in the shallows (Figure 6), the sill radius a, and the azimuthal wave number n by

$$f/\Delta\omega = n(\gamma - s + 1) + \gamma^{1/2}s(s-1)^{-1}(\gamma - s)^{-1/2}, \qquad s = (a\omega/nc_1)^2.$$

The situation is drastically different for waves of period not small compared to a day, and it is best to begin, as in §3, with waves over water of depth $h(x)$ varying only in one direction. Then solutions of (19) of the form (11) are again formally possible, and the horizontal scale L may be chosen so that $m = 1$ and

$$\zeta(x, y, t) = A(x)\exp[iy - i\omega t]. \tag{20}$$

Substitution in (19) gives another Sturm-Liouville equation,

$$\frac{d}{dx}\left(h\frac{dA}{dx}\right) + \left(\omega^2 - f^2 - h - \frac{f}{\omega}\frac{dh}{dx}\right)\cdot A = 0, \tag{21}$$

with coefficient

$$q(x) = (\omega^2 - f^2) - h - (f/\omega)\,dh/dx = \tau^{-2}(1 - \tau^2) - h + \tau\,dh/dx, \tag{22}$$

if the time scale be chosen so that $f = 1$, and

$$\tau = -1/\omega$$

so that $|\tau|$ is the period in units of pendulum days.

Consider now waves over a sea-scarp (Figure 8) for which $dh/dx \geqq 0$ for all x and

$$h(x) \to h_1 < 1 \quad \text{as } x \to -\infty,$$
$$\to h_2 = 1 \quad \text{as } x \to \infty.$$

Then $q(x) \to$ const as $|x| \to \infty$, and from (21),

$$A \sim \exp[l_i x]$$

apart from a constant factor, with

$$l_i^2 = 1 + (\tau^2 - 1)/(h_i\tau^2).$$

In the absence of the Coriolis terms in (21), all the solutions (§3) for such a sea-scarp are progressive waves—some confined to one side of a caustic—tending to plane waves as $x \to \infty$ or $x \to -\infty$ or both. If trapped waves exist, therefore, then they must be due essentially to the Coriolis force.

For such wave trapping, two conditions are necessary [**10**]. First, l_i^2 must be positive, so that no energy is radiated toward $|x| \to \infty$. That

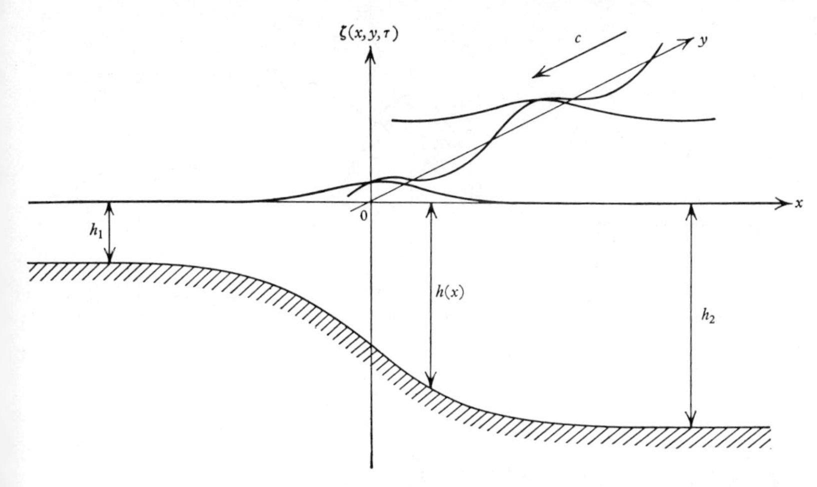

Reprinted with permission of J. Fluid Mech. [**10**, p. 51]

FIGURE 8. Trapped waves being propagated along the transition zone between two regions of uniform depth [**10**].

implies

$$q(x) \to \tau^{-2}(1 - \tau^2) - h_i = -h_i l_i^2 < 0. \tag{23}$$

Secondly, there must be an x-interval on which the solution of (21) is oscillatory, i.e., $q(x) > 0$ or, by (22), $\tau\, dh/dx > \tau^{-2}(\tau^2 - 1) + h \geqq h_1 l_1^2 > 0$, since $h \geqq h_1$ for all x. Since the axes were chosen so that $dh/dx \geqq 0$ for all x, the second condition implies

$$\tau > 0. \tag{24}$$

On the northern hemisphere, where $f > 0$, this means, by (20), that *trapped waves can propagate along a sea-scarp only in the direction* of y *decreasing*, that is, *with the shallows on their right* [**10**].

With the notation $F(x) = h(x) + (\tau^2 - 1)/\tau^2$, we have $q(x) = \tau F' - F$ from (22), and by (23), both F and F' are nonnegative functions. A necessary and sufficient condition for trapping is therefore the existence of a value of τ for which (23) holds and yet $\tau F'$ exceeds F at some x (Figure 9, where $= -q$). As $\tau \to \infty$, from (22), $q \sim -1 - h + \tau\, dh/dx \to +\infty$ at any fixed x where $dh/dx > 0$, and hence the existence of large eigenvalues τ is highly probable.

The bound (24) can be greatly improved [**10**] by arguments characteristic of classical Sturm-Liouville theory. Directly from (21), with d/dx denoted by a prime,

$$\int_{-\infty}^{\infty} (hA')'A'\, dx = \int_{-\infty}^{\infty} \left(\frac{\tau^2 - 1}{\tau^2} + h + \tau h' \right) AA'\, dx$$

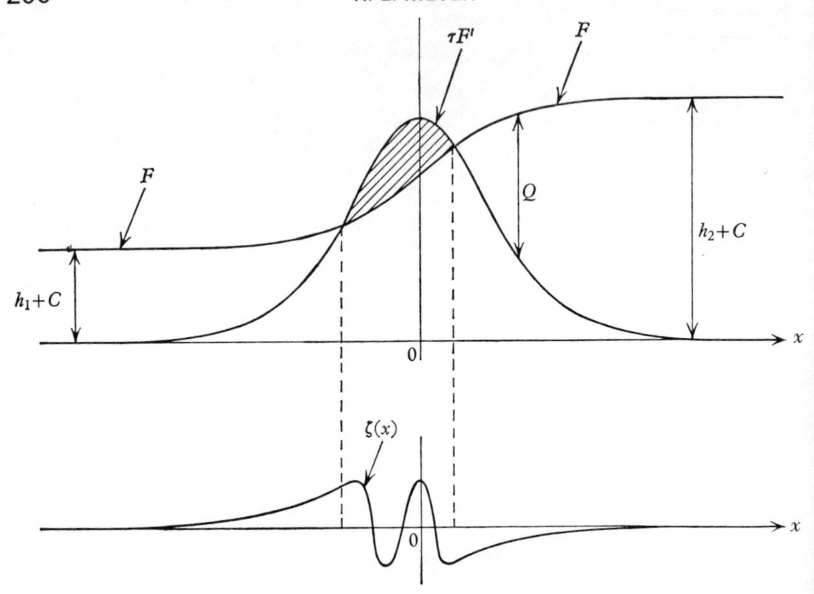

Reprinted with permission of J. Fluid Mech. [**10**, p. 54]

FIGURE 9. Sketch of the functions F, τF and Q for the depth profile of Figure 8. Below is sketched one of the trapped modes ($n = 3$) [**10**].

and the exponential decay of A as $|x| \to \infty$ permits transformation of the right-hand integrand into $-A^2(h' - \tau h'')/2$ by integration by part. In the same way, the left-hand integrand may be transformed into $-hA'A''$ and then into $h'A'^2/2$. The integral identity can therefore be rewritten

$$\tau \int_{-\infty}^{\infty} h'' A^2 \, dx = \int_{-\infty}^{\infty} h'(A'^2 + A^2) \, dx,$$

which shows the left-hand integral to be positive, by (24). Integration by part also gives the trivial relation

$$-\int_{-\infty}^{\infty} h'' A^2 \, dx = \int_{-\infty}^{\infty} 2h' A A' \, dx,$$

and addition now gives

$$(\tau - 1) \int_{-\infty}^{\infty} h'' A^2 \, dx = \int_{-\infty}^{\infty} h'(A + A')^2 \, dx,$$

where also the right-hand integral is necessarily positive for a wave trapped over the sea-scarp. A necessary condition for such trapping is therefore

$$\tau > 1,$$

i.e., *the period must exceed on pendulum day.*

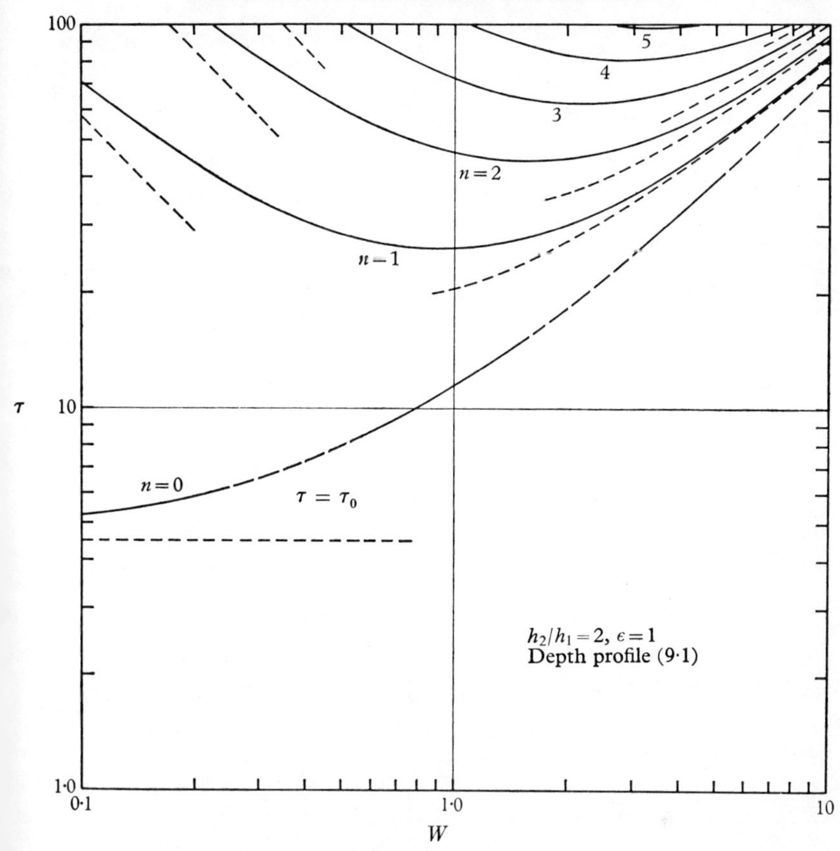

Reprinted with permission of J. Fluid Mech. [**10**, p. 68]

FIGURE 10. The period of the waves (in pendulum-days) as a function of the width W of the transition zone (multiplied by the wave-number m) when the depth profile has the form of Figure 8. Full curves represent the computed values. Broken curves represent asymptotes. For the lowest mode the curve cannot be distinguished from its asymptotes graphically.

By similar arguments it can also be shown [**10**] that $\tau > [\max(h'/h)]^{-1}$ and $\tau > 1/\max h'$ are necessary bounds.

To demonstrate the existence of eigenvalues τ, Longuet-Higgins [**10**] computed them for a tanh-type scarp profile (Figure 8) and for a profile of piecewise constant slope. Figure 10 shows the eigenvalues as functions of the relative scarp width for the former profile, and Figure 11 shows some of the corresponding eigenfunctions $A_n(x)$. Figure 12 shows eigenfunctions for the piecewise linear profile (all four figures are from [**10**]). Furthermore, asymptotic approximations for the eigenvalues and eigenfunctions are given in [**10**] both for general scarp profiles of very gentle slope and for such profiles of very steep slope.

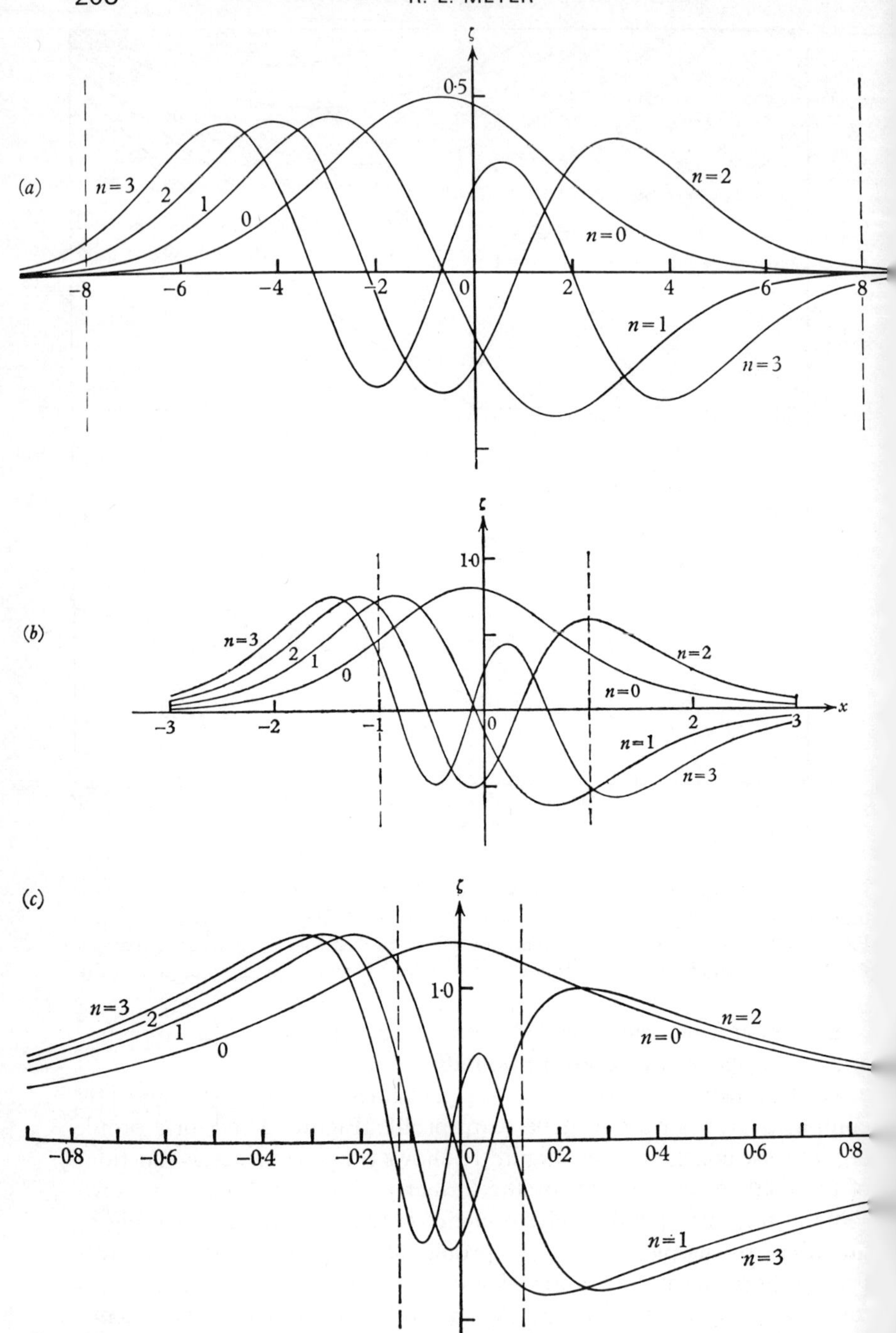

Reprinted with permission of J. Fluid Mech. [**10**, p. 71]

FIGURE 11. Eigenfunctions corresponding to the depth profile of Figure 8. (*a*) $W = 8$ (*b*) $W = 1$, (*c*) $W = \frac{1}{8}$.

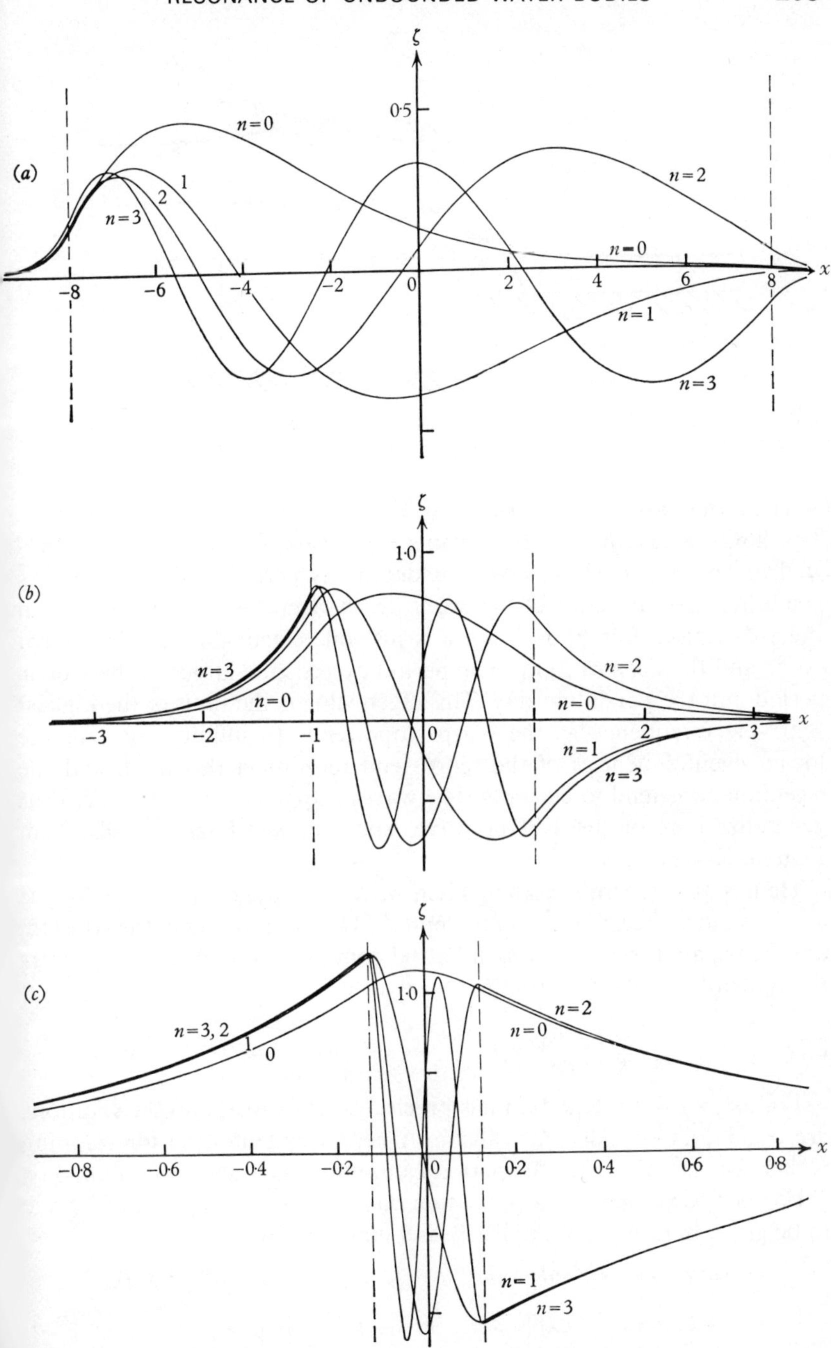

Reprinted with permission of J. Fluid Mech. [**10**, p. 75]

FIGURE 12. Eigenfunctions corresponding to the 'uniform slope' profile. (*a*) $W = 8$, (*b*) $W = 1$, (*c*) $W = \frac{1}{8}$.

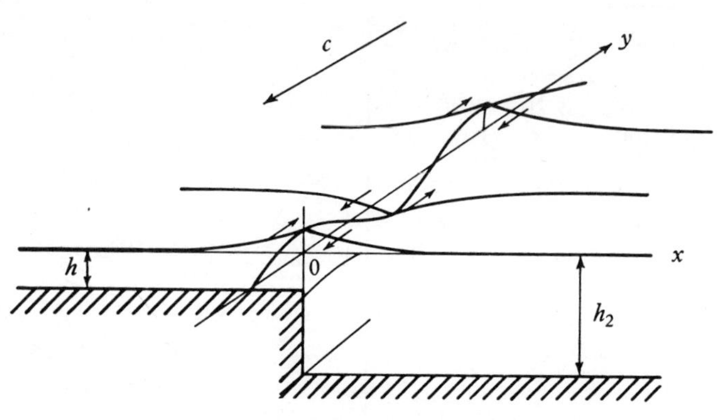

Reprinted with permission of J. Fluid Mech. [**9**, p. 421]

FIGURE 13.

The latter case of the sharp scarp (Figure 13) is treated in detail in [**9**]. The limiting case $h_1 = 0$ is Kelvin's edge wave [**14**] along the vertical wall of an open basin of constant depth. Its period is $\tau_K = L(gH)^{-1/2}$ pendulum days in terms of the water depth H and wave number L^{-1} in the y-direction. For $h_1 > 0$, there is just one eigenvalue [**9**], dependent on τ_K and the deep ratio h_1. The period exceeds the larger of the Kelvin period and the pendulum day. This eigenvalue is the limit of the highest scarp-eigenfrequency as the scarp slope tends to infinity [**10**]. All the lower eigenfrequencies of the scarp tend to zero in this limit, and the eigenfunctions tend to currents. The waves trapped over a scarp are thus generalizations of the Kelvin wave, and Longuet-Higgins calls them Double-Kelvin waves.

He has also generalized the Kelvin wave to a round island standing as a column in an ocean of constant depth h [**11**]. Long-waves of the type (15) are then again formally admissible, and comparison of (9) and (19) shows the Sturm-Liouville equation for them to be

$$\frac{d}{dr}\left(r\frac{dP}{dr}\right) + \left(\frac{\omega^2 - f^2}{h} - \left(\frac{n}{r}\right)^2\right)P = 0. \tag{25}$$

For $\omega^2 > f^2$, this equation has precisely one caustic, and the solutions are oscillatory outside that caustic. A necessary condition for trapping is therefore $\omega^2 \leqq f^2$, i.e., the period cannot be less than a pendulum day.

For periods exceeding a pendulum day, (25) shows any trapped wave to be given in terms of modified Bessel functions by

$$\zeta = K_n(kr)\exp[in\theta - i\omega t], \qquad k = \{(f^2 - \omega^2)/h\}^{1/2} > 0,$$
$$\sim (2kr/\pi)^{-1/2}\exp[-kr + in\theta - i\omega t] \quad \text{as } r \to \infty.$$

The remaining boundary condition for natural modes, that the radial velocity vanishes at the island radius $r = a$, can be satisfied [**11**] only for

$n \geqq 1$ and only if

$$a > [n(n-1)h]^{1/2}/f. \tag{26}$$

If the island radius exceeds this critical radius, there is precisely one eigenvalue ω_n for each positive integer n, and $\omega_n/f < 0$, so that the waves travel around the island clockwise in the northern hemisphere. As $ka/n \to \infty$, they tend, locally, to Kelvin waves [**11**].

In the ocean, (26) can rarely be satisfied for $n > 1$, and for $n = 1$, an eigenperiod greatly exceeding a pendulum day can rarely be expected (but the density stratification of the ocean may invalidate this conclusion) [**11**].

For a period equal to a pendulum day ("inertial waves"), necessary conditions for trapping are clockwise travel in the northern hemisphere and an island of just critical radius $[n(n-1)h]^{1/2}/f$ [**11**].

Mysak [**15**] and Rhines [**16**] have studied still longer waves of period large compared with a day, for certain topographies assisting the trapping mechanism of the Coriolis force. They show that resonances of several days' period are possible at a very large island (modelling Australia) [**15**], and how such slow waves may represent local modifications of Rossby waves [**16**].

6. **Oceanographic resonance approximation.** The more spectacular resonances (§1) are of much shorter period than a day and are due to the influence of the seabed, not the Coriolis force. They need not involve long-wave modes, but Ursell established (§2) that the classical theory of small amplitude wave motion, formulated by equations (1), (2), and (3), is a valid basis for the calculation of their natural modes and eigenfrequencies, even for edge waves close to shore. The work of Shen et al. [**17**] is therefore based on (1) to (3) and on the smallness of seabed slopes in the ocean. Indeed, 1/1000 is a typical seabed slope, and slopes much in excess of 1/100 are quite exceptional [**18**]. Motions in the ocean on which the seabed has an influence therefore depend on a *small parameter* ε *representing a typical seabed slope.*

This is insufficient to define an approximation or even to make the equations nondimensional. For water bodies adjacent to continents or islands, the topography may define a water depth scale, but not generally a horizontal length scale, and it is conventional to choose the wave length for the latter. Neither need be relevant, however, and progress depends on discarding them. Since trapping (§§3, 4) plays a decisive role in the motions under discussion, the most plausibly relevant, horizontal length scale L of the motion is a typical dimension of the region of trapping. The relevant vertical length scale of surface waves is their penetration distance, which is [5] the reciprocal of the wave number K. Both scales are therefore intrinsic scales of the problem, not known in advance (as is not unnatural for an eigenvalue problem).

For surface waves on which the seabed has even a small influence, the local phase velocity is comparable [5] to $(gH)^{1/2}$ when the seabed slope is small, where H denotes again the local water depth. If Ω denotes the frequency, the local wave number K is therefore comparable to $\Omega(gH)^{-1/2}$, which varies greatly with position, near a shore. It has a lower bound of scale $K_0 = \Omega(g\varepsilon L)^{-1/2}$, and surface waves trapped due to the influence of the seabed must therefore be expected to depend on the parameter

$$M = K_0 L = \Omega L^{1/2}(\varepsilon g)^{-1/2}.$$

The proper nondimensional formulation is then given by the transformation

$$\begin{gathered} x = X/L, \qquad y = Y/L, \qquad z = MZ/L, \qquad t = T(g/L)^{1/2}, \\ h(x, y) = MH(X, Y)/L, \\ \Phi(X, Y, Z, T) = (L^3 g)^{1/2} \phi(x, y, z; M, \varepsilon) \exp[-i\omega t], \\ M = \varepsilon^{-1/2}\omega, \end{gathered} \tag{27}$$

which brings (1), (2) and (3), respectively, into the form

$$M^2\phi_{zz} + \phi_{xx} + \phi_{yy} = 0 \quad \text{for } 0 < x, \quad 0 > z > -h, \tag{28}$$

$$M^2\phi_z + h_x\phi_x + h_y\phi_y = 0 \quad \text{for } z = -h(x, y), \tag{29}$$

$$\phi_z = \varepsilon M \phi \quad \text{for } z = 0. \tag{30}$$

In an open water body resonating in the lowest mode, the wave length scale must be comparable to L, but for higher modes, the wave pattern must contain many wave lengths, so that M is large. Most of the spectrum therefore corresponds to large M and the analysis of Shen et al. [17] aims at an asymptotic approximation in the limit

$$M \to \infty, \qquad \varepsilon \to 0, \qquad \varepsilon M \text{ bounded}. \tag{31}$$

It was first a surprise that—as indicated by comparison with the long-wave limit (§§3, 4) and Ursell's special solution (§2)—the limit (31) of (28) to (30) yields correct results for values of M as low as unity. But the magnitude of the eigenvalues M^2 depends on the choice of L, usually aimed at simplicity. For a recirculating wave pattern (§3), the relevant parameter is the average of the local scale ratio over the recirculation path corresponding to one cycle. If this parameter be used in place of $K_0 L$ with the simplest dimension L, then the lowest eigenvalue M^2 tends to exceed 5.

In any case (31) corresponds to neither a long-wave nor a short-wave approximation in a conventional oceanographic sense. In a mathematical sense, however, it is a "short-wave limit." Since M represents the non-dimensional wave number scale, an asymptotic theory of (28)–(30) in the limit (31) stands in the same relation to classical surface wave theory as geometrical optics stands to wave optics.

7. **Ray theory of spectra.** An heuristic asymptotic theory of (28)–(30) based on the analogy to geometrical optics has been developed by Keller [**19**] for the case $\Omega^2 L/g \to \infty$, and it appears equally applicable to the case (31), even if εM and $\Omega^2 L/g$ be small. It is, of course, an analytical version of ray-methods of wave calculation familiar in oceanography. But its power [**20**] is so much greater that little resemblance is apparent, apart from a mockingly similar nomenclature.

The theory assumes that the solution of (28)–(30) has, in the limit (31), an approximation of the form

$$\phi \sim \sum_{j=1}^{N} [A_j(x, y) + O(M^{-1})] \cosh[(z + h)k(x, y)] \exp[iMS_j(x, y)], \tag{32}$$

each term of which is called a wave and is assumed to satisfy (28)–(30) in that limit. That implies

$$(\nabla S)^2 = k^2, \tag{33}$$

$$k \tanh(kh) = \varepsilon M, \tag{34}$$

and an equation for A, where ∇ is again the horizontal gradient and the subscript j has been dropped. Equation (33) is the eiconal equation of geometric optics, and its characteristics are orthogonal trajectories, called *rays*, of the wave fronts $S(x, y) = \text{const}$. As developed to date [**19**], [**20**], [**17**], the theory assumes M to be real, and it thereby concentrates on the eigenfrequencies without seeking direct information on leakage rates (§4); the real spectrum turns out to be of quite enough complexity and interest, for a start.

It should be emphasized that the analysis to be discussed now cannot be understood by interpreting its waves, rays, caustics, etc. in a direct physical sense. These terms are used for attributes, not of water waves, but of an abstract two-dimensional map of the limit (31) of three-dimensional solutions of (28)–(30). The nomenclature derives, in fact, from an analogy with the physics, not of water waves, but of optics.

The equation for $A(x, y)$ shows this amplitude function to be singular at shore lines and at caustics (which are envelopes of rays), indicating that the approximation (32) fails there. But such local failure need not have a major influence on spectral predictions, which will be seen presently to depend only on the wave patterns in the large and on averages over long ray paths. Shen et al. [**17**] proceed on the assumption [**19**] that, if a ray meets a shore line, caustic, or vertical wall, then another ray emerges with the same value of S and A, except that A is changed by a factor $\exp[-i\pi/2]$ at a shore line[2] and caustic. This assumption is supported by a closer study of the caustics [**21**] and by agreement of the results obtained [**17**] with the exact eigenvalues for Ursell's beach (§2).

To determine the spectrum then requires only three steps [**20**], each of which may be illustrated conveniently by its application [**17**] to a beach of monotone depth $H(X)$ (Figure 14) in a channel bounded by vertical walls $y = 0, y = b > 0$. Then H increases from $H(0) = 0$ to $\lim_{x\to\infty} H(X) = H_\infty$,

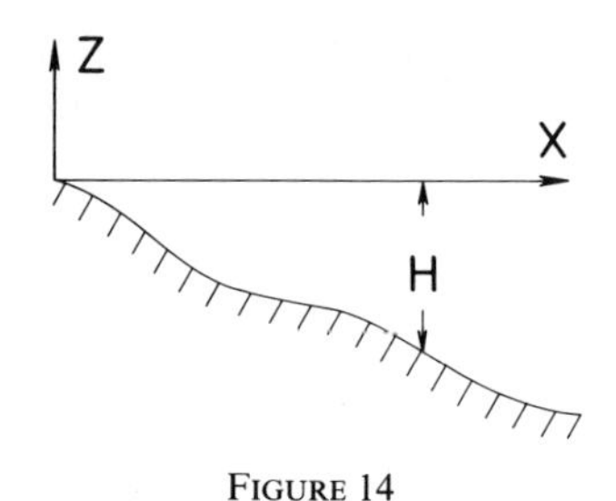

FIGURE 14

and without loss of physical generality, $dH/dX = H'(x)$ exists, and $0 \leqq H'(x) \leqq \varepsilon$, for all $x \geqq 0$.

The first step is to find the pattern of the caustics. They determine the *wave regions*, that is, those subregions of the equilibrium water surface of which every point can be reached by rays associated with a wave. For simplicity, it will be assumed in this section that if a region of trapping occurs, then it is a connected set.

That will be seen presently to hold always for the channel beach where, from (34), $k = k(x)$ decreases monotonically from ∞ at $x = 0$ to $k_\infty > 0$ as $x \to \infty$. (When k is considered as a function of x, y in the following, M

---

[2] This implies again perfect reflection of waves at a shore since $|A|$ is not reduced. It is not difficult to account for a reduction, once complex eigenvalues M^2 are admitted. But Ursell's results (§2) show that inviscid theory with the boundedness condition at the shore predicts the correct *edge wave* resonances. (Its failure is the paradoxical one of underestimating the degree of resonance at cut-off [**4**].) The perfect-reflection condition can therefore ruin only the analysis of *progressive wave* reflection. However, the shore boundary condition does not enter into the spectral conditions (40), (48) for the continuous spectra!

and ε will be assumed fixed.) The eiconal equation (33) then gives readily [**22**]

(35) $$dy/dx = \pm c\,(k^2 - c^2)^{-1/2}$$

for the slope of the rays, where c is an arbitrary constant. Without loss of generality, $0 \leqq c < \infty$, and two cases emerge. For $c < k_\infty$, the slope (35) is real for all $x \geqq 0$, and (33) defines a real-valued phase function

(36) $$S = \int^x k^2(k^2 - c^2)^{-1/2}\,dx$$

with integral taken along a ray, for all $x > 0$. The wave region thus extends over the whole equilibrium surface, and the waves are progressive in the sense of § 3. But for $c > k_\infty$, there is a unique number $a > 0$ such that

(37) $$k(a) = c,$$

and the slope (35) and phase function (36) are imaginary for $x > a$, indicating exponential decay, by (32). Hence $x = a$ is a caustic, the wave region is $0 < x < a$, and these waves are trapped in the sense of §3. Step one is thus seen to be the application of a ray approximation to a general trapping theory, free of the restriction to special forms of solution (§§3, 4).

Starting at some definite wave front $S(x, y) =$ const, the family of rays orthogonal to it, called a normal congruence of rays, may be followed in the sense of increasing S. After several encounters with boundaries of the wave region, the continuation of this normal congruence may again be orthogonal to the original wave front (corresponding to the loose notion of recirculation in §3). Since S has increased, however, it must then be a multivalued function on the wave region, and the same may hold of $A(x, y)$. To avoid confusion, it is necessary to associate each sweep of the rays across the wave region with a separate sheet of this region, and to distinguish the branches S_p of the phase function S defined on different sheets. The sheets are naturally joined to each other by reflection of rays at boundaries of the wave region, and in this way, the sheets form a connected *covering space* of the wave region on which ∇S and ∇A are single-valued. The second step is to determine the topological genus of this covering space, since it largely determines the spectral structure.

For the channel beach, the congruences can be labelled by the value of c in (35). Begin with the trapping case $c > k_\infty$, and fix c. From (35), four sheets of the wave region $0 < x < a$, $0 < y < b$ are possible, which are sketched in Figure 15 (where W, S and C denote respectively wall, shore and caustic). Since $k(x)$ is monotone decreasing, the rays turn toward the shallows, as predicted in §3. It is easy to follow a ray and its reflection at

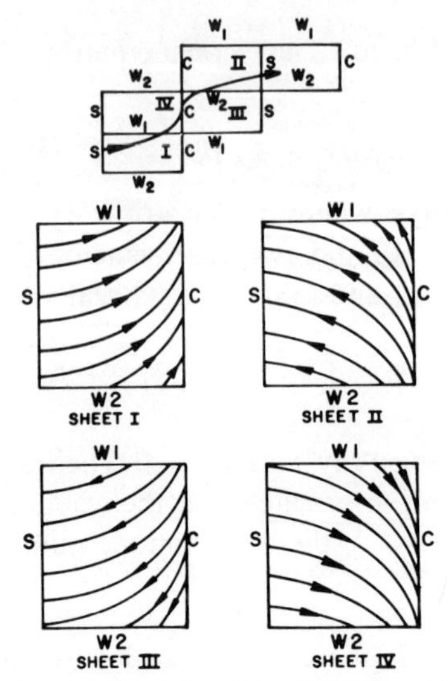

Reprinted with permission of Phys. Fluids [**17**, p. 2297]

FIGURE 15. Four congruences of rays for edge waves near the shore of a channel with monotonely sloping bottom.

boundaries from sheet to sheet (upper part of Figure 15) and to see from it that sheets adjacent in the longshore direction should be joined at both walls, while sheets adjacent in the offshore direction should be joined at shore and caustic. The four sheets are then joined so as to form a cover space equivalent to a torus.

The progressive case $c < k_\infty$ is obtained from that just discussed by letting the caustic distance $a \to \infty$. The torus then degenerates into a cylinder.

Now consider any closed path on the covering space. It corresponds to a closed path in the wave region, which cannot have circulation, if the wave region is simply connected. If not, as for an annulus of trapping, we are not here interested in pure swirling motions, whether steady or periodic in time, and their exclusion implies again that *the potential ϕ must be single-valued on the covering space.* That is the basic spectral condition [**20**]. To formalize it, let Γ_q denote any irreducible, closed path on the covering space, and let m_q denote the total number of times Γ_q crosses a shore line or caustic as it is traced once around. Since $H(x, y)$ and, by (34), $\cosh[k(z + h)]$ are single-valued, (32) shows the increase of a wave potential.

$$\phi = \cosh[k(z + h)] \exp[i(MS - i \log A)],$$

as Γ_q is traced once around in the sense of S increasing, to be given by

$$-i\delta(\log \phi) = M \int_{\Gamma_q} \nabla S \cdot d\mathbf{s} - \tfrac{1}{2} m_q \pi$$

where $d\mathbf{s}$ is the vector element of arc length. The spectral condition therefore implies

$$(38) \qquad M \int_{\Gamma_q} \nabla S \cdot d\mathbf{s} = 2\pi(n_q + \tfrac{1}{4}m_q)$$

for every such path Γ_q, where n_q is any nonnegative integer.[3] The genus of the covering space counts the number of homology classes of irreducible, closed paths, and each class contributes an independent condition (38) distinguished by the value of q. The third step is to evaluate them to determine the eigenvalues M^2.

8. **Coastal and channel spectra.** To apply the third step to the channel beach, begin again with the trapping case $c > k_\infty$ of wave region $0 < x < a$, $0 < y < b$. From (36), the values of S at the same x but on different rays, differ by an amount independent of x, so that $S(x, y) = X(x) + Y(y)$ with $X'^2 = k^2 - \mu^2$, $Y'^2 = \mu^2$, by (33), for constant μ. Comparison with (35) shows $\mu = c$. The four possible branches of S are therefore

$$S_1 = -S_4 = \int_0^x (k^2 - c^2)^{1/2}\, dx + cy,$$

$$S_3 = -S_2 = \int_0^x (k^2 - c^2)^{1/2}\, dx - cy,$$

and by associating them with the sheets of Figure 15 according to subscript, a phase function of gradient continuous on the cover space is found to be obtained.

Since the cover space is a torus, there are two homology classes. For Γ_1, we may choose the path along W_1 (upper part of Figure 15) from S across C back to S; it crosses shore and caustic once each, so $m_1 = 2$. For Γ_2, we may choose the shore line S (upper part of Figure 15) from W_2 back to W_2; no shore or caustic need be crossed, so $m_2 = 0$. Substitution in (38) thus gives immediately

$$(39) \qquad 2M \int_0^a (k^2 - c^2)^{1/2}\, dx = 2\pi(n_1 + 1/2), \qquad n_1 = 0, 1, \ldots,$$

$$(40) \qquad 2Mcb = 2n_2\pi, \qquad n_2 = 1, 2, \ldots .$$

These, together with (37) and (34), are a set of four equations, subject to the inequality $c > k_\infty$, for the four unknowns M, c, a, $k(x)$ in terms of the parameters n_1, n_2. Since $k(x)$ is monotone, continuous, and unbounded,

[3] It turns out to count, in the direction of Γ_q, the number of nodes in a natural mode.

the system has a finite number of solutions for each n_2, giving a countably infinite, discrete spectrum.

The system may be streamlined a little by elimination of c to obtain

$$M = n_2\pi/(bk(a)) < n_2\pi/(bk_\infty),$$

$$n_2 \int_0^a \left[\left(\frac{k(x)}{k(a)}\right)^2 - 1\right]^{1/2} dx = (n_1 + \tfrac{1}{2})b,$$

$$k(x)\tanh[k(x)h(x)] = \varepsilon n_2\pi/(bk(a)),$$

and the last two equations then represent a functional relation determining the caustic distance a. The implicit nature of this relation is characteristic of the theory and is the reason why Keller's ray theory succeeds in making quite simple relations represent very complex spectral structures.

For the progressive case $c < k_\infty$, the cover space is a cylinder, so has only one homology class, of which Γ_2 is a representative path. Hence (38) implies only (40), with $c < k_\infty$, i.e. a continuous spectrum

$$M > n_2\pi/(bk_\infty), \qquad n_2 = 1, 2, \ldots$$

of countable multiplicity; recall that the frequency is $\omega = \varepsilon^{1/2}M$, in units of $(g/L)^{1/2}$. By (34), the cut-off $M_0 = n_2\pi/(bk_\infty)$ is given by

$$\varepsilon M_0^2 = \omega_0^2 = (n_2\pi/b)\tanh[n_2\pi H_\infty/(bL)]$$

if H_∞ exists; if not, $k_\infty = \varepsilon M$ and $\omega_0^2 = n_2\pi/b$ [**17**].

To determine individual, discrete eigenfrequencies, (39) must be solved numerically, except for idealized and laboratory beaches [**17**]. For Ursell's beach (§2), e.g., $H(X) = \varepsilon X$ and integration of (39) gives [**17**]

$$(2n_1 + 1)\varepsilon = \arcsin(M\varepsilon/c),$$

whence by (40), (37) and (34), the frequency and caustic distance are given by

$$\omega^2 = (\pi n_2/b)\sin[(2n_1 + 1)\varepsilon],$$

$$a = b(\pi\varepsilon n_2)^{-1}\tanh^{-1}\{\sin[(2n_1 + 1)\varepsilon]\}$$

with $n_1 = 0, 1, \ldots$ but $\leqq (4\varepsilon)^{-1}\pi - \frac{1}{2}$ and $n_2 = 1, 2, \ldots$. Comparison with (6), (7) shows the subset $n_2 = 1$ to be precisely the small-slope approximation to Ursell's spectrum, even for the lowest resonant frequency ($n_1 = 0$, $n_2 = 1$).

This last observation is of logical importance for the theory in its present state, because no proof of the existence of an asymptotic approximation (32) has been given yet (and we have already noted that it is invalid near shores and caustics). From a purely mathematical point of view, the comparison with Ursell's edge wave spectrum is therefore one of the firm patches from which the ray theory of spectra is extended heuristically.

The dependence of the eigenvalues on two counting parameters will emerge to be typical of simple trapping configurations (§§3, 4). It is helpful to call the part of the spectrum corresponding to a fixed value of n_2 in (40) a spectral component. The results just given then show the spectrum of any monotone channel beach to have a countable infinity of spectral components. Each component consists of a continuous spectrum with cut-off and a finite discrete spectrum below the cut-off. But different spectral components have different cut-offs, and hence most of the discrete eigenvalues are actually embedded in the continuous spectrum (of other components).

The lowest cut-off, and the discrete frequencies below it, are undoubtedly resonant [**4**]. Spectral theory [**24**] indicates that the other cut-offs and discrete eigenfrequencies are similarly resonant within the framework of the linear theory, but experimental checks are still outstanding.

The resonance at a cut-off must be expected to be particularly dangerous for those members of topography families for which that cut-off coincides with a discrete eigenvalue [**4**]. For natural topography families of small seabed slope, there are many such critical topographies, and they are close together, as shown explicitly by (6), (7), so that an asymptotic approximation cannot be expected to pinpoint them precisely. The effect of such crowding of critical topographies on the resonant response at cut-off remains to be studied.

The same analysis covers the basic, continental slope topography, with or without shelf, in which the water depth depends only on distance from the shore [**17**]. This corresponds to the limit $b \to \infty$ of the monotone channel beach, in which the cover space degenerates for $c > k_\infty$ into a cylinder with homology class represented by Γ_1. Hence only (39) applies, together with (34) and (37); that is a set of only three equations for four unknowns, giving M as a continuous function of a for $0 < h(a) < h_\infty$. The spectrum is therefore continuous, and of finite multiplicity because $c > k_\infty$ still bounds n_1. For $c < k_\infty$, on the other hand, there is no caustic, and as $b \to \infty$, the cover space degenerates to a plane so that no restriction on M is left.

The ideal continental coast, therefore, has an entirely continuous spectrum, even though the natural modes include trapped waves of the type of edge waves or shelf waves [**17**]. These trapped waves, therefore, are *nonresonant* modes (observable as forced oscillations, but only of amplitude comparable to that of the excitation). For finite ocean depth, there are secondary cut-offs embedded in the continuous spectrum at which the spectral multiplicity changes, and their degree of resonance is not yet established.

This nonresonant nature of *basic* shelf waves merits emphasis in view of the relatively loose use of the term in the oceanographical literature. Of course, it applies only to straight coasts of indefinite longshore extent. Any lateral boundary conditions at the ends of the slope or shelf change the genus of the cover space, and hence the spectral structure. Similarly, any round island or continent has a discrete spectrum to which a shelf may contribute (§9).

The analysis is quite similar for topographies with nonmonotone depth distribution. For instance, for a submerged mountain range across a channel with flat bed (Figure 16), three types of wave modes are possible. They are progressive waves over the whole channel, progressive waves over half the channel and the crown of the submerged range (confined to one side of a single caustic), and *waves trapped over the submerged range* between two caustics, as in Figure 16. This is in strong contrast to a round

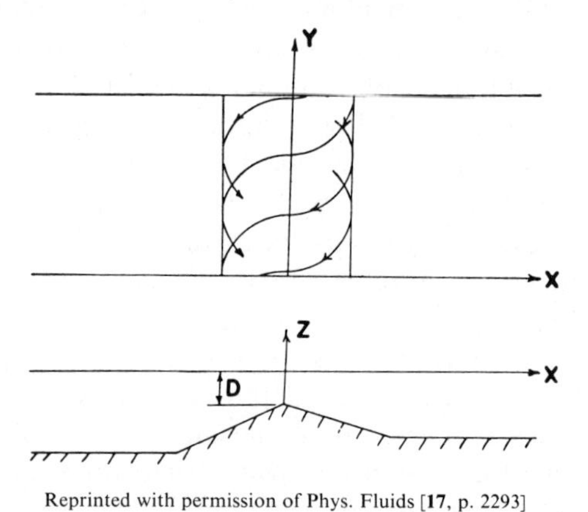

Reprinted with permission of Phys. Fluids [**17**, p. 2293]

FIGURE 16. Ray patterns over a protrusion in a channel.

seamount or reef (§§4, 9), where the water surface over the crown of the reef is a damping disc and where wave trapping can result only from a "hedge." For the submerged mountain range in a channel, the spectrum has again a countable number of components, each of which has again a continuous part, corresponding to progressive modes, with cut-off, and a finite number of discrete eigenvalues below the cut-off, corresponding to trapped modes [**17**].

This result emphasizes how essentially wave resonance in unbounded water bodies due to the influence of the seabed is a three-dimensional effect. For "two-dimensional" topographies, where the water depth depends only on one Cartesian coordinate, lateral boundary conditions—as at the side walls of a channel—are necessary for discrete eigenvalues.

9. **Spectra of round islands and seamounts.** As in §§4, 5 cylindrical coordinates R, θ, Z are convenient. Consistently with (27), let

$$r = R/L, \qquad h(r) = MH(R)/L;$$

then (34) becomes

$$k(r)\tanh[k(r)h(r)] = \varepsilon M, \tag{41}$$

and (33) gives [22]

$$r d\theta/dr = \pm c\,(k^2r^2 - c^2)^{-1/2} \tag{42}$$

for the ray directions, with arbitrary constant $c \geqq 0$. As in §8, $S(r, \theta) = C(r) + D(\theta)$ and from (42),

$$S(r, \theta) = \pm \int^r \{[k(r')]^2 - (c/r')^2\} dr' \pm c\theta. \tag{43}$$

As in §7, the first step is to determine the caustic radii, which correspond to

$$rk(r) = c,$$

by (42). Now, any caustic bounds a wave region, on the cover space of which the caustic circle represents one of the homology classes (corresponding to Γ_2 in §7). The corresponding spectral condition (38) is, by (43), $2\pi Mc = 2\pi n_2$. With

$$rk(r) = p(r, M), \qquad H(R)/R = G(r),$$

a caustic radius must therefore satisfy both

$$p(r, M) = n_2/M \tag{44}$$

and (41), which may be written

$$p\tanh(pMG) = \varepsilon M r. \tag{45}$$

Conversely, a solution $r = a(M)$ of both (44) and (45), for given n_2 and M, is a caustic radius, if M^2 is an eigenvalue in the spectral component n_2.

Elimination of M from (44), (45) gives the *spectral curve* [17] $p = p_s(r)$ of the component n_2 with

$$p_s(r) = \{\varepsilon n_2 r \coth[n_2 G(r)]\}^{1/2}, \tag{46}$$

from which it is easy [17] to read off (Figures 17, 19, 21) which r-intervals would be wave regions $rk = p_s > c = n_2/M$, and which would be damping

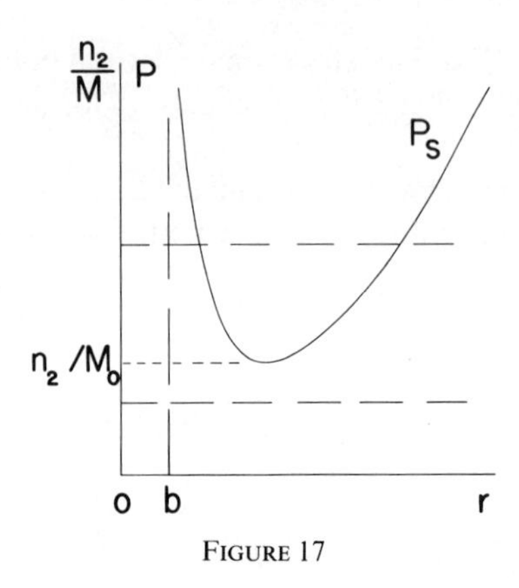

FIGURE 17

regions $p_s < n_2/M$, for a given M, if that were an eigenvalue. The extrema of the spectrum curve define cut-off values of M because they separate M-intervals on which the number of caustics is constant, and hence also the cover space genus and the spectral formulae are constant. At the cut-offs, the genus jumps, and the spectral relations change accordingly.

For any topography, $p_s(r, M) \to \infty$ as $r \to \infty$, by (45). For a *round island*, also $p_s \to \infty$ as r tends to the island radius b, and the spectrum curve must have a minimum. For simplicity, assume it has only one (Figure 17). Then $M_0 = n_2/\min(p_s)$ is the cut-off of the spectral component n_2. For $M > M_0$, this component cannot have a caustic $p_s = n_2/M$ (Figure 17) so that the entire water surface is a progressive wave region, and the only class of closed paths on the cover space can be those represented by the path Γ_2 along the shore. For it, (38) gives again $2\pi Mc = 2\pi n_2$, whence the spectrum is continuous for $M > M_0(n_2)$.

For $M < M_0$, on the other hand, there are always two caustics (Figure 17) bounding a damping annulus $p_s < n_2/M$ which, in turn, separates a trapping annulus around the island from the unbounded, progressive wave region at large r (Figure 18). This agrees entirely with the discussion of §4, and indeed, the long-wave limit ($kh \to 0$) of (45) is $\varepsilon p_s^{-2} = h(r)/r^2$. The ray pattern in the trapping annulus (Figure 18) is homeomorphic to that in the channel (Figure 15), with periodicity in θ now automatic, instead of implied by the channel wall boundary condition. The cover space of the trapping region is therefore again a torus. One homology class is again represented by the path Γ_2 along the shore, and the other, by a radial path Γ_1 from shore to caustic and back. For this, (38) and (43) give [**17**]

$$(47) \qquad 2M \int_b^a (p^2 - c^2)^{1/2} r^{-1}\, dr = (2n_1 + 1)\pi, \qquad n_1 = 0, 1, \dots .$$

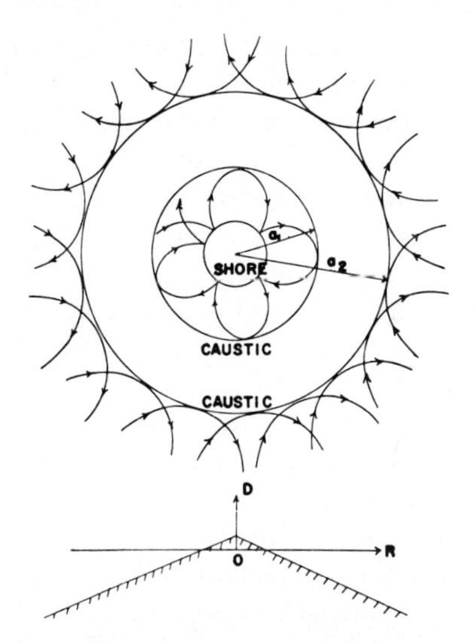

Reprinted with permission of Phys. Fluids [**17**, p. 2294]

FIGURE 18. Ray patterns for trapped waves near an island.

Together with the Γ_2-condition

$$Mc = n_2, \qquad n_2 = 1, 2, \ldots, \tag{48}$$

the caustic condition

$$0 < c = p(a), \tag{49}$$

and (45), this determines a finite set of discrete eigenvalues M^2 in

$$M < M_0(n_2). \tag{50}$$

The structure of the spectrum, then, appears to be the same as for the monotone channel beach, with a countably infinite number of components, each of which has a continuous part, and below its cut-off, a finite discrete part. The total spectrum consists of, as the frequency increases, first a finite number of discrete eigenfrequencies and then a continuous spectrum of multiplicity increasing by and by at secondary cut-offs, with a countable infinity of embedded discrete eigenfrequencies. But this reflects only the structure of the real part of the spectrum to which the analysis has been restricted. The finite width of the damping annulus implies, as in §4, a nonzero imaginary part for the discrete eigenvalues, which represents leakage of energy by radiation to infinity. This leakage is the more serious the narrower the damping annulus, i.e. from Figure 17, the nearer n_2/M is to the minimum of the spectrum curve. For any spectral component, therefore, only the lowest eigenfrequencies can be anticipated to be strongly resonant. The highest ones, close to the cut-off, are effectively nonresonant, and a new investigation appears needed to decide whether a cut-off is resonant under such circumstances.

A further surprise arises from the bound (50) for the discrete spectrum, which implies, as in §8, an upper bound $N_1(n_2)$ on the counting parameter n_1. This bound can be negative for a spectral component, implying that the discrete spectrum of the component is empty. For a conical island, e.g., that occurs [**17**] for $n_2 = 1$ and 2, so that the long-wave end of the discrete spectrum is missing! For components with n_2 not large compared to 3, moreover, $N_1(n_2)$ is relatively small, i.e. all the discrete eigenvalues of such components are close to the respective cut-offs. The surviving, long-wave end of the spectrum is therefore effectively nonresonant. This result may explain the observation that tsunamis are not amplified at small islands. Shen's work [**17**] on resonance was, in fact, prompted by the problem of catastrophic tsunami amplification.

The spectral structure changes completely if, in a wave tank, water be added to *submerge* the island. Then from (46), $p_s(0) = 0$; and if the spectrum curve is monotone (Figure 19), there is precisely one caustic $p_s(r) = n_2/M$

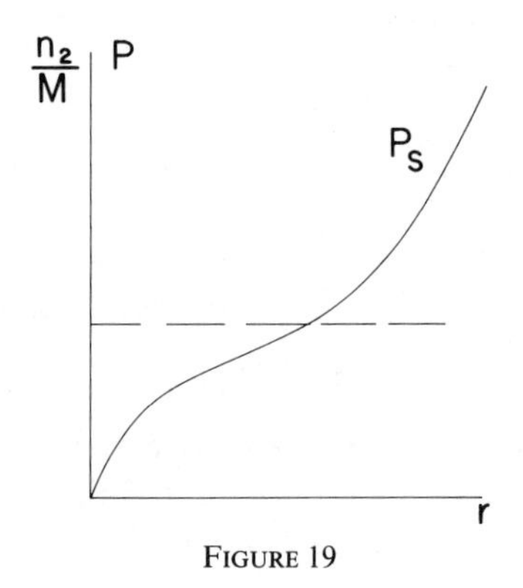

FIGURE 19

for each n_2/M, and (48) is the only spectral condition. Accordingly, the spectrum is entirely continuous, without cut-off, and all natural modes are progressive (Figure 20).

This result, however, depends crucially on the monotonicity of the spectrum curve. For a submerged reef or seamount with nonmonotone spectrum curve (Figure 21), the highest and lowest extremes of this curve define a pair of cut-offs bounding a *gap in the continuous spectrum* of the spectral component. A necessary and sufficient condition for this [**17**] is existence of an r-interval in which $\partial p_s/\partial r < 0$, i.e. by (46), $\sinh[2n_2G(r)] <$

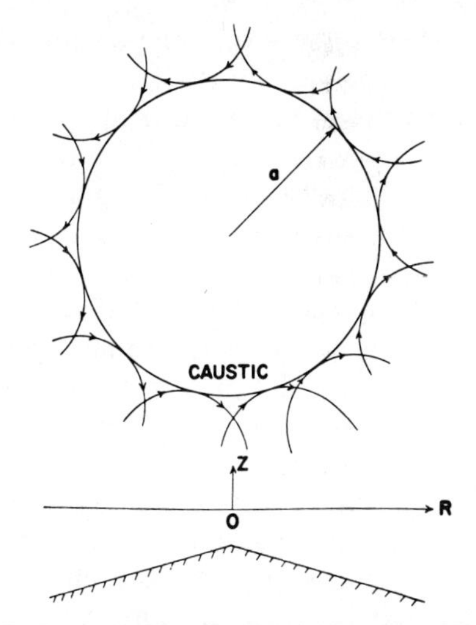

Reprinted with permission of Phys. Fluids [**17**, p. 2295]

FIGURE 20. Ray patterns for trapped waves over a submerged peak.

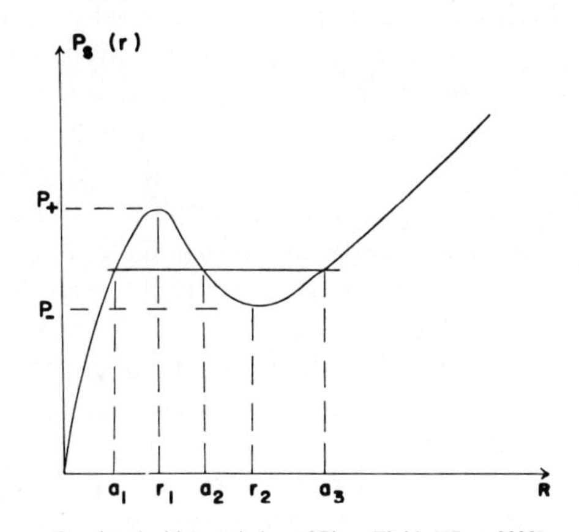

Reprinted with permission of Phys. Fluids [**17**, p. 2303]

FIGURE 21

$2n_2rG'(r)$, or in terms of the dimensional water depth $H(R)$,

$$\frac{R}{H}\frac{dH}{dR} > 1 + \frac{R}{2n_2H}\sinh\left(\frac{2n_2H}{R}\right) > 2. \tag{51}$$

This shows the hedge condition (18) to be both a necessary condition and the long-wave limit of the more general condition. For a given topography, however, the hedge condition (51) cannot be satisfied for all spectral components—rather [**17**], all spectral components with n_2 exceeding a bound n_* are entirely continuous.

To understand the nature of the spectral components with gap in the continuous spectrum, assume for simplicity that the hedge condition (51) is satisfied on only one connected interval (Figure 21). Any mode in the gap must then have three caustics (Figure 21) and a wave pattern as shown in Figure 6. The ray pattern in the trapping annulus is homeomorphic to that around the shore of an island (Figure 18), so the cover space must again be a torus, and the spectral relations must therefore again be (47) and (48), with b and a denoting the smaller pair of caustic radii given by (49). Accordingly, there are discrete eigenfrequencies for $n_1 = 0, 1, 2, \ldots < N_1(n_2)$ in the gaps of the spectral components with $N \leqq n_2 < n_*$, where N is the smallest positive integer for which $N_1(N) \geqq 0$. As the margin decreases, by which (18) can be satisfied by a topography, N increases and n_* decreases [**17**], so that some gaps are usually empty ($n_2 < N$) and a fair margin is, in fact, required for any nonempty gap to exist ($N < n_*$). But even if a spectral component does have a nonempty gap, the number of discrete eigenvalues in it must be appreciable in order to include an eigenvalue with small leakage. A very pronounced shelf topography indeed is therefore necessary for the existence of effectively resonant eigenfrequencies over small seabed slopes.

10. **Conclusion.** The spectral analysis, by the ray-approximation, of still more realistic oceanographical topographies requires the same three steps (§7), and the first two may be more formidable than in the preceding sections. To determine the configuration of caustics and ray patterns may require numerical work, which may then need to be dovetailed with analytical argument, to establish the genus of the cover space for various intervals of parameters. The third step, however, will always benefit from the simple structure of the spectral relations. Since these are integral relations, moreover, the finer details of the seabed topography will have little quantitative influence on the eigenvalues.

An extension of the theory beyond the idealized or laboratory topographies so far considered, therefore, appears entirely practical. It may be premature, however, since it is clear that the theory has forged far ahead of experiment and observation. It has done so heuristically, moreover, and not on a very wide front. The most urgent need is thus for experiments, and the primary aim of this review is to promote them. These columns may seem inappropriate for a call for experiments, but so would the columns of experimental journals seem for the review of a physical subject in which the genus of a cover space is a central notion. The field would appear to call for cooperation between mathematician and experimental scientist, and to offer the most fruitful prospects to such cooperation.

ACKNOWLEDGEMENTS. This work was supported by NSF Grant GP-18641 and by a Water Resources Research grant from the U.S. Department of the Interior under P.L. 88-379.

Figures 6 and 7 are from [**8**]; Figures 8 to 12 from [**10**]; Figure 13 from [**9**]; and Figures 15, 16, 18, 20 and 21 from [**17**].

REFERENCES

1. A. S. Peters, *Water waves over sloping beaches and the solution of a mixed boundary value problem for $\Delta^2\phi - K^2\phi = 0$ in a sector*, Comm. Pure Appl. Math. **5** (1952), 87–108. MR **13,** 789.

2. F. Ursell, *Surface waves on deep water in the presence of a submerged circular cylinder*, I, Proc. Cambridge Philos. Soc. **46** (1950), 141–152. MR **11,** 480.

3. ———, *Trapping modes in the theory of surface waves*, Proc. Cambridge Philos. Soc. **47** (1951), 347–358. MR **12,** 870.

4. ———, *Edge waves on a sloping beach*, Proc. Roy. Soc. London Ser. A **214** (1952), 79–97. MR **14,** 326.

5. J. J. Stoker, *Water waves: The mathematical theory with applications*, Pure and Appl. Math., vol. 4, Interscience, New York, 1957. MR **21** #2438.

6. G. G. Stokes, Rep. British Assoc. **1846,** part 1, 1.

7. C. J. Calvin, Trans. Amer. Geophys. Union **46** (1965), 110.

8. M. S. Longuet-Higgins, J. Fluid Mech. **29** (1967), 781.

9. ———, J. Fluid Mech. **31** (1968), 417.

10. ———, J. Fluid Mech. **34** (1968), 49.

11. ———, J. Fluid Mech. **37** (1969), 773.

12. R. E. Meyer and A. D. Taylor, J. Geophys. Res. **68** (1963), 6443.

13. G. F. D. Duff, *On tidal resonance and tidal barriers in the Bay of Fundy system*, J. Fisheries Res. Board Canad. **27** (1970), 1701.

14. W. Thompson, Proc. Roy. Soc. Edinburgh **10** (1879), 92.

15. L. A. Mysak, J. Marine Res. **25** (1967), 205.

16. P. B. Rhines, J. Fluid Mech. **37** (1969), 161, 191.

17. M. C. Shen, R. E. Meyer and J. B. Keller, Phys. Fluids **11** (1968), 2289.

18. F. P. Shepard, *Submarine geology*, Harper & Row, New York, 1963.

19. J. B. Keller, *Surface waves on water of non-uniform depth*, J. Fluid Mech. **4** (1958), 607–614. MR **21** #1070.

20. J. B. Keller and S. I. Rubinov, Ann. Phys. **9** (1960), 24.

21. R. N. Buchal and J. B. Keller, *Boundary layer problems in diffraction theory*, Comm. Pure Appl. Math. **13** (1960), 85–114. MR **22** #10549.

22. R. Courant, *Methods of mathematical physics*. Vol. 2, Interscience, New York, 1962. MR **25** #4216.

23. C. C. Lautenbacher, *Gravity wave refraction by islands*, J. Fluid Mech. **41** (1970), 655–672.

24. N. I. Ahiezer and I. M. Glazman, *Theory of linear operators in Hilbert space*. Vol. 2, Ungar, New York, 1963.

25. A. C. Vastano and R. O. Reid, J. Marine Res. **25** (1967), 129.

UNIVERSITY OF WISCONSIN

On Planetary Atmospheres and Interiors

Raymond Hide

PREFACE

A comprehensive discussion of planetary atmospheres would treat their origin and evolution, the chemical composition of their main gaseous components, the nature and distribution of large particles in suspension, the mean temperature and pressure at the lower boundary and the nature of that boundary, the vertical and horizontal distributions of density, temperature and pressure, the nature of small-scale turbulent motions and the nature of energy sources and large-scale organized motions; a comprehensive discussion of planetary interiors would deal with a list of comparable or even greater length. These lectures will discuss a limited selection of topics from these general areas and are organized under the separate headings: "Laboratory experiments on baroclinic waves and the theory of the global circulation of the Earth's atmosphere" (Lecture I); "Dynamics of the atmospheres of the major planets" (Lecture II); "Jupiter's rotation, magnetism, internal dynamics and structure" (Lecture III); "Planetary magnetic fields" (Lecture IV); and "The core-mantle interface and the Earth's magnetism, gravitation and rotation" (Lecture V).

Because a great deal of this material was presented by means of slides and ciné films, the preparation of a set of notes which the lectures would follow closely was not practicable. These supplementary notes, in which the subject matter is summarized and reviewed and extensive lists of references are given, were prepared as background material. I must thank the American Meteorological Society, the Royal Meteorological

---

AMS 1970 *subject classifications*. Primary 86-02, 86A25, 86A05, 86A35; Secondary 76W05, 76E20.

Society, and the publishers of the *Journal of Fluid Mechanics* (Cambridge University Press), *Magnetism and the Cosmos* (Oliver and Boyd, Ltd., Edinburgh), *Nature* (MacMillan and Co. Ltd., London), *Planetary and Space Science* (Pergamon Press, Ltd., London) and *Planetary Surfaces and Interiors* (Academic Press Inc., London) for granting permission to reproduce extracts from various articles cited explicitly in these notes.

I. LABORATORY EXPERIMENTS ON BAROCLINIC WAVES AND THE THEORY OF THE GLOBAL CIRCULATION OF THE EARTH'S ATMOSPHERE[1]

Summary. Laboratory investigations of thermal convection due to an impressed horizontal temperature gradient in a rotating fluid of low viscosity and thermal conductivity, involving precise determinations of the principal spatial and temporal characteristics of the fields of temperature and flow velocity over a wide range of accurately specified and carefully controlled conditions, have led to the discovery and partial elucidation of four basic types of flow of varying degrees of spatial and temporal irregularity. The experiments have rendered feasible crucial investigations of effects due to systematic departures from axial symmetry in the impressed conditions and detailed comparisons of a few laboratory flows with the global circulation of the atmosphere. They have led to advances in the theory of baroclinic waves and to interesting numerical work concerned with the study of fine details of the simpler laboratory flows, indicating what might be accomplished by such means in the investigation of more complicated and geophysically realistic flows when very large and fast computers become available for this type of research. But above all, by making possible the separation of essential theoretical considerations from minor and irrelevant ones, the experiments provide a context in which the global circulation of the atmospheres of the Earth and the other planets can be studied in a truly quantitative and scientific way. (Various extensions of Eady's theory of baroclinic instability are presented in the Appendices.)

1. Introduction. If the central scientific problem concerning the global circulation of the Earth's atmosphere is (as one leading investigator has put it) that of "predicting from the laws of classical physics that the atmosphere is necessarily organized as it is," then research toward a solution must, of necessity, include systematic quantitative investigations of many different but related fluid-dynamical systems, of which the atmosphere is but one very complex example. This family of systems could comprise other natural systems, notably the atmospheres of other planets

[1] Supplementary notes for Lecture I, based on a recent article by the lecturer [**81**].

and the oceans, laboratory systems (see Appendix A), as well as mathematical or numerical models. What is evidently required is a thorough understanding of thermal convection due to an impressed horizontal temperature gradient in a rotating fluid of low viscosity and low thermal conductivity.

Were not the formulation and analysis of mathematical or numerical models fraught with the serious and often insuperable technical difficulties encountered in most realistic theoretical studies in fluid dynamics, it would be unnecessary to look to other systems. Studies of the atmospheres of other planets and of the oceans—exciting and important areas of scientific inquiry in their own right, in which encouraging progress is now being made—have not yet advanced to a useful stage in the present context. Thus it is fortunate, but not entirely fortuitous, that research carried out by fluid dynamicists on the hydrodynamics of rapidly rotating fluids includes laboratory experiments that are relevant to the theory of the global atmospheric circulation.

Of these laboratory studies, the most relevant in the first instance have been controlled and reproducible experiments on thermal convection in systems characterized by steadiness and axial symmetry in the shape of the bounding surfaces and in the distribution of applied heating and cooling (see §§2–5). Precise determination of the principal spatial and temporal characteristics of the fields of temperature and flow velocity over a wide range of accurately specified and carefully controlled experimental conditions (i.e. rate of rotation, amplitude and form of the impressed temperature distribution, dimensions and shapes of the boundary surfaces and physical properties of the convecting fluid) led to the discovery of four fundamentally different free types of flow (see Figure 1), only one of which is symmetrical about the axis of rotation. The delineation of empirical criteria for the occurrence of transitions from axisymmetric flow to nonaxisymmetric flow led to the early identification of the form of principal "external dimensionless parameters" in terms of which the experimental conditions should be specified. The first steps have now been taken towards a satisfactory theoretical interpretation of these empirical criteria, of certain statistical and hysteresis effects and of the experimentally-determined dependence on the external dimensionless parameters of internal dimensionless parameters such as the Rossby, Burger and Nusselt numbers and the dominant azimuthal wavenumber (measures of the r.m.s. zonal velocity, vertical stability, convective heat transfer and departures from axial symmetry of the general flow pattern, respectively). Theoretical work along these lines promises to add greatly to our understanding of incipient and fully-developed baroclinic waves, quasi-geostrophic detached thermal boundary layers, ageostrophic viscous boundary layers at free and rigid bounding surfaces and various types of

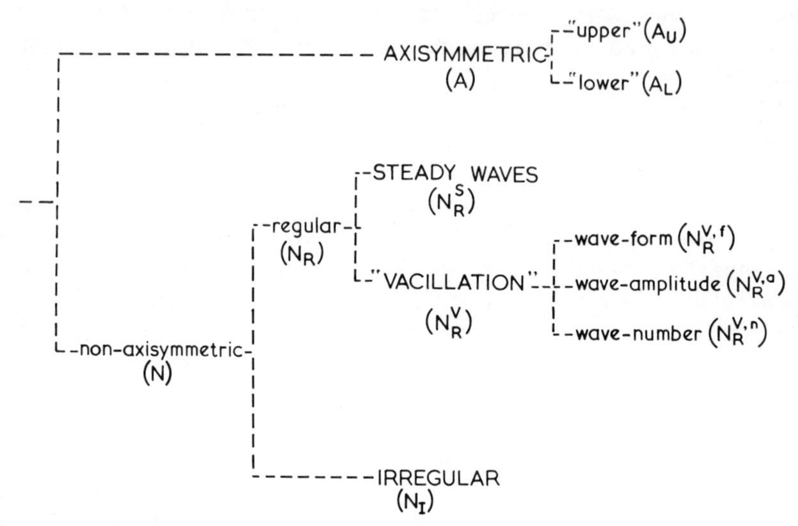

FIGURE 1. Thermal convection in a rotating fluid subject to axisymmetric differential heating and cooling [**81**].

interactions between these processes, and should in due course bring about further advances in geophysical and astrophysical fluid dynamics.

The laboratory experiments have rendered feasible crucial investigations of effects due to systematic departures from axial symmetry in the impressed conditions as well as useful detailed comparisons with a few laboratory flows of the global circulation of the Earth's atmosphere. They have led to interesting numerical work concerned with the reproduction, by means of a computer, of some of the simpler laboratory flows, carried out in order to describe the fields of motion in great detail or to test numerical schemes. But above all, so far as their meteorological significance is concerned, the experiments (to paraphrase remarks of Lorenz [**52**], by indicating the flow patterns that can occur and the conditions favourable to each, have made possible the separation of essential from minor and irrelevant considerations in the theory of the global atmospheric circulation. They show, for instance, that while condensation of water vapour may yet play an essential role in the tropics, it appears to be no more than a modifying influence in temperate latitudes, because hydrodynamical phenomena found in the atmosphere, including even cyclones, jet streams and fronts, also occur in the laboratory apparatus where there is no analogue of the condensation process. Similar remarks apply to topographic features, which were intentionally omitted in the experiments. The so-called "beta-effect"—the tendency for the relative vorticity to decrease in northward flow and increase in southward flow because of the variation with latitude of the Coriolis parameter—now appears to play a lesser role than had once been assumed. Certainly a numerical weather forecast would fail if the beta-effect were disregarded, but the beta-effect does not seem to be required for the production of typical atmospheric systems. The experiments have emphasized the necessity for truly quantitative considerations of planetary atmospheres. These considerations must, at the very least, be sufficient in the first instance to place the Earth's atmosphere in one of the free nonaxisymmetric regimes of thermal convection discovered in the laboratory work.

2. Free types of flow. Consider a vertical annulus of liquid bounded by rigid cylindrical surfaces in $r = a$ and $r = b$, where $b > a$, by a lower surface $z = z_l(r, \phi)$ and an upper surface $z = z_u(r, \phi)$ which may be either rigid or free ((r, ϕ, z) being cylindrical polar coordinates with the z axis vertical), which rotates uniformly with angular velocity $\mathbf{\Omega} = (0, 0, \Omega)$ about the axis of symmetry. If the liquid contains no sources of heat and the temperature is kept uniform and steady over the bounding surfaces then after a sufficient lapse of time—typically of order $d/(\nu\Omega)^{1/2}$ where d is the average value of $(z_u - z_l)$ and ν is the coefficient of kinematical viscosity—following the setting up of the system, the motion of the liquid will be that of solid body rotation with angular velocity $\mathbf{\Omega}$. If on the other hand internal heat sources and/or the heating of certain areas of the bounding surfaces and the cooling of others produce horizontal temperature gradients within the liquid, then solid body rotation is impossible.

Denote by $\boldsymbol{u}$ the Eulerian velocity relative to the rotating frame of the hydrodynamical motion (thermal convection) that then ensues and by p the corresponding dynamic pressure. Axisymmetric flow, for which $\boldsymbol{u}$ and p are independent of ϕ, can occur if z_u, z_l and the distribution of heating and cooling are independent of ϕ. Differential heating produces meridional circulation, with warm fluid rising and cold fluid sinking, and the action of Coriolis forces on this basic circulation produces azimuthal flow, so that the stream-lines have the form of spirals.

Whether or not axisymmetric flow occurs in practice will depend on its stability to small, adventitious and ϕ-dependent perturbations. The horizontal temperature gradient associated with the vertical shear of aximuthal flow (thermal wind) makes for instability, but this is opposed by the vertical temperature gradient produced and maintained by the upward heat transfer associated with the meridional flow. In the first annulus experiments (summarized in [**30**], [**31**] and described in full in [**29**], [**32**])—in which the two bounding cylinders were held at different temperatures T_a and T_b and no internal heat sources were present—the upper surface was free and effectively parallel to the flat rigid horizontal lower surface and primary effects due to viscosity were, by design negligible. These experiments demonstrated and subsequent experiments (see [**39**])—in which effects due to internal heating, rigid upper bounding surface, sloping upper and lower surfaces, vanishingly small inner cylinder, et cetera, were investigated—confirmed that the general character of the flow depends largely on the value of the external dimensionless parameter.

$$\Theta \equiv gd|\Delta\rho|/\bar{\rho}\Omega^2(b - a)^2. \tag{1}$$

Here $\boldsymbol{g} = (0, 0, -g)$ is the acceleration of gravity, $\bar{\rho}$ is the mean density of the liquid and $\Delta\rho$ is a measure of the impressed density contrast produced by the applied differential heating; in the case of a wall-heated annulus, for example, $\Delta\rho = |\rho(T_a) - \rho(T_b)|$, where $\rho(T)$ is the density of the liquid at temperature T. Thus, the flow is axisymmetric or nonaxisymmetric according as

$$\Theta \gtrless \Theta_R \tag{2}$$

where Θ_R is a certain critical value of Θ which is typically of the order of, but greater than, unity and relatively insensitive to the other parameters (see below). The character of the nonaxisymmetric flow found when $\Theta < \Theta_R$ depends on whether

$$\Theta \gtrless \Theta_I \tag{3}$$

where Θ_I, typically less than 10^{-1}, is another critical value of Θ whose complicated dependence on the other parameters has not yet been fully elucidated. When $\Theta_R > \Theta > \Theta_I$ the flow is spatially and temporally regular; it then comprises fully-developed baroclinic waves, with their associated upper-level jet stream, that are either steady (apart from a possible uniform drift in the azimuthal direction) or undergo regular periodic fluctuations ("vacillation," see Figure 1 and Appendix B) in form, amplitude, wavenumber or some combination of these properties. Otherwise, when $\Theta < \Theta_I$, the flow, though roughly wavelike, exhibits complicated, irregular and nonperiodic variations in both space and time.

Here, perhaps, is the place to mention an erroneous conjecture—that steady waves and vacillation must be directly associated with the presence of an inner cylinder and with the particular way in which heating and cooling were applied in the first annulus experiments, notably at the bounding cylinders—which seems to have gained fairly widespread acceptance (see Davies [**13**], Lorenz [**52**]), presumably because these regular nonaxisymmetric flows were observed neither in the "open dishpan" work of Fultz et al. [**26**], Fultz [**24**], Faller [**19**] nor in the much earlier laboratory work of Vettin [**74**] and Exner [**18**] on similar systems.

The first annulus experiments were designed so as to avoid primary effects due to viscosity, but a systematic study of viscous effects was subsequently undertaken by Fowlis and Hide [**22**] who found $\mathscr{T}^{-1/2}$, where

(4a) $$\mathscr{T} \equiv 4\Omega^2(b-a)^5/\nu^2 d,$$

to be the appropriate external dimensionless parameter in terms of which the coefficient of kinematical viscosity ν should be measured (when $(b-a) \ll d$). In a diagram with $\mathscr{T}$ as abscissa and Θ as ordinate (see schematic diagram given in Figure 2; for full quantitative details see Fowlis and Hide [**22**, Figures 3 and 6–10]), axisymmetric flow is found outside an anvil-shaped region whose upper boundary, by definition $\Theta = \Theta_R$ (see Equation (2)), lies below $\Theta = 4.0$ (see Hide [**37**], Hide and Mason [**39**]), the value to which Θ_R apparently tends when $\mathscr{T}$ is very large ($>10^9$). The flow is axisymmetric for all $\mathscr{T} < \mathscr{T}_p$, where $\mathscr{T}_p$ is the value of $\mathscr{T}$ at the point of the anvil, given by

(4b) $$\mathscr{T}_p = (1.85 \pm 0.08) \times 10^5 \quad \text{(standard error)}$$

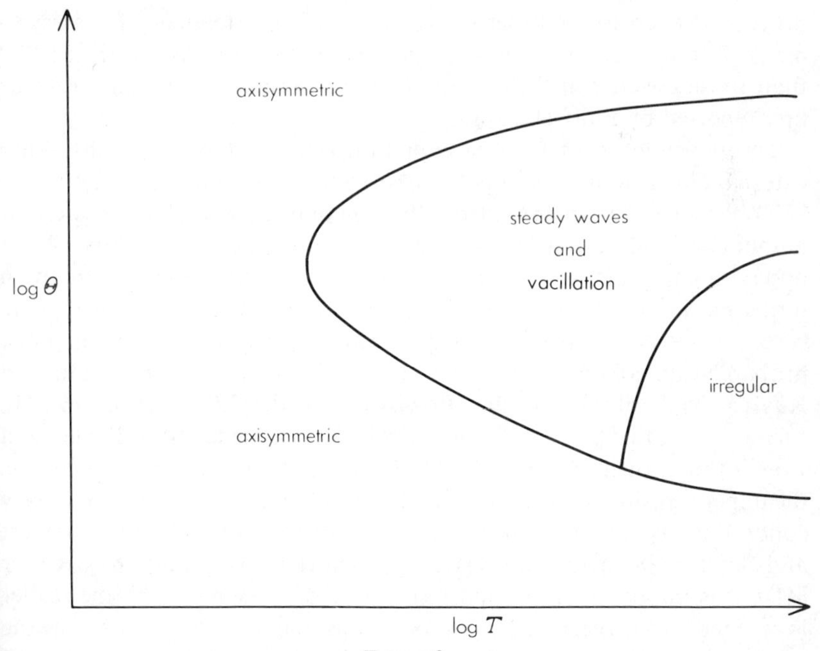

FIGURE 2

in the case of wall-heating (cf. Equation (D18) below). The corresponding lower boundary of the anvil-shaped region is a line $\Theta = \Theta_L$ sloping downwards from left to right. Over the range $\mathscr{T}_p < \mathscr{T} < 2 \times 10^7$ the equation of the line is given by

(5a)
$$\log_{10}(\Theta_L \nu/\kappa) = (5.05 \pm 0.30) - (0.864 \pm 0.043) \log_{10}(\mathscr{T} d/(b - a))$$
(standard errors),

where κ is the coefficient of thermal diffusivity; the position of the lower boundary when $\mathscr{T} > 2 \times 10^7$ has not yet been fully investigated, but indications are that the curve then becomes more nearly parallel to the Θ axis.

It has been found convenient to refer to the region above the anvil-shaped area in the Θ versus $\mathscr{T}$ diagram as the "upper (axi-) symmetric regime" and to the region below as the "lower (axi-) symmetric regime" (see Figure 1). The latter regime, by Equation (5a), arises when $|\Delta\rho|$ is so small that

(5b)
$$g|\Delta\rho|(b - a)^2 d/\bar{\rho}\kappa\nu < (3.5 \pm 2.1) \times 10^4[4\Omega^2(b - a)^4/\nu^2]^{0.136 \pm 0.043}$$

(if $\mathscr{T}_p < \mathscr{T} < 2 \times 10^7$); in these circumstances the agencies responsible for baroclinic instability cannot overcome damping due to transport processes (viscosity and thermal conduction). The tendency for axisymmetric flow to occur at much smaller values of Θ (typically less than 10^{-2}) than those covered in the original annulus experiments (see above) was first reported by Fultz et al. [**26**].

The dependence of Θ_I (see Equation (3)) on $\mathscr{T}$, ν/κ and the other external dimensionless parameters (see Tables 1 and 2 of Fowlis and Hide [**22**]) has also been investigated, albeit incompletely, and a considerable amount of information has been obtained on a variety of properties of nonaxisymmetric flows, some of which can be summarized succinctly in terms of simple and theoretically-suggestive empirical relationships, in certain cases necessarily involving statistical quantities such as the most probable value of the wavenumber (Hide [**29**]–[**32**], Smith [**68**], Fultz and Kaylor [**25**], Fultz et al. [**26**], Fowlis and Hide [**22**], Uryu et al. [**73**], Pfeffer and Chiang [**56**], Ketchum [**44**], Koschmieder [**45**], Pfeffer and Fowlis [**58**], Fowlis and Pfeffer [**23**], Hide and Mason [**39**]). The effect on the upper transition of variations in the mechanical and thermal boundary conditions has been the subject of a number of useful experiments (Lambert and Snyder [**48**], Hide and Mason [**39**], Snyder and Youtz [**69**], Kaiser [**41**]). It is impossible to do full justice to this work here, and the reader is referred to the original papers for details. Suffice it to remark that the work generally confirms that, notwithstanding the large number of external dimensionless parameters required to specify the system exactly, when $\mathscr{T} \gg \mathscr{T}_p$ it is upon the values of Θ and (to a lesser extent) $\mathscr{T}$ that the principal characteristics of the flow largely depend. Thus, it is possible with the aid of only rudimentary theoretical considerations to apply the results of the experiments with some confidence to large-scale geophysical or astrophysical systems, and for meteorologists to exploit detailed comparisons of laboratory flows with corresponding flows in the atmosphere, along lines exemplified by the pioneering work of Riehl and Fultz [**63**], [**64**] (see also Fultz et al. [**26**], Pfeffer and colleagues [**57**], Coté [**12**], Elsberry [**17**]). It is significant, for example, that Θ for the Earth's atmosphere is of the order of but somewhat less than unity, so that it is no longer surprising that—in spite of the fantastic differences in horizontal dimensions (amounting to a factor of 10^8), in aspect ratio $d/(b-a)$ (amounting to 10^{-3}) and also in other parameters—some of the laboratory flows we have described bear a striking resemblance to the global atmospheric circulation (see Appendix C and Hide [**35**], especially Figure 10).

3. Digression on techniques. Neither the scope nor the limitations of the laboratory work can be appreciated fully without reference to techniques of measurement and observation, upon which we shall therefore briefly digress before going on to complete this review of experimental results and their theoretical interpretation.

The results outlined in §2 were obtained largely by exploiting the feasibility of varying in the laboratory parameters such as Ω, d, $\Delta\rho$, b, a and ν over wide ranges and in a controlled manner. It was sufficient in most of the experiments to determine no more than the broad characteristics of the pattern of flow at the top surface, which was done with the aid of quite simple flow-visualization techniques involving the use of dyes (e.g. fluorescein) or tiny reflecting particles (e.g. aluminium powder) either in suspension or floating on the top surface.

In principle the horizontal velocity field is readily found at any level by illuminating the suspended particles with a flat beam of light aimed at that level and obtaining streak photographs of the horizontal trajectories of illuminated particles by exposing, usually for several seconds, the film in a camera mounted on the rotating apparatus. Streak photography has been applied mainly to the study of top-surface flow patterns. It has not yet been used extensively for obtaining three-dimensional velocity fields (although visual work with flat horizontal and vertical light beams has proved instructive in this connection), but the technique increases in feasibility as methods of automatic data processing improve and ought therefore to be considered more seriously in future work. Typical flow speeds fall within a range that precludes the application of most available anemometer techniques, but progress with the use of thermistors for this purpose has been reported recently (Fowlis [**21**]).

Three-dimensional temperature fields are comparatively easy to investigate and, following the use in the first annulus experiments of single thermocouples and of simple thermocouple arrays, numerous studies have now been reported, in some cases using very complex arrays in conjunction with sophisticated data acquisition systems (Ketchum [**44**], Pfeffer and Fowlis [**58**], Fowlis and Pfeffer [**23**], Kaiser et al. [**42**]). Thermistors have also been used in recent work (Fowlis [**21**]), but Toepler-schlieren techniques and various chemical indicators have not yet been exploited.

Accurate direct determinations of total heat transfer and of local heat transfer over limited areas of the bounding surfaces are also comparatively easy to make, provided that adequate precautions are taken to eliminate spurious heat losses (Hide [**29**]–[**32**], Bowden [**4**], Bowden and Eden [**6**], Uryu et al. [**73**]), but preliminary attempts to measure net and local momentum transfer directly have not yet proved successful.

High-speed electronic computers will attain in due course the speed and size required to repeat and extend the laboratory experiments by direct integration of the hydrodynamical equations. From the detailed three-dimensional velocity and temperature fields that will be forthcoming from such studies it will be possible to answer questions of the kind that many dynamical meteorologists—preoccupied as they usually are with atmospheric problems of incredible complexity—tend to put to the laboratory workers. The numerical work of Piacsek [**60**], Quon [**62**], and Williams [**75**]–[**78**] provides a foretaste of what the use of computers in this connection might accomplish in the coming years.

4. Axisymmetric flows. The structure of the temperature and velocity fields and other characteristics of the simplest of the four basic flow types described in §2—axisymmetric flow—have been studied in a number of experimental (Hide [**29**]–[**32**], Smith [**68**], Fultz et al. [**26**], Bowden [**4**], Bowden and Eden [**5**], Ketchum [**44**], Kaiser [**41**], Kaiser et al. [**42**], Hide and Mason [**39**]), theoretical (Robinson [**65**], Hide [**36**], Hunter [**37**], McIntyre [**53**], Brindley [**8**], see also Kreith [**46**]) and numerical (Piacsek [**60**], Quon [**62**], Williams [**75**]–[**78**]) investigations. These studies not only lead to insight into the hydrodynamical processes taking place, especially in the complicated boundary layers present on the side-walls of the system and in the free shear layers that appear, under certain circumstances, in the main body of the fluid, but they constitute a necessary preliminary to the stability analyses required to account for the occurrence of non-axisymmetric flow when $\Theta_L < \Theta < \Theta_R$ (see Equations (2) and (5)).

Introduce three "internal dimensionless parameters" N, σ_z and σ_r (see Hide [**37**]). N, the Nusselt number, is defined as the total rate of heat transfer by convection and conduction (and radiation) divided by the rate at which conduction (and radiation) alone would transfer heat if the fluid were replaced by a solid with the same thermal properties. σ_z is defined as the spatial average value of the vertical temperature gradient, $\partial T/\partial z$, divided by $|\Delta T|/d$, where $|\Delta T|$ is a measure of the impressed horizontal temperature contrast in the system, equal to $|T_a - T_b|$ in the case of wall-heating, and σ_r is defined as the spatial average value of the magnitude of the radial temperature gradient, $\partial T/\partial r$, outside the side-wall boundary layers divided by $|\Delta T|/(b - a)$.

Two further "internal dimensionless parameters" of theoretical significance (see Equation (C23)) are:

$$R \equiv g\alpha\sigma_r|\Delta T|d/4\Omega^2(b - a)^2 \tag{6}$$

and

$$B \equiv g\alpha\sigma_z|\Delta T|d/4\Omega^2(b - a)^2 \tag{7}$$

where α is the thermal coefficient of cubical expansion. (Because α depends on T it is preferable in practice to define σ_z, σ_r, R and B in terms of density rather than temperature.) By the thermal wind equation, R, a thermal Rossby number, is the mean azimuthal flow divided by $2\Omega(b - a)$, which is typically much less than unity, even when $\Theta \simeq 1$ (cf. Equation (2) and see Hide [**37**], especially the footnote on p. 61). B is the dimensionless measure of the vertical stability of the fluid that appears in the theory of baroclinic instability (see Appendix C; cf. Phillips [**59**]).

In the theory of the thermal structure and heat transfer, the external dimensionless parameter

$$\Pi \equiv g\alpha|\Delta T|\nu^{1/2}/8\kappa\Omega^{3/2} \tag{8}$$

plays a crucial role (see Hide [**37**], Equation (1.5)). Π has the character of a Péclét number. When $\Pi \ll \Pi_c$ (where Π_c is sensitive to the upper-surface boundary condition) thermal conduction is the dominant heat transfer process and convection can be treated as a small perturbation. Under these circumstances it is possible to determine from the equations of motion, etc., exact expressions for the small departures of N and σ_r from unity and σ_z from zero for axisymmetric flow (Robinson [**65**], Hunter [**40**], Hide [**37**]). Unfortunately, when Π is large, as in the case of the laboratory experiments, convection is the dominant heat transfer process and therefore the governing equations are nonlinear and cannot be solved analytically. Hide [**37**], without attempting to solve the equations, advanced heuristic arguments leading to the following approximate expressions for σ_z, σ_r and N:

(9a) $$\sigma_z \doteqdot 0.67 \qquad (\text{when } \Pi \gg \Pi_c)$$

and

(9b, c) $$\sigma_r \doteqdot 0.33 \qquad \text{and} \qquad N = (1 + 2\Pi/3)(1 + 3/\Pi)/3$$
$$(\text{when } \Pi \gg \Pi_c \text{ and } (b - a)\Omega^{1/2}/\nu^{1/2} > \Pi \gg 1).$$

The expression for σ_z agrees remarkably well and the expressions for σ_r and N moderately well with laboratory and numerical investigations (see Hide [37], Ketchum [**44**], Kaiser et al. [42], Hide and Mason [39]). Improvements in the expressions given by Equations (9b, c) will require, in the first instance, a theory of the detailed structure of the side-wall boundary layers [**53**].

(a) STABILITY OF AXISYMMETRIC FLOW. In his theoretical treatment of baroclinic instability Eady [**16**] (see Appendix C) effectively considered the following problem. A rapidly-rotating inviscid fluid subject to steady horizontal and vertical temperature gradients, $\partial T_0/\partial r$ and $\partial T_0/\partial z$, occupies an annular region $-\frac{1}{2}d \leqq z \leqq \frac{1}{2}d$ and $a \leqq r \leqq b$, the sign of $\partial T_0/\partial z$ being such that the density stratification is stable.

Relative to a frame of reference rotating with the angular velocity ($\mathbf{\Omega} = (0, 0, \Omega)$) of the bounding surfaces there is a basic axisymmetric hydrodynamical flow which is entirely azimuthal when the "thermal Rossby number,"

(10) $$R \equiv gd\alpha(\partial T_0/\partial r)/4\Omega^2(b - a)$$

(cf. Equation (6)), tends to zero. According to the thermal wind equation (see Equation (C23)) this zonal flow varies with z at a rate proportional to $\partial T_0/\partial r$.

It is important to emphasize the simplifying assumptions implicit in Eady's baroclinic instability model before comparing predictions based on the model with results of the laboratory experiments. These assumptions are:

(i) The width of the annulus is much less than its mean radius (i.e. $(b - a) \ll \frac{1}{2}(b + a)$), so that geometrical effects due to curvature can be neglected.

(ii) $\Omega^2 r$ is so very much less than g that geopotential surfaces do not depart significantly from z = constant.

(iii) Viscous effects are negligible, so that ν can be set equal to zero.

(iv) Thermal conduction (and radiation) are negligible, so that κ can be set equal to zero.

(v) The basic steady horizontal density gradient is independent of position, so that the vertical shear of the basic azimuthal flow is uniform.

(vi) The basic steady vertical density gradient is negative (i.e. stabilizing) and independent of position.

(vii) The thermal coefficient of cubical expansion, α, is independent of temperature, T.

(viii) The basic flow is quasi-geostrophic (i.e. $R \ll 1$, see Equation (10)).

(ix) The slope of the isotherms in the basic state is of order R multiplied by $d/(b - a)$, i.e

$$[(b - a)/d]\,|(\partial T_0/\partial r)/(\partial T_0/\partial z)| = O(R). \tag{11}$$

(x) The amplitude of the wave-like disturbance is infinitesimal, so that linear perturbation theory applies.

(xi) The growth rate of the disturbance is sufficiently slow for quasi-geostrophic balance to hold throughout its development.

Under these assumptions baroclinic waves of wavenumber m (around the annulus) should grow at the expense of the basic state provided that

$$B \equiv g\alpha(\partial T_0/\partial z)\, d^2/4\Omega^2(b - a)^2 \tag{12}$$

(cf. Equation (7)) is less than a certain critical value B_m (say), where B_m depends only on the quantity

$$\Sigma \equiv m(b - a)/\pi(b + a), \tag{13}$$

the ratio of $(b - a)$ to the wavelength $\pi(b + a)/m$. B_m increases monotonically with decreasing Σ, attaining the value 0.581 when $m = 0$ (see Equation (C60)). Hence, if

$$B > 0.581 \tag{14}$$

then the basic axisymmetric state is stable to nonaxisymmetric disturbances of the form considered.

Lorenz [**49**] was the first to suggest that the upper symmetrical regime found in the laboratory experiments arises when the vertical stability in the axisymmetric state is sufficiently high to suppress baroclinic instabilities. It is instructive to compare the stability criterion given by Equation (14) with the experimental criterion for the transition from the upper-symmetric regime to the nonaxisymmetric regime when $\mathscr{T}$ is very large. By Equations (2), (7) and (9a), if $\Theta_R \doteqdot 4$ (see §2) the corresponding value of B is roughly 0.7, which falls midway between the theoretical value, 0.581, and a direct experimental determination of 0.8 based on Ketchum's [**44**] measurements of the internal temperature field in a very large annulus. (The small discrepancy between the two sets of experimental values might be more apparent than real, since the former is based largely on visual observations and the latter on thermocouple traces.)

The comparison of the results of a theoretical study by Davies [**13**] with Eady's analysis indicates that the neglect of curvature effects should not be too serious (see Assumption (i)). Quantitative errors are to be expected from the neglect of centrifugal effects (see Assumption (ii)) but the experimental result that the principal effect of changing the sign of the impressed temperature gradient is the reversal of the sense of flow indicates that centrifugal effects are qualitatively unimportant.

Assumption (iii), that viscous effects can be neglected, is certainly not valid at the lowest values of $\mathscr{T}$ (see Equation (4)) covered by the experiments. It may be shown (see Appendix C) that the inclusion of almost any kind of frictional effects in Eady's baroclinic instability theory can account qualitatively for the general anvil shape of the transition curve in the Θ versus $\mathscr{T}$ diagram, in keeping with the work of Brindley [7], who took friction into account in the boundary layers at the rigid lower surface and free upper surface (see Hide [**33**], [**34**]), Barcilon [**1**], who considered the case of a rigid rather than free upper surface, and of Kuo [**47**], Lorenz [**50**], and Merilees [**54**], who treated two-layer models and were therefore obliged to introduce friction in a somewhat artificial way (cf. Davies [**14**]). Unfortunately no theory capable of accounting for the critical value of $\mathscr{T}_p$ and the shape of the lower transition curve, where $\Theta = \Theta_L$ (see Equations (4b) and (5)), has yet been advanced. Future theories of the transition at low values of $\mathscr{T}$ and Θ (see Fowlis and Hide [**22**]) should take side-wall friction and internal friction into account, as well as thermal conduction (cf. Assumption (iv)), the neglect of which is probably not too bad a supposition in the theory of the *upper* transition.

Assumptions (v) to (viii) are certainly not strictly valid. Temperature measurements within the convecting system show that $\partial T_0/\partial z$ and $\partial T_0/\partial r$ vary strongly with z and r (see Hide [**29**]–[**32**], Smith [**68**], Bowden and Eden [**5**], Ketchum [**44**], Kaiser et al. [**42**]), and in the work of Hide and Mason [**39**], very big variations in $\partial T_0/\partial r$ were deliberately introduced by the use of internal heating; the corresponding azimuthal velocity fields possess nonuniform shear with respect to both r and z. Pedlosky [**55**] has given a formal theory of the effect on Eady's stability criterion of small departures from linearity of the vertical gradient and from zero of the horizontal gradient of the basic azimuthal velocity, but the theory is evidently vitiated by an error [**79**].

Assumption (ix) implies that the essential heat balance is between horizontal and vertical advection (see Hide [**36**]); when the slope of the isotherms exceeds the value given by Equation (11) the differential equation satisfied by the pressure field (cf. Equation (C49)) is no longer separable and serious mathematical difficulties arise in the theory. In the experiments the isothermal slope is indeed significantly greater than $Rd/(b - a)$, but it is just possible that the small contribution of thermal conduction might, when $d/(b - a)$ is typically greater than 2 or 3 as in many of the experiments, play a significant role in diminishing any discrepancy between theory and experiment expected on this particular count. Experiments with small values of $d/(b - a)$, of which there have been comparatively few to date, are now being carried out to clarify this and related points.

The good agreement between theory and experiment in regard to the onset of baroclinic instability when $\mathcal{T} \to \infty$ indicates that finite amplitude effects are not too important in the theory of the transition (cf. Assumption (x)). (Such effects are, of course, important in the theory of the development of baroclinic waves when they are no longer small, as evinced by the poor agreement between the theoretical value of m for the most rapidly growing incipient wave and the observed value of m for the fully-grown wave, see §5 below.) The observed growth rates are much less than Ω, in keeping with Assumption (xi).

(b) RADIAL BARRIER EXPERIMENTS. The flow due to an axisymmetric and z-independent arrangement of fluid sources and sinks in an annulus of barotropic fluid is strongly affected by rotation (Hide [**35**], [**38**]). Though purely radial when $\Omega = 0$, the flow develops azimuthal components when $\Omega \neq 0$, and when Ω is large, this azimuthal motion becomes so pronounced that it inhibits radial flow outside boundary layers near $r = a$, $r = b$, $z = z_u$ and $z = z_l$. However (provided that $z_u - z_l$ is constant), the insertion of a thin rigid radial barrier blocking the whole of the cross-section of the annulus has the remarkable but predictable and verifiable effect (Hide [**38**]) of changing $\boldsymbol{u}$ in the main body of the fluid to that found when $\Omega = 0$. The pressure field, of course, is not unaffected by Ω; associated with the radial motion is an azimuthal pressure gradient $\partial p/\partial \phi$ proportional to $2\Omega(\boldsymbol{u})_r$, supported by the pressure difference between opposite sides of the barrier.

While the foregoing considerations apply strictly to barotropic fluids, they led the writer to suggest that the effect of a radial barrier on the axisymmetric regime of thermal convection in a rotating fluid annulus might be to increase the value of N and decrease the value of σ_z (see Hide [**37**], especially footnote on p. 65) to their nonrotating values if the flow were to remain stable. These conjectures were subsequently confirmed by the work of Bowden [**4**] and Bowden and Eden [**6**] on heat transfer (which also included an investigation of the form of instabilities that arise in the new system at sufficiently large values of Ω, see also Bless [**3**]), and by the work of Kester [**43**] on the effect of the radial barrier on σ_z. (Kester also investigated the dependence of σ_z on the width of the gap between the $z = z_u$ and the top of a partial barrier extending down to $z = z_l$.) Further experiments with radial barriers will be of great interest in their own right and might bear on the theory of the thermohaline circulation of the oceans.

5. Nonaxisymmetric flows. Nonaxisymmetric flows have been studied in a variety of experimental (Hide [**29**], [**30**], Smith [**68**], Riehl and Fultz [**64**], Fultz et al. [**26**], Fultz and Kaylor [**25**], Bowden [**4**], Fowlis and Hide [**22**], Hide [**35**], Uryu et al. [**73**], Pfeffer and Chiang [**56**], Ketchum [**44**], Koschmieder [**45**], Pfeffer and Fowlis [**58**], Fowlis and Pfeffer [**23**], Hide and Mason [**39**]), theoretical (Rogers [**66**], [**67**], Davies [**13**], Lorenz [**50**], [**51**]), and numerical (Williams [**78**]) investigations, but, owing to their complex structure, especially the irregular flows (see Figure 1), what has been accomplished to date is slight in comparison with what will be required to gain a satisfactory detailed description of the spatial and temporal characteristics of these flows and to elucidate the underlying hydrodynamical processes.

The sheer complexity of most of the experimental results and the paucity of quantitative theory with which the results can be compared renders impossible the task of summarizing these results in an interesting way. The reader is referred, therefore, to the original papers and to forthcoming publications reporting experimental determinations of quantities of the kind that meteorologists find useful in their work on the global atmospheric circulation (e.g. available potential energy, et cetera, see Lorenz [**52**], Pfeffer et al. [**57**], Dutton and Johnson [**15**], Ketchum [**44**]). Only a few comparatively straightforward results will be discussed in what follows in the remainder of this section.

(a) Steady Flows. The most extensive investigations of steady waves to date have been concerned with the dependence of m, the number of waves, on the external dimensionless parameters. It has been found ([**29**]–[**32**]) that m, which takes only discrete integral values, is not uniquely determined by the external dimensionless parameters Θ, $\mathscr{T}$, et cetera (see Equations (1) and (4)) but only statistically so, a property of the waves that is undoubtedly related to the hysteresis effects exhibited by transitions between wavenumbers when the external dimensionless parameters are varied very slowly [**26**].

Denote by $\hat{m}$ the most probable value of m, as determined by repeatedly destroying the wave pattern by vigorous stirring and allowing the pattern to reform. When $\mathscr{T}$ exceeds about $2\mathscr{T}_p$, for fixed values of $(b - a)/\frac{1}{2}(b + a)$ the quantity $\hat{m}$ depends largely on Θ; $\hat{m}$ increases with decreasing Θ in a manner that agrees qualitatively, but not quantitatively, with Eady's baroclinic instability theory (see §4 and Appendix C). The range of m depends on $(b - a)/\frac{1}{2}(b + a)$ in a very simple way; within the limitations imposed on m by the fact that it must be an integer the highest and lowest values of m, namely $m_{\max}$ and $m_{\min}$, satisfy the empirical formulae:

$$m_{\min} \doteqdot 0.25\pi(b + a)/(b - a), \qquad m_{\max} \doteqdot 0.7\pi(b + a)/(b - a) \tag{15}$$

(cf. Equation (13)). Complications, which have not yet been fully investigated, arise when $\mathcal{T} < 2\mathcal{T}_p$; for details see [**22**], [**26**].

Certain broad characteristics of the steady wave flow pattern can be understood by treating the jet stream found in the experiments as a quasi-geostrophic detached thermal boundary layer (Hide [**29**]–[**32**], Hide and Mason [**39**]). Rogers [**66**]–[**67**] and Davies [**14**] have examined the theory of the structure of such boundary layers, a very difficult problem, with some qualitative success, but values of N, the heat transfer coefficient (see §4), based on Davies's theoretical work are much less than empirical values of that quantity (Hide [**29**]–[**32**], Bowden [**4**], Hide and Mason [**39**]) which typically range from 3 to 9.

According to these heat transfer experiments (see also Hide [**37**]), N shows no systematic dependence on Ω (in contrast to the axisymmetric regime, for which N decreases with increasing Ω, see Equation (9c)). At the transition from axisymmetric flow to steady waves, N undergoes an abrupt increase of about 20 percent. Within the steady waves regime N satisfies the empirical law

$$N \doteqdot C[g\alpha|\Delta T|(b - a)^3/\kappa\nu]^{1/4} \tag{16}$$

where C is nearly independent of the other parameters and equal to 0.164 ± 0.004. For a nonrotating annulus N satisfies a similar equation but with $C = 0.203 \pm 0.010$.

Measurements of the rate at which steady waves drift relative to the rotating apparatus are consistent with the hypothesis that (when $z_u - z_l$ is constant) this rate is equal to the mean azimuthal flow velocity, a result which agrees with baroclinic instability theory, see Equation (C59).

(b) Vacillation and Irregular Flows. Of the basic flow types found in the experiments, vacillation and irregular flow are undoubtedly the most interesting to meteorologists, and a considerable amount of effort, especially by Pfeffer and his colleagues, is now being directed towards the detailed description of these phenomena and to making comparisons with the atmosphere and with numerical models of the kind investigated by Phillips, Leith, Mintz, Smagorinsky and others (see Lorenz [**52**]). Lorenz's [**51**] theory of nonlinear effects in baroclinic instability provides valuable insight into the processes underlying regular and irregular time-dependent flows, but as the theory is qualitative, useful quantitative comparisons with experiments cannot be made.

The conditions under which the different types of vacillation (see Figure 1 and Appendix B) occur have not yet been fully delineated (but see Hide [**29**]–[**32**], Pfeffer and Chiang [**56**], Pfeffer and Fowlis [**58**], Fowlis and Pfeffer [**23**]). Wave-form vacillation occurs largely when $m = m_{\max}$ as a transitional phenomenon between steady waves and irregular flows and wave-amplitude vacillation is probably most pronounced near the upper transition, where $m = m_{\min}$.

Of obvious importance is the dependence of the period of vacillation on other parameters. Under certain conditions at least this period seems to be equal to the wave-drift period multiplied by the ratio of two small integers, indicating that vacillation might be a manifestation of interactions between different modes (Hide [**29**]–[**32**]).

The dominant wavenumber of irregular flow is equal to $m_{\max}$, but the complexity of the flow is clearly due to the presence of many other modes. If, as seems likely, the virtual independence of N on Ω within the steady-waves regime is due to the tendency for m to increase with increasing Ω (thus giving the system a "degree of freedom" with which to counteract the inhibiting influence of Coriolis forces on radial heat transfer), then in the irregular flow regime, where the dominant wavenumber is fairly insensitive to Ω, N should decrease with increasing Ω; such behaviour is consistent with the results of Bowden's [**4**] heat transfer experiments.

(c) EFFECTS DUE TO NONAXISYMMETRIC BOUNDARY CONDITIONS. In carrying out experiments using systems that are thermally and mechanically symmetrical about the axis of rotation, it is necessary of course to design the apparatus in such a way as to avoid spurious effects due to departures from axial symmetry in the boundary conditions. Only a few, often brief and mostly unpublished, careful studies have been carried out on effects due to obstacles or azimuthal temperature gradients on the bounding surfaces. It is known, for instance (see Hide [**35**]), that a sizeable solid object placed in the annulus will deform steady baroclinic waves locally without otherwise changing their properties (cf. Fultz and Spence [**27**]) and that (according to Fowlis and Pfeffer) a slight increase in the apparent viscosity of the working liquid is the principal effect of the presence within the fluid of a complicated array of probes (notwithstanding the dramatic effects of a full radial barrier, as outlined in §4). Thus, systematic investigations of effects due to deliberately introduced and accurately specified departures from axial symmetry in the mechanical and thermal boundary conditions have not yet been taken very far. In the light of present knowledge of the behaviour of axisymmetric systems and with the aid of modern techniques (see §3), quite sophisticated quantitative investigations of nonaxisymmetric systems are now feasible. Such experiments could be of both intrinsic and meteorological importance.

Appendix A : Models. As the use of the term "model" has occasionally led to needless confusion about the role and scientific objectives of laboratory experiments in relation to problems in geophysical fluid dynamics, the meaning of the word deserves a brief discussion. In everyday life a "model" is "an imitation of something on a smaller scale," "imitation" being "that which is produced as a copy, or counterfeit," but this definition is not very useful here. Theoreticians make extensive use of the term to mean "a conceptual idealization of an actual (usually physical) system constructed with the objective of determining through the analysis of the governing mathematical equations the behaviour of the idealized system, in the hope that the analysis will shed light on the behaviour of the actual system," but neither of the two well-known dictionaries of mathematical terms consulted by the lecturer attempted a definition of the word. Van Nostrand's *Scientific Encyclopedia* (1968) proposes that a "model" is a "small-scale reproduction of the prototype in all the factors which are pertinent to the investigation."

Some of the laboratory experiments might be considered models in the spirit of the last definition if the principal objective of the work is the discovery of the pertinent factors and not the manufacture of a finished product satisfying the definition to the letter; such a model would deservedly end up in a dark corner of a museum. But most of the experiments come closer to the mathematical type of model, with the techniques of the physics laboratory replacing those of mathematical or numerical analysis.

The idea of a laboratory model is closely linked with dimensional and similarity analysis, with which physicists have long been familiar and which meteorologists are now gradually exploiting in their studies. A cogent account of dimensional analysis in the context of fluid dynamics is given in Birkhoff's [**2**] important book, but even there it is assumed that the term "model" is understood by the reader when the author remarks that

> the use of models to study fluid dynamics has an appeal for everyone endowed with natural curiosity.... And yet in few departments of the physical sciences is there a wider gap between (scientific research) and engineering practice than in the use of models to study hydrodynamical phenomena. (Scientists) tend to gloss over uncomfortable facts that do not fit nicely into simple logical theory, whereas engineers, constantly faced with reality ... are usually too engrossed with technical special problems to enter the arena of (scientific) controversy. It is easier to pay lip service to current theories, relying on experience and judgment for the solution of design problems.

Appendix B : Vacillation. The term "vacillation" is now established in the literature and geophysicists occasionally wonder how it first came to be used.

During the course of the first annulus experiments, the writer was largely concerned initially with the study of steady waves and especially with the determination of the conditions under which such waves can occur. The procedure followed in a typical experiment involved first setting at pre-determined values the various quantities (e.g. Ω, d, T_a, T_b, etc.) required to specify the impressed experimental conditions and then waiting until effects due to the starting-up processes had disappeared before carrying out systematic measurements of various quantities. Judging when this final state had been reached was quite easy in the cases of axisymmetric flow and steady waves, but even within what eventually came to be called the "steady-waves regime," under certain conditions (which have not yet been fully elucidated) the flow pattern exhibited regular periodic time-variations. Typical of these variations were pulsations in amplitude (which, at their most pronounced, were accompanied by changes in wavenumber from one cycle to the next), the regular progression around the wave pattern of a sizeable distortion (often amounting to the complete splitting of one of the waves) and wavering of the shape of the flow pattern. To the most pronounced form of wavering, which was found to occur near the transition to irregular flow, the descriptive name "vacillation" was given. (It was, of course, found necessary to establish experimentally that vacillation and the irregular flow were real phenomena associated with the higher values of Ω used in experiments, not spurious effects.)

Through usage and by consensus "vacillation" now includes not only the pronounced wavering phenomenon to which the name was originally applied, but also all other flows characterized by regular periodic temporal variations. Thus, it subsequently became necessary to subdivide vacillation phenomena into several different types, each characterized by the nature of the most prominent variations taking place (see Figure 1). When in due course further experimental and theoretical work has led to a thorough understanding of these phenomena it will be possible to abandon the term "vacillation" in favour of more appropriate terms descriptive of the underlying hydrodynamical processes.

Appendix C: On the theory of baroclinic instability in a liquid of low viscosity. Of the original theoretical treatments of baroclinic instability in an inviscid fluid (Charney [**11**], Eady [**16**], Fjortoft [**20**], Sutcliffe [**72**]) and subsequent investigations (see [**9**], [**79**]–[**81**] for an extensive list of references) the problem considered by Eady and defined in §4 seems the most relevant to the experiments at high values of $\mathscr{T}$ (i.e. $\gg \mathscr{T}_p$ (see Equation (4b)). Extensions of Eady's work to the case of a nearly inviscid fluid (Brindley [7], Barcilon [**1**], Hide [**81**]) have not yet developed to the point of accounting quantitatively for the transition between axisymmetric and nonaxisymmetric flow when $\mathscr{T} \not\gg \mathscr{T}_p$, but they are instructive in several respects and they indicate the next steps required in the theory (see Appendix D). Eady's theory and the aforementioned extensions will be outlined in what follows, together with a discussion of the effects due to the presence of sloping upper and/or lower bounding surfaces. The simplifying assumptions underlying Eady's theory and listed in §4(a), will be referred to frequently in what follows as Assumption (i), et cetera.

(a) EQUATIONS OF THE PROBLEM. In ordinary units the equations of motion, continuity and heat flow of an incompressible Boussinesq fluid, for which the departure ρ of the density $(\bar{\rho} + \rho)$ from the mean density $\bar{\rho}$ is small and satisfies

$$\rho = -\bar{\rho}\alpha(T - \overline{T}), \tag{C1}$$

$\overline{T}$ being the mean temperature and α the thermal coefficient of cubical expansion, taken as uniform (see Assumption (vii)), are the following:

$$\partial \boldsymbol{u}/\partial t + (\boldsymbol{u} \cdot \nabla)\boldsymbol{u} + 2\boldsymbol{\Omega} \times \boldsymbol{u} = -(1/\bar{\rho})\nabla(p + \bar{p}) + \boldsymbol{g}(\rho + \bar{\rho})/\bar{\rho} + \nu\nabla^2\boldsymbol{u}, \tag{C2}$$

assuming that ν is independent of position,

$$\nabla \cdot \boldsymbol{u} = 0 \tag{C3}$$

and

$$\partial\rho/\partial t + \boldsymbol{u} \cdot \nabla\rho = \kappa\nabla^2\rho \tag{C4}$$

where t denotes time. It will be convenient to suppose that $\boldsymbol{g} = (0, 0, -g)$ (see Assumption (ii)) so that $\bar{p}$, by definition, satisfies

$$\partial\bar{p}/\partial z + g\bar{\rho} = 0. \tag{C5}$$

(b) BOUNDARY LAYERS. Introduce a local Cartesian system of co-ordinates (x, y, z) related, under Assumption (i), to the cylindrical polar co-ordinates introduced in §2 as follows:

$$(x, y, z) = (\tfrac{1}{2}(b + a)\phi, -(r - \tfrac{1}{2}(b + a)), z). \tag{C6}$$

Divide the cross-sectional area of the annulus into an interior region, occupying

$$z_l + \delta_l < z < z_u - \delta_u \quad \text{and} \quad -\tfrac{1}{2}(b - a) + \delta_b < y < \tfrac{1}{2}(b - a) - \delta_a,$$

thin side-wall boundary layers of thickness δ_a and δ_b and thin end-wall boundary layers of thickness δ_l and δ_u.

In this Appendix we consider those tractable cases when $\kappa = 0$ everywhere (see Assumption (iv)), $\nu = 0$ in the interior region and dissipative effects in the end-wall boundary layers are so much more important than in the side-wall boundary layers that it is possible to set $\delta_a = \delta_b = 0$ (cf. Appendix D). Clearly no model in which $\kappa = 0$ can account for the "lower transition" (see Equation (5)), which depends on κ. Less obvious (see below) is the upshot of the calculations to follow, namely that the neglect of internal and side-wall friction is an oversimplification which will have to be rectified in order to account for the level of dissipation implied by Equation (4b) for $\mathscr{T}_{\mathrm{p}}$.

(c) INTERIOR FLOW. Denote by $\boldsymbol{u}_0$, ρ_0 and p_0 the values of $\boldsymbol{u}$, ρ and p in the basic axisymmetric state, whose stability we are considering; $\boldsymbol{u}_0$, ρ_0 and p_0 are independent of x and t. Assume that ρ_0 varies linearly with y and z; thus

(C7) $$\rho_0 = y\partial\rho_0/\partial y + z\partial\rho_0/\partial z$$

where $\partial\rho_0/\partial y$ and $\partial\rho_0/\partial z$ are constants (see Assumptions (v) and (vi)). If $\boldsymbol{u} = (u, v, w)$ and $\boldsymbol{u}_0 = (u_0, v_0, w_0)$ then

(C8) $$v_0 = w_0 = 0,$$

(C9) $$2\Omega\frac{\partial u_0}{\partial z} = \frac{g}{\bar{\rho}}\frac{\partial\rho_0}{\partial y},$$

(C10) $$\partial p_0/\partial z = -g\rho_0, \qquad \partial p_0/\partial y = -2\Omega\bar{\rho}u_0$$

are solutions of Equations (C2)–(C4) that are compatible with the boundary conditions (see Equations (C19) and (C20) below). By Equation (C9)

(C11) $$u_0 = \left(\frac{g}{2\Omega\bar{\rho}}\frac{\partial\rho_0}{\partial y}\right)z + \bar{u}_0 = u_0(z)$$

where $\bar{u}_0 \equiv u_0\ (z = 0)$.

If we denote by $\boldsymbol{u}_0(z) + \boldsymbol{u}_1(x, y, z, t)$, $p_0(y, z) + p_1(x, y, z, t)$ and $\rho_0(y, z) + \rho_1(x, y, z, t)$ the perturbed values of $\boldsymbol{u}$, p and ρ and treat $\boldsymbol{u}_1$, p_1 and ρ_1 as small quantities compared with $\boldsymbol{u}_0$, p_0 and ρ_0 (see Assumption (x)), then by Equations (C2) to (C4) and (C8) to (C11),

(C12) $$\frac{du_1}{dt} + w_1\,Du_0 - 2\Omega v_1 = -\frac{1}{\bar{\rho}}\frac{\partial p_1}{\partial x},$$

(C13) $$\frac{dv_1}{dt} + 2\Omega u_1 = -\frac{1}{\bar{\rho}}\frac{\partial p_1}{\partial y},$$

(C14) $$\frac{dw_1}{dt} = -\frac{1}{\bar{\rho}}\frac{\partial p_1}{\partial z} - \frac{\rho_1}{\bar{\rho}}g,$$

(C15) $$\partial u_1/\partial x + \partial v_1/\partial y + \partial w_1/\partial z = 0,$$

and

(C16) $$d\rho_1/dt + v_1\, \partial\rho_0/\partial y + w_1\, \partial\rho_0/\partial z = 0,$$

where

(C17) $$d/dt \equiv \partial/\partial t + u_0(z)\, \partial/\partial x \qquad \text{and} \qquad D \equiv d/dz.$$

(d) BOUNDARY CONDITIONS. As there can be no flow normal to the rigid surfaces in $y = \pm\frac{1}{2}(b - a)$ and, by hypothesis, $\delta_a = \delta_b = 0$, the "interior" flow must satisfy

(C18) $$v_1 = 0 \qquad \text{when} \quad y = \pm\tfrac{1}{2}(b - a).$$

At a rigid surface in $z = z_u$ or z_l both the tangential as well as the normal components of $\boldsymbol{u}$ must vanish, while at a free surface the tangential component of the stress must vanish. Continuity of the normal component of stress is the other boundary condition that must be satisfied at a free surface, but it is often convenient to follow Lord Rayleigh's treatment of the Bénard convection problem and require, instead, that the normal component of $\boldsymbol{u}$ should vanish at a free surface (see Chandrasekhar [**10**]). It is a straightforward application of the theory of Ekman boundary layer suction (see Prandtl [**61**], Hide [**33**], Greenspan [**28**]) to show that these requirements lead to the following boundary conditions on the "interior flow":

$$w_1 = \Gamma_l^* v_1 + \tfrac{1}{2}(\nu/\Omega)^{1/2}(\partial v_1/\partial x - \partial u_1/\partial y)$$
$$\text{on} \quad z = -\tfrac{1}{2}d \quad \text{(lower surface rigid)}$$

(C19) or

$$w_1 = \Gamma_l^* v_1 - \tfrac{1}{2}(\nu/\Omega)(\partial/\partial z)(\partial v_1/\partial x - \partial u_1/\partial y)$$
$$\text{on} \quad z = -\tfrac{1}{2}d \quad \text{(lower surface free)}$$

and

$$w_1 = \Gamma_u^* v_1 - \tfrac{1}{2}(\nu/\Omega)^{1/2}(\partial v_1/\partial x - \partial u_1/\partial y)$$
$$\text{on} \quad z = \tfrac{1}{2}d \quad \text{(upper surface rigid)}$$

(C20) or

$$w_1 = \Gamma_u^* v_1 - \tfrac{1}{2}(\nu/\Omega)(\partial/\partial z)(\partial v_1/\partial x - \partial u_1/\partial y)$$
$$\text{on} \quad z = \tfrac{1}{2}d \quad \text{(upper surface free).}$$

Here it is assumed that the slopes of the upper and lower surfaces are uniform and small, that is to say

(C21) $$z_l = -\tfrac{1}{2}d + \Gamma_l^* y, \qquad z_u = \tfrac{1}{2}d + \Gamma_u^* y,$$

where Γ_u^* and Γ_l^* are constants that are very much less than $(b - a)/d$. The problem solved by Eady corresponds to the case $\Gamma_u^* = \Gamma_l^* = \nu = 0$; Brindley's problem is the case $\Gamma_u^* = \Gamma_l^* = 0$ but $\nu \neq 0$ with free upper surface and rigid lower surface, and Barcilon's problem the case $\Gamma_u^* = \Gamma_l^* = 0$ but $\nu \neq 0$ with both upper and lower surfaces rigid. (Surface tension affects the structure of free-surface boundary layers but disappears from the boundary layer suction formula (Hide [**34**]).)

(e) Dimensionless Parameters. At this point it is convenient (1) to scale the foregoing equations by changing to the following units: $(b - a)$ for horizontal distance, d for vertical distance,

(C22) $$U_0 \equiv gd(\partial\rho_0/\partial y)/2\Omega\bar{\rho}$$

for horizontal velocity (assuming that $\partial\rho_0/\partial y > 0$, cf. Equation (C9)), $U_0 d/(b - a)$ for vertical velocity, $(b - a)/U_0$ for time, $2\Omega U_0\bar{\rho}(b - a)$ for pressure and $\bar{\rho}(b - a)2\Omega U_0/gd$ for ρ (the difference between the actual density and the mean density), and (2) to introduce the dimensionless parameters:

(C23) $$R \equiv gd(\partial\rho_0/\partial y)/4\Omega^2\bar{\rho}(b - a) \quad \text{and}$$
$$B \equiv -gd^2(\partial\rho_0/\partial z)/4\Omega^2\bar{\rho}(b - a)^2$$

(both essentially positive, see Equations (10) and (12)),

(C24) $$\Gamma_l \equiv \Gamma_l^*(b - a)/d \qquad \text{and} \qquad \Gamma_u \equiv \Gamma_u^*(b - a)/d$$

(cf. Equation (C21)), and

(C25) $$\mathscr{E} \equiv \tfrac{1}{2}(\nu/\Omega d^2)^{1/2}$$

(cf. Equations (C21) and (4a)).

The scaled Equations (C7) to (C11) governing the basic flow are:

(C26) $$\rho_0 = y - Bz/R,$$

(C27) $$v_0 = w_0 = 0,$$

(C28) $$Du_0 = \partial\rho_0/\partial y = 1,$$

(C29) $$\partial p_0/\partial z = -\rho_0, \quad \partial p_0/\partial y = -u_0;$$

and the corresponding equations for the perturbed flow (cf. Equations (C12) to (C17)) and the boundary conditions (cf. Equations (C18) to (C21)) are:

(C30) $$R\left[\frac{du_1}{dt} + w_1\right] - v_1 = -\frac{\partial p_1}{\partial x},$$

(C31) $$R\frac{dv_1}{dt} + u_1 = -\frac{\partial p_1}{\partial y},$$

(C32) $$R\frac{dw_1}{dt} = -\frac{\partial p_1}{\partial z} + \rho_1,$$

(C33) $$\partial u_1/\partial x + \partial v_1/\partial y + \partial w_1/\partial z = 0,$$

and

(C34) $$d\rho_1/dt + v_1 - Bw_1/R = 0,$$

where

(C35) $$d/dt \equiv \partial/\partial t + (z + \bar{u}_0)\partial/\partial x, \qquad D \equiv d/dz,$$

and

(C36) $$v_1 = 0 \quad \text{when} \quad y = \pm\tfrac{1}{2},$$

(C37) $$w_1 = \Gamma_l v_1 + \mathscr{E}(\partial v_1/\partial x - \partial u_1/\partial y) \quad \text{on} \quad z = -\tfrac{1}{2} \quad \text{(lower surface rigid)}$$
or
$$w_1 = \Gamma_l v_1 - 2\mathscr{E}^2\partial(\partial v_1/\partial x - \partial u_1/\partial y)/\partial z \quad \text{on} \quad z = -\tfrac{1}{2} \quad \text{(lower surface free)}$$

and

(C38) $$w_1 = \Gamma_u v_1 - \mathscr{E}(\partial v_1/\partial x - \partial u_1/\partial y) \quad \text{on} \quad z = \tfrac{1}{2} \quad \text{(upper surface rigid)}$$
or
$$w_1 = \Gamma_u v_1 - 2\mathscr{E}^2\partial(\partial v_1/\partial x - \partial u_1/\partial y)/\partial z \quad \text{on} \quad z = \tfrac{1}{2} \quad \text{(upper surface free).}$$

It follows directly from Equations (C30), (C32) and (C33) that

(C39) $$\partial w_1/\partial z = R\{d(\partial v_1/\partial x - \partial u_1/\partial y)/dt - \partial w_1/\partial y\}.$$

This equation shows that for quasi-geostrophic perturbations [i.e. those having "growth times" not less than unity in order of magnitude in the dimensionless units (or $(b - a)/U_0$ in ordinary units) in which case $d/dt = O(R^0)$ see Assumption (xi)] and to which attention will be confined in the remainder of this appendix,

(C40) $$w_1 = O(R).$$

(f) Expansion in Rossby Number Series. As, by hypothesis, the basic (as well as the perturbed) flow is quasi-geostrophic, $R \ll 1$ (see Equation (C23)). Introduce the following series expansions (cf. Stern [70]):

$$\text{(C41)} \quad \begin{aligned} u_1 &= u_{(0)} + u_{(1)}R + u_{(2)}R^2 \ldots, \quad v_1 = v_{(0)} + v_{(1)}R + v_{(2)}R^2 \ldots \\ w_1 &= w_{(0)}R + w_{(1)}R^2 + w_{(2)}R^3 \ldots, \quad p_1 = p_{(0)} + p_{(1)}R + p_{(2)}R^2 \ldots \end{aligned}$$

and

$$\rho_1 = \rho_{(0)} + \rho_{(1)}R + \rho_{(2)}R^2 \ldots$$

(cf. Equation (C39)) where the coefficients are of order unity, and suppose, in accordance with Equation (11), that

$$\text{(C42)} \qquad B = O(R^0).$$

Substitute these expansions in Equations (C30) to (C34) and show, by equating terms of order R^0, that:

$$\text{(C43)} \quad \begin{aligned} &v_{(0)} = \partial p_{(0)}/\partial x, \qquad u_{(0)} = -\partial p_{(0)}/\partial y, \qquad \partial p_{(0)}/\partial z = -\rho_{(0)}, \\ &\partial u_{(0)}/\partial x + \partial v_{(0)}/\partial y = 0, \qquad d\rho_{(0)}/dt + v_{(0)} - Bw_{(0)} = 0, \end{aligned}$$

the corresponding boundary conditions (see Equations (C36) to (C38)) being:

$$\text{(C44)} \qquad v_{(0)} = 0 \quad \text{on} \quad y = \pm\tfrac{1}{2}$$

$$\text{(C45)} \quad \begin{aligned} &w_{(0)} = [\Gamma_l v_{(0)} + \mathscr{E}(\partial v_{(0)}/\partial x - \partial u_{(0)}/\partial y)_{\text{J}}/R \\ &\qquad\qquad \text{on} \quad z = -\tfrac{1}{2} \quad \text{(lower surface rigid)} \\ \text{or} \quad &w_{(0)} = [\Gamma_l v_{(0)} - 2\mathscr{E}^2(\partial(\partial v_{(0)}/\partial x - \partial u_{(0)}/\partial y)/\partial z)]/R \\ &\qquad\qquad \text{on} \quad z = -\tfrac{1}{2} \quad \text{(lower surface free)} \end{aligned}$$

and

$$\text{(C46)} \quad \begin{aligned} &w_{(0)} = [\Gamma_u v_{(0)} - \mathscr{E}(\partial v_{(0)}/\partial x - \partial u_{(0)}/\partial y)]/R \\ &\qquad\qquad \text{on} \quad z = \tfrac{1}{2} \quad \text{(upper surface rigid)} \\ \text{or} \quad &w_{(0)} = [\Gamma_u v_{(0)} - 2\mathscr{E}^2(\partial(\partial v_{(0)}/\partial x - \partial u_{(0)}/\partial y)/\partial z)]/R \\ &\qquad\qquad \text{on} \quad z = \tfrac{1}{2} \quad \text{(upper surface free).} \end{aligned}$$

By Equation (C43) the zeroth-order solution corresponds to strictly geostrophic flow in hydrostatic equilibrium in the vertical, and it is necessary, therefore, to go to first order in R in order to discuss the stability problem. Thus, by Equations (C30) to (C34) and (C41) to (C42):

$$\text{(C47)} \quad \begin{aligned} &du_{(0)}/dt - v_{(1)} = -\partial p_{(1)}/\partial x, \qquad dv_{(0)}/dt + u_{(1)} = -\partial p_{(1)}/\partial y, \\ &\partial p_{(1)}/\partial z = -\rho_{(1)}, \qquad \partial u_{(1)}/\partial x + \partial v_{(1)}/\partial y + \partial w_{(0)}/\partial z = 0, \\ &d\rho_{(1)}/dt + v_{(1)} - Bw_{(1)} = 0. \end{aligned}$$

Eliminate $u_{(1)}$, $v_{(1)}$ and $\rho_{(1)}$ between Equation (C47) and find

$$\text{(C48)} \qquad \frac{\partial w_{(0)}}{\partial z} = \frac{d}{dt}\left(\frac{\partial^2}{\partial x^2} + \frac{\partial^2}{\partial y^2}\right)p_{(0)}$$

which, when combined with the results of eliminating $u_{(0)}$, $v_{(0)}$ and $\rho_{(0)}$ between Equation (C43), namely

$$\text{(C49)} \qquad \frac{\partial w_{(0)}}{\partial z} = -\frac{1}{B}\frac{d}{dt}\frac{\partial^2 p_{(0)}}{\partial z^2},$$

leads to the partial differential equation for $p_{(0)}$:

$$\text{(C50)} \qquad \left(\frac{\partial}{\partial t} + u_0\frac{\partial}{\partial x}\right)\left\{\frac{\partial^2}{\partial x^2} + \frac{\partial^2}{\partial y^2} + \frac{1}{B}\frac{\partial^2}{\partial z^2}\right\}p_{(0)} = 0.$$

If

$$\text{(C51)} \qquad p_{(0)}(x, y, z, t) = Z(z)\, e^{ikx}\, e^{-ikct} \cos n\pi y,$$

where k is real and positive, then Equation (C44) (cf. the first of Equations (C43)) and (C50) are satisfied if n is an odd integer 1, 3, 5 . . . , and

$$\text{(C52)} \qquad [c - (\bar{u}_0 + z)][(k^2 + n^2\pi^2) - B^{-1}D^2]Z = 0.$$

As $c \neq (\bar{u}_0 + z)$ in general, Z must satisfy

$$\text{(C53)} \qquad (D^2 - \gamma^2)Z = 0,$$

where

$$\text{(C54)} \qquad \gamma^2 \equiv B\bar{k}^2 \qquad \text{and} \qquad \bar{k}^2 \equiv k^2 + n^2\pi^2.$$

Hence

$$\text{(C55)} \qquad Z = Ke^{\gamma z} + L\, e^{-\gamma z},$$

where K and L are constants whose values can be determined from the boundary conditions expressed by Equations (C45) and (C46). These boundary conditions, when expressed in terms of Z, are the following:

$$\text{(C56a)} \qquad (\hat{c} + \tfrac{1}{2})DZ + [1 + \mathscr{E}\bar{k}^2B/ikR - \Gamma_l B/R]Z = 0$$
on $z = -\frac{1}{2}$ (lower surface rigid),

or

$$\text{(C56b)} \qquad (\hat{c} + \tfrac{1}{2} - 2\mathscr{E}^2B\bar{k}^2/ikR)DZ + (1 - \Gamma_l B/R)Z = 0$$
on $z = -\frac{1}{2}$ (lower surface free),

and

$$\text{(C57a)} \qquad (\hat{c} - \tfrac{1}{2}DZ + (1 - \mathscr{E}\bar{k}^2B/ikR - \Gamma_u B/R)Z = 0$$
on $z = \frac{1}{2}$ (upper surface rigid)

or

$$\text{(C57b)} \quad (\hat{c} - \tfrac{1}{2} - 2\mathscr{E}^2 B\bar{k}^2/ikR)DZ + (1 - \Gamma_u B/R)Z = 0 \quad \text{on} \quad z = \tfrac{1}{2} \text{ (upper surface free)}$$

where

$$\text{(C58)} \qquad \hat{c} \equiv c - \bar{u}_0 \equiv \hat{c}_R + i\hat{c}_I$$

(cf. Equations (C11) and (C35)); $\hat{c}$ is the complex phase speed of the disturbance relative to the mean basic flow.

Equations (C55) to (C58) suffice to determine $\hat{c}$ as a function of $B, R, k, n, \mathscr{E}, \Gamma_u$ and Γ_1. As the disturbance varies with time as $\exp(-ik(\hat{c} + \bar{u}_0)t)$, the system is unstable if for any mode the growth rate

$$\text{(C59)} \qquad k\hat{c}_I > 0;$$

otherwise the system is stable.

(g) Some Special Cases. When $\mathscr{E} = \Gamma_u = \Gamma_1 = 0$ we have the Eady problem, for which

$$\text{(C60)} \qquad \hat{c}^2 + [1 + \gamma^2/4 - \gamma \coth \gamma]/\gamma^2 = 0.$$

$\hat{c}$ is either real or pure imaginary according as $\gamma \gtrless 2.399$, so that, by Equations (C54) and (C59), the system is stable when

$$\text{(C61)} \qquad B > (2.399/\pi)^2 = 0.581$$

(cf. Equation (14)). A simple physical interpretation of this result is presented in Appendix D below. The unstable disturbances that arise when $B < 0.581$ drift with the mean velocity of the basic flow, $\bar{u}_0$.

When $\mathscr{E} = 0$ but $\Gamma_u \neq 0$ and $\Gamma_1 \neq 0$, we have

$$\text{(C62)} \qquad \begin{aligned} \hat{c}^2 + \hat{c}(Q_u - Q_l)\coth \gamma/\gamma - \{1 + \gamma^2/4 + Q_u Q_l - (Q_u + Q_l) \\ - \gamma[1 - \tfrac{1}{2}(Q_u + Q_l)] \coth \gamma\}/\gamma^2 = 0. \end{aligned}$$

Here

$$\text{(C63)} \qquad Q_u \equiv \Gamma_u B/R = \Gamma_u^*/\tan \vartheta, \; Q_l \equiv \Gamma_l B/R = \Gamma_l^*/\tan \vartheta$$

where

$$\text{(C64)} \qquad \tan \vartheta \equiv -(\partial \rho_0/\partial y)/(\partial \rho_0/\partial z)$$

(in ordinary units), ϑ being the inclination to the horizontal of the surfaces of equal density (temperature) in the basic state. The detailed discussion of this case will be presented elsewhere when experiments now nearing completion are written up for publication.

When $\Gamma_u = \Gamma_1 = 0$ but $\mathscr{E} \neq 0$ and both upper and lower surfaces are rigid we have Barcilon's problem, for which

$$\hat{c}^2 + 2i\hat{c}F \coth\gamma/\gamma - \{1 + \gamma^2/4 - \gamma\coth\gamma + F^2\}/\gamma^2 = 0, \tag{C65}$$

where

$$F \equiv \mathscr{E}\gamma^2/kR = \mathscr{E}(k^2 + n^2\pi^2)d/k(b - a)\tan\vartheta. \tag{C66}$$

$\hat{c}_I \leqq 0$, corresponding to stability, for all B when

$$F^2 \geqq 0.096 \tag{C67}$$

and outside a finite range of B when $F^2 < 0.096$; otherwise, i.e. within the finite range of B when $F^2 < 0.096$, the quantity $\hat{c}_I > 0$, corresponding to instability. When $k = n\pi$, the quantity $(k^2 + n^2\pi)/k$ has its maximum value $2\pi n$. Denote by $\mathscr{E}_p$ and B_p the corresponding values of $\mathscr{E}$ and B when, in addition, $\hat{c}_I = 0$. By Equation (C67)

$$\mathscr{E}_p > 0.31(b - a)\tan\vartheta/2\pi n d \tag{C68}$$

(which, when substituted in Equation (C65), leads to an expression for B_p which will not be written down here). All modes are stable when $\mathscr{E} \geqq \mathscr{E}_p$.

When $\Gamma_u = \Gamma_l = 0$ but $\mathscr{E} \neq 0$ and the lower surface is rigid but the upper surface is free we have Brindley's problem, for which

$$\begin{aligned}\hat{c}^2 + i\hat{c}F(\coth\gamma/\gamma + 2\mathscr{E}) + \{1 + \gamma^2/4 - \gamma(1 - 2\mathscr{E}F^2)\coth\gamma\}/\gamma^2 \\ - iF[\tfrac{1}{2}\coth\gamma(1 + 4\mathscr{E}) - (1 + \mathscr{E}\gamma^2)/\gamma]/\gamma = 0.\end{aligned} \tag{C69}$$

$\hat{c}_I > 0$ (instability) or $\hat{c}_I \leqq 0$ (stability) according as:

$$\begin{aligned}[1 + \tfrac{1}{4}\gamma^2 - \gamma(1 - 2\mathscr{E}F^2)\coth\gamma][\coth\gamma + 2\mathscr{E}\gamma]^2 \\ \lesseqgtr [\tfrac{1}{2}(1 + 4\mathscr{E})\coth\gamma - (1 + \mathscr{E}\gamma^2)]^2.\end{aligned} \tag{C70}$$

When

$$\mathscr{E}F^2 > 0.030 \tag{C71}$$

we have stability for all B (cf. Equation (C67)); in place of Equation (C68) (see also Equation (C75)) we have

$$\mathscr{E}_p > [0.17(b - a)\tan\vartheta/2\pi n d]^{2/3} \tag{C72}$$

as the expression for the critical value of $\mathscr{E}$ above which all modes are stable.

Finally, consider the case when $\Gamma_u = \Gamma_l = 0$, $\mathscr{E} \neq 0$ and both upper and lower surfaces are free; then

$$(\hat{c} + 2i\mathscr{E}F)^2 = \{1 + \gamma^2/4 - \gamma\coth\gamma\}/\gamma^2. \tag{C73}$$

When

$$\mathscr{E}^2F^2 > 0.021 \tag{C74}$$

we have stability for all B, but when $\mathscr{E}^2F^2 < 0.021$ we have stability or instability according as B is greater or less than a certain critical value; unlike the Brindley and Barcilon cases there is no "lower axisymmetric regime" when both upper and lower surfaces are free. The critical value of $\mathscr{E}$ (cf. Equations (C68) and (C72)) is given by

$$\mathscr{E}_p > [0.021(b - a)\tan\vartheta/2\pi nd]^{1/2}. \tag{C75}$$

According to Equations (C68), (C72) and (C75), of the three cases considered viscous damping is greatest when both upper and lower surfaces are rigid and least when both are free. But it is readily shown (see [**22**], cf. [**41**]), that even in the rigid/rigid case the amount of damping falls short of that implied by the empirical criterion expressed by Equation (4). Therefore, as anticipated above, in the experiments viscous effects in sidewall boundary layers and in the interior region were not negligible in comparison with those arising in Ekman layers on the endwalls (see Equation (D18) below).

Appendix D: The essence of baroclinic instability.

(a) INVISCID SYSTEMS. As the basic vertical density gradient is negative, and therefore stabilizing, the potential energy of the basic state can only be converted into kinetic energy of baroclinic waves when the angle of inclination θ to the horizontal of surfaces containing typical trajectories of individual fluid elements is less than the slope of the surfaces of equal density, but greater than zero. Hence, for instability we must require (in ordinary units) that

$$0 < \theta < (-\partial\rho_0/\partial y)/(\partial\rho_0/\partial z). \tag{D1}$$

Only when θ satisfies this criterion is it possible for relatively dense fluid elements to sink and relatively light fluid elements to rise, a process recognized in the use of the terms "sloping" or "slantwise" convection.

Since, by Assumptions (viii) and (xi) of §4, $(-\partial\rho_0/\partial y)/(\partial\rho_0/\partial z) \ll 1$, it is plausible that for the mode of maximum instability

$$\theta = \theta' = \varepsilon(-\partial\rho_0/\partial y)/(\partial\rho_0/\partial z), \tag{D2}$$

where ε is a positive constant neither greater nor much less than 0.5. The corresponding "e-folding" time, τ', will, in the absence of friction, be roughly equal to the time taken for a freely moving particle to slide a distance $[(b - a)^2 + (\pi/k)^2]^{1/2}$ down a slope inclined at an angle

$$\theta'(b-a)/[(b-a)^2 + (\pi/k)^2]^{1/2}$$

to the horizontal under the influence of reduced gravity $|g\bar{\rho}^{-1}(b-a)(\partial\rho_0/\partial y)|$; whence

$$\text{(D3)} \qquad \tau' \doteqdot \left[\frac{2\bar{\rho}}{\varepsilon g}\left(\frac{\partial\rho_0/\partial z}{(\partial\rho_0/\partial y)^2}\right)\right]^{1/2} [1 + (\pi/k(b-a))^2]^{1/2}$$

(see Equations (C23), (D9), (D10)). This approximate expression for τ' is in satisfactory agreement with exact values based on Equation (C60).

In order to demonstrate that the instability criterion expressed by Equation (D1) is equivalent to that found in the exact analysis of the Eady problem (see Equation (C61)), it is necessary to express θ in terms of the other parameters. First we take the z component of the vorticity equation (equivalent to eliminating p_1 between Equations (C12) and (C13) making use of Equation (C15)) and thus show that

$$\text{(D4)} \qquad \frac{\partial w_1}{\partial z} = -\frac{1}{2\Omega}\left[\left(\frac{\partial}{\partial t} + u_0\frac{\partial}{\partial x}\right)\left(\frac{\partial u_1}{\partial y} - \frac{\partial v_1}{\partial x}\right) + \frac{du_0}{dz}\frac{\partial w}{\partial y}\right]$$

(cf. Equation (C39)). Now, for quasi-geostrophic instabilities (see Assumption (xi) of §4, also Equation (C40)) $\partial u_1/\partial x \doteqdot -\partial v_1/\partial y$ (see the fourth of Equations (C43)) and the dominant term in the square brackets on the r.h.s. of Equation (D4) is $u_0\partial(\partial u_1/\partial y - \partial v_1/\partial x)/\partial x$. Therefore, Equation (D4) simplifies to

$$\text{(D5)} \qquad \partial w_1/\partial z \doteqdot (u_0/2\Omega)(\partial^2 v_1/\partial x^2 + \partial^2 v_1/\partial y^2).$$

Denote by (V_1, W_1) the r.m.s. average values of (v_1, w_1). We can find a relationship between W_1 and V_1 by replacing $\partial w_1/\partial z$ in Equation (D5) by $\pi W_1/d$ (remembering that we are dealing with the case when the upper and lower bounding surfaces are horizontal) and $(\partial^2 v_1/\partial x^2 + \partial^2 v_1/\partial y^2)$ by $k^2 + \pi^2/(b-a)^2)V_1$, and by taking $|gd(\partial\rho_0/\partial y)/4\Omega\bar{\rho}|$ as a measure of the average value of $|u_0|$ (see Equation (C22)). Thus we find that

$$\text{(D6)} \qquad \frac{W_1}{V_1} \doteqdot \left|\frac{gd^2(\partial\rho_0/\partial y)\pi}{8\Omega^2(b-a)^2}\right|\left\{1 + \frac{k^2(b-a)^2}{\pi^2}\right\} = \frac{\pi R d}{2(b-a)}\left\{1 + \frac{k^2(b-a)^2}{\pi^2}\right\}.$$

The instability criterion sought can be obtained by combining the last equation with Equation (D1), remembering that by definition

$$\text{(D7)} \qquad \theta = W_1/V_1.$$

Thus, baroclinic instability arises when

$$\text{(D8)} \qquad 0 < B(1 + k^2(b-a)^2/\pi^2) < 0.63$$

but not otherwise, which is in perfect qualitative and excellent quantitative agreement with the exact criterion given in Appendix C (cf. Equation (C61)).

We can combine Equations (D6), (D7) and (D2) and find for the wave-number of the mode of maximum instability the equation

$$\text{(D9)} \qquad k' \doteqdot (\pi/(b-a))[2\varepsilon/\pi B - 1]^{1/2},$$

which agrees satisfactorily with the exact theory. By Equations (D3) and (D9),

$$\text{(D10)} \qquad \tau' \doteqdot (\Omega^{-1}/R)[B/(2\varepsilon - \pi B)]^{1/2}.$$

(b) Viscous Effects. According to Appendix C and Equations (4b) and (5), in the annulus experiments the total rate of energy dissipation by viscosity exceeds that attributable to the end-wall boundary layers. It is of interest, therefore, to calculate the effect on baroclinic instability of viscous friction in the main body of the fluid (cf. Equations (C68), (C72) and (C75)).

The inclusion of viscosity leads to an additional term

$$-(\nu/2\Omega)\nabla^2(\partial u_1/\partial y - \partial v_1/\partial x)$$

on the r.h.s. of Equation (D5) and therefore to an additional factor on the r.h.s. of Equation (D6), which then takes the form

$$\text{(D11)} \quad \frac{W_1}{V_1} \doteqdot \frac{\pi R d}{2(b-a)}\left(1 + \frac{k^2(b-a)^2}{\pi^2}\right)\left(1 + \frac{2\pi^2(1 + k^2(b-a)^2/\pi^2)}{\mathcal{T}^{1/2}k(b-a)^{1/2}d^{1/2}R}\right),$$

where $\mathcal{T} \equiv 4\Omega^2(b-a)^5/\nu^2 d$ (see Equation (4a)). It follows from Equations (D1), (D7) and (D11) that in place of Equation (D8) we have

$$\text{(D12)} \quad 0 < B\left(1 + \frac{k^2(b-a)^2}{\pi^2}\right)\left(1 + \frac{2\pi^2(1 + k^2(b-a)^2/\pi^2)}{\mathcal{T}^{1/2}k(b-a)^{1/2}d^{1/2}R}\right) < 0.63.$$

The raising of the value of θ, the inclination of surfaces containing typical trajectories of individual fluid elements, is not the only effect of viscosity, so that when $\mathcal{T}$ is not infinite, Equation (D12), though necessary, is not a sufficient condition for the occurrence of baroclinic instability. We must also consider energy dissipation by viscosity and require that this be less than that released by buoyancy forces. Thus, we must require that

$$\text{(D13)} \qquad \tau'_v > \tau'$$

where τ'_v is a time-constant associated with viscous dissipation and τ' is the "e-folding" time of the mode of maximum instability in the absence of viscosity (cf. Equation (D3)).

An approximate expression for τ'_v when viscous boundary layers are ignored (see above) and $(b-a) \ll d$ is

(D14) $$\tau'_v \doteqdot (b-a)^2/\nu\pi^2(1 + (k'(b-a)/\pi)^2).$$

By Equations (D11), (D7) and (D2), the wavenumber k' of the mode of maximum instability is modified by viscosity and satisfies

(D15) $$\frac{2\varepsilon}{\pi B} = \left(1 + \left(\frac{k'(b-a)}{\pi}\right)^2\right)\left(\frac{1 + 2\pi^2(1 + (k'(b-a)/\pi)^2)}{\mathcal{T}^{1/2}k'(b-a)^{1/2}d^{1/2}R}\right)$$

(cf. Equation (D9)). Equations (D14), (D15) and (D3) suffice to determine the wavenumber k' and the "e-folding" time $(1/\tau' - 1/\tau'_v)^{-1}$ of the mode of maximum instability.

It is of interest to make on the basis of Equations (D3), (D13), (D14) and (D15) a rough calculation of $\mathcal{T}_{\text{crit}}$, the value of $\mathcal{T}$ below which viscosity completely inhibits the growth of all baroclinic waves, and compare $\mathcal{T}_{\text{crit}}$ with the empirical quantity $\mathcal{T}_p$ given by Equation (4b). For simplicity we shall ignore the effect of viscosity on k' so that, by Equations (D14) and (D15),

(D16) $$\tau'_v \doteqdot (b-a)^2 B/2\pi\nu\varepsilon.$$

By Equation (D3) (cf. Equation (D10)), the criterion expressed by Equation (D13) for baroclinic waves to grow is satisfied when

(D17) $$\mathcal{T}^{1/2} > \frac{4\pi\varepsilon d^{1/2}\cot\vartheta}{(2\varepsilon - \pi B)^{1/2}B^{1.5}(b-a)^{1/2}},$$

where $\cot\vartheta = -(\partial\rho_0/\partial z)/(\partial\rho_0/\partial y)$, see Equation (C64). Now the maximum value of $B^3(2\varepsilon - \pi B)$ is $27\varepsilon^4/16\pi^3$ and occurs when $B = 3\varepsilon/2\pi$, so that the last equation cannot be satisfied when

(D18) $$\mathcal{T} < \frac{256\pi^5}{27\varepsilon^2}\left(\frac{d}{b-a}\right)\left(\frac{\partial\rho_0/\partial z}{\partial\rho_0/\partial y}\right) = \mathcal{T}_{\text{crit}} \text{ (say).}$$

When $|\varepsilon(b-a)^{1/2}(\partial\rho_0/\partial y)/(\partial\rho_0/\partial z)|$ has the reasonable value of 0.1, $\mathcal{T}_{\text{crit}}$ is approximately numerically equal to $\mathcal{T}_p$ (cf. Equations (4b) and (D18)).

References, Lecture I

1. V. Barcilon, *Role of the Ekman layers in the stability of the symmetric regime obtained in a rotating annulus*, J. Atmospheric Sci. **21** (1964), 291–299. MR **29** #1835.

2. G. Birkhoff, *Hydrodynamics: a study in logic, fact and similitude*, 2nd ed., Princeton Univ. Press, Princeton, N.J., 1960. MR **22** #12919.

3. S. J. Bless, *The effect of a radial barrier on thermally driven motions in a rotating fluid annulus*, B.S. Thesis, M.I.T., Cambridge, Mass., 1965.

4. M. Bowden, *An experimental investigation of heat transfer in a rotating fluid*, Ph.D. Thesis, University of Durham, Newcastle-upon-Tyne, 1961.

5. M. Bowden and H. F. Eden, *Thermal convection in a rotating fluid annulus: temperature, heat flow and flow field observations in the upper symmetric regime*, J. Atmospheric Sci. **22** (1965), 185–195.

6. ———, *Effect of a radial barrier on thermal convection in a rotating fluid annulus*, J. Geophys. Res. **73** (1968), 6887–6896.

7. J. Brindley, *Stability of flow in a rotating viscous incompressible fluid subjected to differential heating*, Philos. Trans. Roy. Soc. London Ser. A **253** (1960), 1–25. MR **22** #10441.

8. ———, *Symmetric flow in a differentially-heated rotating annulus of fluid*, Technical Report #11, Geophys. Fluid Dynamics Inst., Florida State University, Tallahassee, Fla., 1968.

9. John A. Brown, Jr., *Numerical investigation of hydrodynamic instability and energy conversions in the quasi-geostrophic atmosphere*, J. Atmospheric Sci., **26** (1969), 366–375.

10. S. Chandrasekhar, *Hydrodynamic and hydromagnetic stability*, Clarendon Press, Oxford, 1961. MR **23** #B1270.

11. J. G. Charney, *The dynamics of long waves in a baroclinic westerly current*, J. Meteorol. **4** (1947), 135–162. MR **9**, 163.

12. Owen R. Coté, *Dye mixing processes in the periodic waves generated in a heated rotating annulus: Is this an analog to atmospheric mixing* (Abstract), Trans. Amer. Geophys. Union **49** (1968), 181.

13. T. V. Davies, *The forced flow due to heating of a rotating liquid*, Philos. Trans. Roy. Soc. London, Ser. A **249** (1956), 27–64. MR **17**, 1149.

14. ———, *On the forced motion due to heating of a deep rotating liquid in an annulus*, J. Fluid Mech. **5** (1959), 593–621. MR **21** #4679.

15. John A. Dutton and Donald R. Johnson, *A theory of available potential energy and a variational approach to atmospheric energetics*, Advances in Geophysics **12** (1967), 333–436.

16. E. T. Eady, *Long waves and cyclone waves*, Tellus **1** (1949), no. 3, 33–52. MR **13**, 86.

17. R. L. Elsberry, *A high rotation general circulation model experiment with cyclic time changes*, Atmos. Sci. Paper #134, Colorado State University, Fort Collins, Col., 1968.

18. F. M. Exner, *Über die Bildung von Windhosen und Zyklonen*, S.-B. Akad. Wiss. Wien. Abt. IIa **132** (1923), 1–16.

19. A. J. Faller, *A demonstration of fronts and frontal waves in atmospheric models*, J. Meteorol. **13** (1956), 1–4.

20. R. Fjörtoft, *Application of integral theorems in deriving criteria of stability for laminar flows and for the baroclinic circular vortex*, Geofys. Publ. Norske Vid.-Akad. Oslo, **17** (1950), no. 6, 1–52. MR **14**, 815.

21. W. W. Fowlis, *Techniques for fast and precise measurement of fluid temperatures and flow speeds using multi-probe thermistor assemblies*, Technical Report #10, Geophys. Fluid Dynamics Inst., Florida State University, Tallahassee, Fla., 1968.

22. W. W. Fowlis and R. Hide, *Thermal convection in a rotating fluid annulus: effect of viscosity on the transition between axisymmetric and non-axisymmetric flow regimes*, J. Atmospheric Sci. **22** (1965), 541–558.

23. W. W. Fowlis and R. L. Pfeffer, *Characteristics of amplitude vacillation in a rotating differentially-heated fluid determined by a multi-probe technique*, J. Atmospheric Sci. **26** (1969), 100–108.

24. D. Fultz, *Developments in controlled experiments on large-scale geophysical problems* Advances in Geophysics **7** (1961), 1–103.

25. D. Fultz and R. Kaylor, *The propagation of frequency in experimental waves in a rotating annular ring*, Rossby Memorial Volume, B. Bolin (editor), Rockefeller Univ. Press New York, 1959.

26. D. Fultz, R. R. Long, G. V. Owens, W. Bowan, R. Kaylor and J. Weil, *Studies of thermal convection in a rotating cylinder with some implications for large-scale atmospheric motions*, Meteorol. Monographs, **4**, Amer. Meteorol. Soc., Boston, Mass., 1959.

27. D. Fultz and T. Spence, *Preliminary experiments on baroclinic westerly flow over a north-south ridge*, Proc. Sympos. on Mountain Met., Atmos. Sci. Paper #122. Colorado State University, Fort Collins, Col., 1967.

28. H. P. Greenspan, *The theory of rotating fluids*, Cambridge Univ. Press, New York, 1968.

29. R. Hide, *Some experiments on thermal convection in a rotating liquid*, Ph.D. Thesis, Cambridge University, Cambridge, 1953.

30. ———, *Some experiments on thermal convection in a rotating liquid*, Quart. J. R. Met. Soc. **79** (1953), 161.

31. ———, "Fluid motion in the Earth's core and some experiments on thermal convection in a rotating liquid," *Fluid models in geophysics*, edited by R. R. Long, U.S. Government Printing Office, Washington, D.C., 1953, pp. 101–116.

32. ———, *An experimental study of thermal convection in a rotating liquid*, Philos. Trans. Roy. Soc. London Ser. A **250** (1958), 442–478.

33. ———, *The viscous boundary layer at the free surface of a rotating baroclinic fluid*, Tellus **16** (1964), 523–529.

34. ———, *The viscous boundary layer at the free surface of a rotating baroclinic fluid: effects due to temperature dependence of surface tension*, Tellus **17** (1965), 440–442.

35. ———, *On the dynamics of rotating fluids and related topics in geophysical fluid dynamics*, Bull. Amer. Meteorol. Soc. **47** (1966), 873–885.

36. ———, *On the vertical stability of a rotating fluid subject to a horizontal temperature gradient*, J. Atmospheric Sci. **24** (1967), 6–9.

37. ———, *Theory of axisymmetric thermal convection in a rotating fluid annulus*, Phys. Fluids **10** (1967), 56–68.

38. ———, *On source-sink flows in a rotating fluid*, J. Fluid Mech. **32** (1968), 737–764.

39. R. Hide and P. J. Mason, *Baroclinic waves in a rotating fluid subject to internal heating*, Philos. Trans. Roy. Soc. London Ser. A **268** (1970), 201–232.

40. C. Hunter, *The axisymmetric flow in a rotating annulus due to a horizontally applied temperature gradient*, J. Fluid Mech. **27** (1967), 753–778.

41. J. A. C. Kaiser, *Rotating deep annulus convection: wave instabilities, vertical stratification and associated parameters and thermal properties of the upper symmetrical regime*, Ph.D. Thesis, University of Chicago, Chicago, Ill., 1969. See also Tellus **21** (1969), 789–805; **22** (1970), 275–287.

42. J. A. C. Kaiser, J. Weil and D. Fultz, *Measured temperature fields and parameter summary for the upper symmetric regime in a rotating annulus of baroclinic fluid*, Hydrodynamics Lab. Report. HL-4-E-69, Dept. Geophys. Sci., University of Chicago, Ill., 1969.

43. J. E. Kester, *Thermal convection in a rotating annulus of liquid: nature of the transition from an unobstructed annulus to one with a total radial wall*, B.S. Thesis, M.I.T., Cambridge, Mass., 1966.

44. C. B. Ketchum, *An experimental study of baroclinic instability of a rotating baroclinic liquid*, Ph.D. Thesis, M.I.T., Cambridge, Mass., 1968.

45. E. L. Koschmieder, *Convection in a rotating laterally heated annulus*, 1968. (Unpublished report.)

46. F. Kreith, *Convection heat transfer in rotating systems*, Advances in Heat Transfer **5** (1968), 129–251.

47. H. L. Kuo, *Further studies of thermally driven motions in a rotating fluid*, J. Meteorol. **14** (1957), 553–558.

48. R. B. Lambert and H. A. Snyder, *Experiments on the effects of horizontal shear and change of aspect ratio on convective flow in a rotating annulus*, J. Geophys. Res. **71** (1966), 5225–5234.

49. E. N. Lorenz, "A proposed explanation for the existence of two regimes of flow in a rotating symmetrically heated cylindrical vessel," *Fluid models in geophysics*, Edited by R. R. Long, U.S. Government Printing Office, Washington, D.C., 1953, pp. 73–80.

50. ———, *Simplified dynamic equations applied to the rotating basin experiment*, J. Atmospheric Sci. **19** (1962), 39–51.

51. ———, *The mechanics of vacillation*, J. Atmospheric Sci. **20** (1963), 448–464.

52. ———, *The nature and theory of the general circulation of the atmosphere*, W.M.O. Geneva, Publ. #218, 1967.

53. M. E. McIntyre, *The axisymmetric convective regime for a rigidly-bounded rotating annulus*, J. Fluid Mech. **32** (1968), 625–655.

54. P. E. Merilees, *On the transition from axisymmetric to non-axisymmetric flow in a rotating annulus*, J. Atmospheric. Sci. **25** (1968), 1003–1014.

55. J. Pedlosky, *On the stability of baroclinic flows as a functional of the velocity profile*, J. Atmospheric. Sci. **22** (1965), 137–145.

56. R. L. Pfeffer and Y. Chiang, *Two kinds of vacillation in rotating laboratory experiments*, Mon. Weath. Rev. **95** (1967), no. 2, 75–82.

57. R. L. Pfeffer and Colleagues, *A new concept of available potential energy*, Final Report U.S.W.B., Grant WBG 45, Report #66–1, Dept. Met., Florida State University, Tallahassee, Fla., 1965.

58. R. L. Pfeffer and W. W. Fowlis, *Wave-dispersion in a rotating, differentially heated cylindrical annulus of fluid*, J. Atmospheric. Sci **25** (1968), 361–371.

59. N. A. Phillips, *Geostrophic motion*, Rev. Geophys. **1** (1963), 123–173.

60. S. A. Piacsek, *Thermal convection in a rotating annulus of liquid: numerical studies of the axisymmetric regime of flow*, Ph.D. Thesis, M.I.T., Cambridge, Mass., 1966.

61. L. Prandtl, *Essentials of fluid dynamics*, Blackie and Sons, London, 1952.

62. C. Quon, *Numerical studies of the upper-symmetric regime in a rotating fluid annulus*, Ph.D. Thesis, Cambridge University, Cambridge, 1967.

63. H. Riehl and D. Fultz, *Jet streams and long waves in a steady rotating dishpan experiment: structure and circulation*, Quart. J. R. Met. Soc. **83** (1957), 215–231.

64. ———, *The general circulation in a steady rotating dishpan experiment*, Quart. J.R. Met. Soc. **84** (1958), 389–417.

65. A. R. Robinson, *The symmetric state of a rotating fluid differentially heated in the horizontal*, J. Fluid Mech. **6** (1959), 599–620.

66. R. H. Rogers, *The structure of the jet stream in a rotating fluid with a horizontal temperature gradient*, J. Fluid Mech. **5** (1959), 41–59. MR **20** #5616.

67. ———, *The effect of viscosity near the cylindrical boundaries of a rotating fluid with a horizontal temperature gradient*, J. Fluid Mech. **14** (1962), 25–41.

68. A. R. Smith, *The effect of rotation on the flow of a baroclinic liquid*, Ph.D. Thesis, Cambridge University, Cambridge, 1958.

69. H. A. Snyder and E. M. Youtz, *Transient response of a differentially-heated rotating annulus*, J. Atmospheric. Sci. **26** (1969), 96–99.

70. M. E. Stern, *Eady's theory of baroclinic instability*, Proc. Geophys. Fluid Dynamics Summer School, Woods Hole Oceanographic Inst., 1960.

71. P. H. Stone, S. Hess, R. Hadlock and P. Ray, *Preliminary results of experiments with symmetric baroclinic instabilities*, J. Atmospheric Sci. **26** (1969), 997–1001.

72. R. C. Sutcliffe, *The quasi-geostrophic advective wave in a baroclinic zonal current*, Quart. J. R. Met. Soc. **77** (1951), 226–234.

73. M. Uryu, K. Ukaji and R. Sawada, *Transition of flow patterns and heat transport in a rotating fluid*, 1966. (Unpublished report.)

74. F. Vettin, *Meteorologische Untersuchungen*, Ann. Phys. Lpz. (2) **100** (1857), 99–100.

75. G. P. Williams, *Thermal convection in a rotating fluid annulus*. 1: *The basic axisymmetric flow*, J. Atmospheric. Sci. **24** (1967), 144–161.

76. ———, *Thermal convection in a rotating fluid annulus*. 2: *Classes of axisymmetric flow*, J. Atmospheric. Sci. **24** (1967), 162–174.

77. ———, *Thermal convection in a rotating fluid annulus*. 3: *Suppression of the frictional constraint on lateral boundaries*, J. Atmospheric. Sci. **25** (1968), 1034–1045.

78. ———, *Numerical integration of the three-dimensional Navier-Stokes equations for incompressible flow*, J. Fluid Mech. **37** (1969), 727–750.

79. M. E. McIntyre, *On the non-separable baroclinic parallel flow problem*, J. Fluid Mech. **40** (1970), 273–306.

80. P. G. Drazin, *Non-linear baroclinic instability of a continuous zonal flow*, Quart. J. Roy. Meteorol. Soc. **96** (1970), 667–676.

81. R. Hide, "Some laboratory experiments on free thermal convection in a rotating fluid subject to a horizontal temperature gradient and their relation to the theory of the global atmospheric circulation," In: *The global circulation of the atmosphere*, G. A. Corby (editor), Roy. Meteorol. Soc., London, 1969, pp. 196–221.

II. Dynamics of the Atmospheres of the Major Planets[1]

Summary. Attempts to interpret, in terms of basic hydrodynamical processes, (1) the Great Red Spot and other less persistent and generally smaller spots on Jupiter's visible surface and (2) the banded appearance and complicated and striking variation of rotation rate with latitude, including the equatorial jets, of Jupiter and Saturn have led to advances in the study of the dynamics of the atmosphere of the major planets. The dynamical influence of Coriolis forces on relative motions in these atmospheres is much more pronounced than in the case of the Earth's atmosphere, though effects due to vertical density stratification are probably much less important. Possible complications arise because (1) the major planets rotate hypersonically with respect to the speed of sound in their atmospheres, and (2) the electrical conductivity of the lower reaches of these atmospheres might be sufficiently large for magnetohydrodynamic processes to occur there. If, as has been suggested, these processes produce, or at least modify, Jupiter's magnetic field, then future research on the dynamics of the atmospheres of the major planets should include attempts to detect the magnetic fields of Saturn, Uranus and Neptune, and to determine the configuration of the magnetic field in the vicinity of the visible surface of Jupiter, carried out in conjunction with attempts to measure the electrical properties of the outer layers of the planets and systematic theoretical studies of the hydrodynamics and magnetohydrodynamics of hypersonically rotating fluids.

---

[1] Supplementary notes for Lecture (II) based on recent articles by the lecturer [76], [77].

1. Introduction. The first two papers in the oldest scientific journal still in publication ([**5**], [**39**]) describe observations pertaining to large-scale motions in Jupiter's atmosphere, so that the topic discussed in this lecture is by no means "newfangled." At the same time, however, exciting new results can be expected during the coming decade when space probes make their stupendous journeys to the outer solar system and signal back observations of the atmospheres of the major planets that are well beyond the reach of ground-based equipment. Making sense of these observations will present challenging theoretical problems and attempting to interpret and reconcile certain well-established results based on existing observations might be the best way that theoreticians can prepare for this challenge.

Evidently, it is by no means unlikely that dynamical processes in the lower reaches of Jupiter's atmosphere, well below the visible cloud level, are responsible for a significant part of the magnetic field of Jupiter implied by certain radio-astronomical observations. From this it follows that the determination of magnetic field strength and configuration at levels ranging from above the visible surface to well below that surface, together with appropriate measurements of electrical conductivity, are investigations that should be considered seriously when formulating a program of experiments concerned with the dynamics of the atmospheres of the major planets. A realistic program of related theoretical research should include studies of the magnetohydrodynamics (hydromagnetics) as well as ordinary hydrodynamics of hypersonically rotating fluids, an almost untouched area of fluid dynamics.

The basis of these remarks will be outlined in §§2, 3 and 4, in the context of a systematic (but simplified for the sake of exposition) scale analysis which brings out the importance not only of the familiar (to meteorologists) Rossby, Ekman and Burger numbers (see Phillips [**52**]), but also of the magnetic Reynolds number, magnetic Rossby number, and rotational Mach number.

A comprehensive review (which this article is not) would have to deal in detail with certain major observational features, whose rational interpretation in terms of basic fluid dynamical processes lies at the heart of the subject under discussion. The features are (1) the Great Red Spot and the many other, less persistent and generally smaller, spots on Jupiter's visible surface, (2) the banded appearance, and complicated and striking variation of rotation rate with latitude, especially the equatorial jets, of the visible surfaces of Jupiter and Saturn (see Peek [**51**], Robinson [**57**], Alexander [**1**]) and (3) Jupiter's magnetic field, with its characteristic rotation period (see Warwick [**72**], Douglas and Smith [**11**], Olsen and Smith [**48**]). (This list is by no means exhaustive and final, of course; important additions can surely be expected in the future as instruments improve and new techniques are introduced. Owing to the great distances between the Earth and the outer planets, we have at the present time no knowledge of the scales of motion on Neptune and indications of planetary scale banded structure that are vague in the case of Uranus but rather better for Saturn.) In the time and space available it will be possible to comment only briefly on these phenomena, relying to some extent on previous discussions (Hide [**27**], [**31**]). The bibliography, however, is fairly extensive.

Owing to their inaccessibility, neither the thickness nor the chemical composition and physical properties and conditions (including the nature and strength of energy sources) prevailing in these fluid layers are known with any degree of precision. Hence, any approach to the interpretation of the aforementioned observational features must, of necessity, be indirect. Typically, the procedure to be followed involves (1) making an educated guess at the type of hydrodynamical process that might account for a particular observational feature; (2) then, if necessary, carrying out appropriate theoretical studies, possibly supplemented by laboratory experiments, of the general physical conditions under which the process can occur; and finally (3) investigating in the wider context of the general body of knowledge about the planet the plausibility or otherwise that such physical conditions could prevail in the fluid layer in question.

This procedure, if applied successfully to the major observational features, will in due course lead not only to an understanding of the dynamical processes taking place, but also to information about the interiors of the major planets that may be obtainable in no other way. For example (see Hide [**25**], [**26**], [**27**], [**31**], also §3), Jupiter's Great Red Spot and the equatorial jets on Jupiter and Saturn have been tentatively interpreted in terms of hydrodynamical processes characteristic of rapidly rotating fluids, considered in the context of their implications with regard to (1) angular momentum transfer between different parts of Jupiter, and the interpretation of the rotation about the axis of the planet of sources of decametric and decimetric radiation, (2) the strength of Jupiter's internal magnetic field, (3) the depth and thermal structure of and energy sources within the lower atmosphere and other regions of Jupiter's interior, and (4) the relative thicknesses of the atmospheres of Jupiter and Saturn. These implications are self-consistent, compatible with the limited amount of accurate information (beyond pressure and density as a function of radius) about the internal structure of the major planets that is forthcoming from theoretical studies along more traditional lines (DeMarcus [**9**], Wildt [**75**], Öpik, [**49**], Peebles [**50**], Smoluchowski [**63**], Hubbard [**40**]), and most of them are quite novel.

The central theoretical difficulty in all dynamical studies of motions of planetary atmospheres is that of understanding the interaction between motions on different scales. Current deficiencies in our knowledge of the scales of motions present in the atmospheres of the major planets and of the details of certain conspicuous features (e.g., the edges of the strong equatorial jets on Jupiter and Saturn, and Jupiter's Great Red Spot) will not be remedied until high-resolution photographs and thermal maps have been obtained.

2. Basic equations, scale analysis and energy sources.

(a) BASIC EQUATIONS. The basic equations governing the flow of an electrically-conducting fluid in the presence of a magnetic field, when referred to a frame which rotates with angular velocity Ω relative to an inertial frame (see Cowling [**8**], Ferraro and Plumpton [**14**], Hide and Roberts [**38**], Alfvén and Fälthammar [**2**], Shercliff [**61**], Roberts [**56**]) are as follows:

momentum

$$\rho[\partial \boldsymbol{u}/\partial t + (\boldsymbol{u}\cdot\nabla)\boldsymbol{u} + 2\boldsymbol{\Omega}\times\boldsymbol{u}] = -\nabla p + \rho g + \nu\rho\nabla^2\boldsymbol{u} + \boldsymbol{j}\times\boldsymbol{B}, \tag{2.1}$$

continuity

$$\partial\rho/\partial t + \nabla\cdot(\rho u) = 0, \tag{2.2}$$

state

$$\rho = \rho(p, T, \text{chemical composition}), \tag{2.3}$$

heat flow

$$\partial T/\partial t + (\boldsymbol{u}\cdot\nabla)T = k\nabla^2 T + Q, \tag{2.4}$$

Ampère's law

$$\nabla\times\boldsymbol{B} = \mu\boldsymbol{j}, \tag{2.5}$$

Faraday's law

$$\nabla\times\boldsymbol{E} = -\partial\boldsymbol{B}/\partial t, \tag{2.6}$$

Ohm's law

$$\boldsymbol{j} = (\boldsymbol{E} + \boldsymbol{u}\times\boldsymbol{B})/\eta, \tag{2.7}$$

and Gauss's law

$$\nabla\cdot\boldsymbol{B} = 0. \tag{2.8}$$

Rationalized MKS units are used throughout. Here $\boldsymbol{u}$ denotes the Eulerian flow velocity, ρ density, T temperature, p pressure, $\boldsymbol{g}$ the acceleration of gravity and centrifugal forces, $\boldsymbol{j}$ current density, $\boldsymbol{E}$ electric field, $\boldsymbol{B}$ magnetic field, and t time. The parameters ν, κ, μ and η denote, respectively, the coefficients of kinematical viscosity, thermometric diffusivity, magnetic permeability and electrical resistivity, assumed here, for simplicity, to be independent of position; and Q represents effects due to adiabatic compression and to any heat sources, associated with radiative or chemical processes, mechanical friction, phase changes (see DeMarcus and Wildt [**10**]), dissipation of electromagnetic energy, etc., that might be present in the fluid.

Combining Equations (2.2) and (2.5)–(2.8), thus eliminating $\boldsymbol{j}$ and $\boldsymbol{E}$ from the equations of electromagnetism, we find

(2.9) $$\partial \boldsymbol{B}/\partial t - \nabla \times (\boldsymbol{u} \times \boldsymbol{B}) = \lambda \nabla^2 \boldsymbol{B},$$

where

(2.10) $$\lambda \equiv \eta/\mu \, [\mathrm{m}^2 \, \mathrm{sec}^{-1}],$$

which has the same dimensions as ν and κ. Equations (2.1)–(2.5) and (2.9) form a complete set.

(b) SCALE ANALYSIS. We denote by ΔL, ΔU, $\Delta\rho$, ΔT, ΔB and $\Delta\tau$, respectively, a typical length scale, speed of horizontal motion relative to a system rotating with angular velocity $\boldsymbol{\Omega}$, density variation, temperature variation, magnetic field variation, and interval of time over which significant changes in flow pattern occur; we further define the following dimensionless parameters:

(2.11) $$\mathscr{Y} \equiv \Delta U/(\Omega \Delta L),$$

a Rossby number;

(2.12) $$\mathscr{F} \equiv \Delta L/(\Delta\tau \Delta U),$$

a measure of unsteadiness in the flow;

(2.13) $$\mathscr{E} \equiv \nu/[\Omega(\Delta L)^2],$$

an Ekman number based on an eddy coefficient of viscosity representing effects due to turbulence on scales much less than ΔL;

(2.14) $$\mathscr{B}^2 \equiv g\Delta\rho/(\rho\Omega^2\Delta L),$$

a buoyancy parameter (equal to a Burger number when $\Delta\rho$ is interpreted as the vertical contrast of potential density, positive if the lapse rate is subadiabatic, and $(\Delta L)^{-1}$ as a typical vertical dimension divided by the square of a typical horizontal dimension, the corresponding Richardson number being equal to the Burger number divided by $\mathscr{Y}^2$);

(2.15) $$\mathscr{Y}_m \equiv \Delta V/(\Omega \Delta L),$$

a "magnetic Rossby number" based on the Alfvén speed

$$\Delta V \equiv \Delta B/(\mu\rho)^{1/2} \tag{2.16}$$

[cf. Equations (4.13) and (A13)];

$$\mathscr{M} \equiv \Omega \Delta L/c, \tag{2.17}$$

a "rotational Mach number" (see Hide [**26**]) where c is the speed of compressional waves in the fluid;

$$\mathscr{R}_m \equiv \Delta U \Delta L/\lambda, \tag{2.18}$$

a "magnetic Reynolds number" (see Equations (4.3) and (2.10)); and

$$\mathscr{P} \equiv \Delta U \Delta L/\kappa, \tag{2.19}$$

a Péclét number based on an eddy coefficient of thermal diffusivity representing small scale turbulence, etc. (cf. Equation (2.13)).

If we measure length, time, density variations, temperature variations, and magnetic field strength in terms of ΔL, $\Delta L/\Delta U$, $\Delta\rho$, ΔT and ΔB, respectively, then the equations of motion and continuity become:

$$\begin{aligned} &2\boldsymbol{\Omega} \times \rho \boldsymbol{u} + \nabla p - (\mathscr{B}^2/\mathscr{Y})\boldsymbol{g}\rho \\ &= -\mathscr{Y}\rho[\mathscr{F}\partial \boldsymbol{u}/\partial t + (\boldsymbol{u}\cdot\nabla)\boldsymbol{u}] + \mathscr{E}\rho\nabla \boldsymbol{u}^2 + (\mathscr{Y}_m^2/\mathscr{Y})(\nabla \times \boldsymbol{B}) \times \boldsymbol{B}/\mu, \end{aligned} \tag{2.20}$$

and

$$\nabla\cdot(\rho\boldsymbol{u}) = -\mathscr{M}^2\mathscr{Y}\mathscr{F}\partial p/\partial t, \tag{2.21}$$

(cf. Equations (2.1), (2.2), (2.3) and (2.5); also Equations (2.22)–(2.25)), where the variables, operators, $\boldsymbol{\Omega}$, $\boldsymbol{g}$, Q, etc., are dimensionless and (apart from geometrical factors which, though important, will not be considered in detail here) of order unity. A useful vorticity equation is readily found by operating on (2.20) with *curl* and making use of (2.21); this equation (cf. (5.11) of Hide [**29**]) reduces to the well-known thermal wind equation

$$2(\boldsymbol{\Omega}\cdot\nabla)(\rho\boldsymbol{u}) = (\mathscr{B}^2/\mathscr{Y})\boldsymbol{g} \times \nabla\rho \tag{2.22}$$[2]

in the geostrophic limit, when $\mathscr{Y}$, $\mathscr{Y}\mathscr{F}$, $\mathscr{Y}_m^2/\mathscr{Y}$, $\mathscr{E}$ and $\mathscr{M}\mathscr{Y}\mathscr{F}$ tend to zero. When, in addition, baroclinic effects are absent (i.e., $\mathscr{B}^2/\mathscr{Y} = 0$), we have the celebrated Proudman-Taylor Theorem

$$2(\boldsymbol{\Omega}\cdot\nabla)\boldsymbol{u} = 0. \tag{2.23}$$

The corresponding dimensionless forms of the equations of electromagnetism and heat transfer (Equations (2.9) and (2.4)) are, respectively,

---

[2] See Appendix to Lecture III, pages 305–307.

$$\mathscr{F}\partial \boldsymbol{B}/\partial t - \nabla \times (\boldsymbol{u} \times \boldsymbol{B}) = \mathscr{R}_m^{-1}\nabla^2 \boldsymbol{B}, \tag{2.24}$$

$$\mathscr{F}\partial T/\partial t + \boldsymbol{u}\cdot\nabla T = \mathscr{P}^{-1}\nabla^2 T + Q. \tag{2.25}$$

Numerical estimates of some of the foregoing dimensionless parameters, and some of the interesting, if tentative, conclusions to which they have led (largely concerning Jupiter, the one major planet for which appropriate observations are available) will be considered in the following sections. We conclude the present section with a few remarks on driving mechanisms and kinetic energy sources.

(c) KINETIC ENERGY SOURCES. "Motions in the atmospheres of the major planets derive their kinetic energy not only from solar heating but also from internal sources. Öpik [**49**], for example, believes that the rate at which internal sources supply thermal energy to Jupiter's atmosphere is between 0.8 and 1.6 times that associated with solar radiation, the corresponding quantity for Saturn being slightly higher. For the sake of comparison, the gravitational energy of the planet divided by the solar energy intercepted by the planet in 4×10^9 years is about 50 in the case of Jupiter and 20 in the case of Saturn (the corresponding quantity for the Earth is about 3×10^{-2}). It is conceivable, therefore, that the internal sources are due to gravitational energy release." (From Hide [**27**].) Low[3] makes the same point graphically when he points out that the implied collapse rates are about 1 mm year^{-1} (see also Smoluchowski [**63**]).

The pressure forces that give rise to motions in the *Earth's* atmosphere result entirely from the effect of gravity on the horizontal, almost north-south, density gradient produced by differential solar heating. In the theory of these motions, whose *raison d'être* is the horizontal transfer of heat from equatorial to polar regions, the Burger number plays a key role (see Equation (2.14)). It is significant not only that the Burger number is positive for the Earth's atmosphere, but also that it is of the order of but somewhat less than unity. In these circumstances, an efficient horizontal heat transfer mechanism, involving nonaxisymmetric baroclinic waves, is possible.

Owing to their great depths and internal energy sources, the Burger numbers for the atmospheres of the major planets should be very much less than unity and might even be negative (Hide [**27**]). Future investigations should include systematic theoretical studies of systems characterized by small positive and negative Burger numbers over a wide range of the

[3] Paper presented at the Third Arizona Conference on Planetary Atmospheres, 1969 (see [**76**]).

other parameters of the problem (including $\mathscr{R}_m$, $\mathscr{Y}_m$ and $\mathscr{M}$; see Equations (2.18), (2.15) and (2.17) and §4). It goes almost without saying that direct information about the thermal structure of the atmospheres of the major planets will be of crucial dynamical importance.

3. Rotational effects. The rotation of the planet Jupiter on its axis (Campani [**5**], Peek [**51**]) approximately once every 10 hours is by far the most important dynamical influence on atmospheric motions. If we take for nonequatorial regions $\Delta U \sim 1 \text{ m sec}^{-1}$, $\Delta L \sim 10^7$ m, and $\Omega \sim 10^{-4} \text{ sec}^{-1}$, then the Rossby number [Equation (2.1)] $\mathscr{Y} < 10^{-3}$ (Hide [**25**]). It has been argued that $\nu < 10^2 \text{ m}^2 \text{ sec}^{-1}$ (Hide [**26**]) in which case the Ekman number (Equation (2.3)) $\mathscr{E} < 10^{-8}$. The Great Red Spot (Hide [**25**]), the banded structure (Wasiutyǹski [**73**], Stone [**69**], Ingersoll and Cuzzi [**42**]), the equatorial jets (Schoenberg [**60**], Hess [**23**], Lorenz [**45**], Hide [**27**], Gierasch and Stone [**16**]) and certain dynamical properties of features such as the South Tropical Disturbance (Hide [**26**], see below) are probably direct consequences of, among other things, these very small values of $\mathscr{Y}$ and $\mathscr{E}$.

(a) THE GREAT RED SPOT. This most prominent feature of Jupiter's atmosphere (Hooke [**39**], Peek [**51**], Focas [**15**], Smith and Tombaugh [**62**], Solberg [**64**], Chapman [**6**], Reese and Smith [**55**]) has not yet received a completely satisfactory explanation, although the "Taylor column" hypothesis proposed by Hide [**24**] seems capable of accounting for many of its observed properties (see Sagan [**59**], cf. DeMarcus and Wildt [**10**]).

For the purpose of exposition, the simplest conceivable model of Jupiter's atmosphere was used in making the original proposal (cf. Equation (2.23)). In that and in subsequent work, the assumptions underlying the simple model were examined (Hide [**26**]); the implications of the model with regard to the depth, thermal structure and other properties of Jupiter's atmosphere (Hide [**26**], [**27**], Sagan [**59**]) and to the dynamics and magnetohydrodynamics of Jupiter's interior (Hide [**26**], [**27**], [**31**], Runcorn [**58**]) were considered; and laboratory experiments and related theoretical investigations were carried out (Hide and Ibbetson [**35**], Hide [**30**], Titman [**71**]; see also Hide et al. [**36**], Ingersoll [**41**], cf. Greenspan [**21**]). It would appear from this work that the Taylor column hypothesis continues to have a useful role to play.

One of the problems arising is concerned with hydrodynamical effects due to spatial variations in density (cf. Equation (2.22)), associated with inhomogeneities in temperature (and possibly chemical composition; see Equation (2.3)). This problem has not yet been solved in requisite mathematical detail and further research is certainly needed. Nevertheless, tentative considerations of certain simple hydrodynamical models, characterized, among other things, by very large values of the Péclét number $\mathscr{P}$ (see Equations (2.19) and (2.25)), indicate that the conditions under which the effects of density inhomogeneities in Jupiter's atmosphere would permit "Taylor columns" to form are not so special as to be unlikely.[4]

What appears on the face of it to be exactly the opposite conclusion has been stated by Stone and Baker [**70**] in their recent discussion of the same problem. It would seem, however, that their main argument is based on a linearized theory of heat and momentum transfer that fails to take advective processes properly into account, unless κ has the incredibly high value of 10^6 m^2 sec^{-1} (in which case $\mathscr{P} < 1$; see Equations (2.19) and (2.25)) and ν has a comparable value (in which case the Ekman number $\mathscr{E}$—see Equations (2.13) and (2.20)—and not the Rossby number $\mathscr{Y}$—see Equation (2.11)—is the leading measure of ageostrophic effects). Their conclusion is not upheld when effects due to horizontal advection are included in the analysis. Their arguments apply only to the Jacobs-type Taylor column (whose relationship to other types I have discussed elsewhere; see Hide et al. [**36**]) in which ageostrophic effects are due entirely to viscosity; but, so far as I know, no one has ever suggested that such an unlikely phenomenon could occur in Jupiter's atmosphere. Indeed, the Jacobs-type Taylor column is hard to produce even on the small scale of the laboratory, owing to the difficulty of rendering other ageostrophic effects negligible in comparison with those due to viscosity.

(b) BANDED APPEARANCE OF JUPITER (AND SATURN). It is generally accepted that the banded appearance of Jupiter and Saturn reflects the influence of rotation on atmospheric motions, but no theoretical model that bears close physical scrutiny and comparison with the observations has yet been proposed.

Wasiutyński [**73**] seems to have carried out the first theoretical investigation of this problem, in terms of a model characterized, among other things, by the assumption that, owing to internal heat sources, the lapse rate in Jupiter's atmosphere is superadiabatic (i.e., a negative Burger number; see §2). More recently, Stone [**69**] introduced a different model characterized, among other things, by the assumptions that differential solar heating is largely responsible for Jupiter's atmospheric motions (cf. §2), that the basic flow is mainly zonal (thermal wind; see Equation (2.22)), that the lapse rate is subadiabatic, and that the corresponding Richardson number Ri lies between 0.25 and 0.95.

---

[4] See pages 305–307.

It is known from important work of Stone [**68**] extending earlier work on the stability of axisymmetric stratified shear flows, that within this Richardson number range the most unstable disturbances are baroclinic, axisymmetric, and grow very rapidly, with typical growth times of the order of Ω^{-1} sec. Stone [**69**] conjectures that Jupiter's banded structure reflects the presence of instabilities of this type after they have stopped growing, supposing that the main properties of the fully developed disturbances resemble those of the incipient instabilities. As the theory of fully developed disturbances has not yet been worked out, it is impossible to make a useful quantitative test of Stone's suggestion, which would seem, among other things, to imply horizontal temperature variations over Jupiter's visible disk that are considerably greater than those observed (see Wildey et al. [**74**]). One attractive feature of Stone's suggestion is its ability to account in terms of baroclinic instabilities for both nonaxisymmetric as well as axisymmetric disturbances within the limited range of Richardson number required for his idea to work. However, if $\mathrm{Ri} < 0.25$, then the nonaxisymmetric disturbances could be due to barotropic instabilities, as seems to be indicated by a recent analysis of observations by Ingersoll and Cuzzi [**42**].

Gierasch and Stone [**16**], apparently unaware of previous theoretical work on Jupiter's equatorial jet, cite as evidence in favour of Stone's explanation of the banded structure their claim that the axisymmetric disturbances invoked are capable of advecting zonal momentum toward the warmer parts of the fluid and thus provide a mechanism for driving the jet (see below). However, there are difficulties with these arguments. They are qualitative; only the directions and not the magnitude of the rates of momentum and heat transport have been assessed. Advective processes in directions perpendicular to the rotation axis are impossible when the flow is both geostrophic and axisymmetric (Hide [**32**]), for the simple reason that such processes are associated with motions perpendicular to the rotation axis, which in geostrophic flow must be supported by azimuthal gradients of pressure; such motions vanish when the pressure field is axisymmetric.

It follows, then, that the advective processes invoked by Stone [**69**] and by Gierasch and Stone [**16**] must be associated with departures from geostrophy. While these departures are quite large for the rapidly growing disturbances on which these authors base their analyses, they will be very much less for fully developed quasi-steady motions unless very strong horizontal shears develop in certain regions. It might be possible to find a reasonably efficient horizontal heat transfer mechanism by invoking strong shears, but when the spherical geometry of the system is taken properly into account, the qualitative mechanism proposed by Gierasch and Stone [16] for producing and maintaining the zonal momentum of the equatorial jet would seem to be physically impossible (see below).

(c) EQUATORIAL JETS. The existence of equatorial jets in the fluid layers of rapidly rotating planets seems fairly general; Jupiter and Saturn exhibit such currents and on Earth we have the Cromwell current in the ocean and the Berson current in the lower stratosphere. These jets, which are the subject of a great deal of current theoretical research, are not yet fully understood.

If the reasonable assumption is made that the essential vorticity balance is between horizontal advection of relative vorticity and effects due to the variation of Coriolis parameter with latitude, then $(UR/2\Omega)^{1/2}$ is an approximate expression for the latitudinal width of a jet of typical flow speed U relative to a rotating planet of radius R; the expression agrees satisfactorily with observations (Hide [27]).

These jets in all probability represent sinks for energy and angular momentum originating at higher latitudes and advected horizontally toward the equator. Considerations of the general properties of thermally driven motions in a rotating spherical shell of fluid of outer radius R and ratio of total angular momentum to moment of inertia equal to $\mathbf{\Omega}$ indicate that in order to account for a westerly (i.e., faster than the basic rotation) equatorial jet without appealing to sinking motions from higher levels in the atmosphere, effects due to local azimuthal pressure gradients cannot be neglected, as these gradients provide the only forces (in the absence of magnetohydrodynamic effects) capable of increasing the angular momentum per unit mass of an individual fluid element to a value $> \Omega R^2$. With certain forms of flow pattern angular momentum is transferred upgradient, an effect which, so far as the equations governing the zonally-averaged flow are concerned, can be represented by a pseudo-viscous term in which the "coefficient of viscosity" is negative (see Jeffreys [43], Lorenz [46], Starr [66]).

The fluid motions invoked by Schoenberg [60] (see Hess [23]) and by Gierasch and Stone [16] in their attempts to account for Jupiter's equatorial jet without invoking either a strong source of angular momentum deep within the atmosphere (which would conflict with the observed value of the "radio period"; see §4) or sinking motions in the high atmosphere above the jet, are characterized by symmetry about the axis of rotation, the concomitant pressure gradients having no azimuthal component. According to the foregoing arguments, only easterly jets can be accounted for in this way; the equatorial jets of Jupiter and Saturn are westerly (see [27] and [34]).

(d) AN EFFECT DUE TO HYPERSONIC ROTATION(?). The time scales $\Delta\tau$ on which nonpermanent planetary-scale features undergo significant changes are such that $\mathscr{F}\mathscr{Y} < 1$ (see Equation (2.20)); indeed, these time scales can be incredibly long in comparison with the time required for ordinary planetary waves to disperse the kinetic energy of a typical disturbance (see Hide [**26**]), but not in comparison with the time required for effects arising in viscous boundary layers (see Equation (A30)) to produce significant dissipation, $D/(\nu\Omega)^{1/2}$ sec, assuming that the depth D of the atmosphere is $\ll 10^6$ m (Hide [**25**], [**26**], [**31**], Goldreich and Soter [**19**]) and that $\nu < 10^2 \text{ m}^2 \text{ sec}^{-1}$. A completely satisfactory explanation of this discrepancy has not yet been given, but the following tentative lines along which an explanation might be sought were indicated several years ago by Hide [**26**].

The usual theory of ordinary planetary waves (cf. Equation (3.2)) applies to the case when rotational Mach number $\mathscr{M}$ (see Equations (2.17) and (2.21)) is zero, whereas $\mathscr{M}$ is as high as 10 for Jupiter and not much less for the other major planets, all of which rotate at hypersonic speeds with respect to the speed of sound in their outer, cooler layers (Hide [**26**], Golitsyn and Dikii [**20**]). When $\mathscr{M} \neq 0$, the dispersion relationship for ordinary planetary waves has the form

$$\omega \doteqdot -\beta k(1 - \omega^2/k^2c^2)/[k^2 + l^2 + (f^2 - \omega^2)/c^2], \tag{3.1}$$

where ω denotes angular frequency, k and l the east-west and north-south wavenumbers, respectively, f the Coriolis parameter (see Equation (A12)), β the rate of change of the Coriolis parameter with latitude, and c the speed of sound. This reduces when $\mathscr{M} = 0$ to the well-known "Rossby-Hauwitz" formula, namely

$$\omega \doteqdot -\beta k/(k^2 + l^2). \tag{3.2}$$

According to Hide [**26**], typical periods associated with some of the oscillatory phenomena reported by Jupiter observers (Peek [**51**]; see also Solberg [**65**]) are, at several months, comparable with those expected on the basis of Equation (3.1), but are considerably longer than those based on (3.2). Moreover, planetary waves in hypersonically rotating fluids lose their dispersive properties, as may be shown by setting $k^2 + l^2 \ll f^2/c^2$ and $\omega^2 \ll k^2c^2$ in (3.1), according to which ω/k, the phase velocity, is then the same as the group velocity, $\partial\omega/\partial k$. It was in terms of this result that Hide proposed an explanation of the great duration of Jupiter's South Tropical Disturbance, which first arose in 1901 and gave way in 1939 to three prominent white spots (see Peek [**51**]) which can be seen at the present day.

4. Magnetohydrodynamic effects. Radio-astronomical data (Warwick [72]) indicate that the strength of the magnetic field in the vicinity of Jupiter's visible surface is about 5×10^{-3} Webers (Wb) m^{-2} (50 gauss, about 100 times that of the magnetic field at the surface of the Earth). This field is due, presumably, to electric currents circulating in regions of the interior of the planet where the electrical resistivity η is not too high. If the field is not primordial in origin, then fluid motions in these regions are the most likely cause of the electric currents.[5]

Hide [27], [28], [31] has suggested (1) that D, the depth of the fluid layer underlying Jupiter's visible surface, $\gtrsim 10^7$ m, and (2) that, in consequence, the magnetic Reynolds number $\mathscr{R}_m$ (see Equation (2.18)) might attain moderately high values in the lower reaches of the layer, sufficient for the fluid motions there to generate, by "motional induction" (see Equations (2.25), (2.9) and (2.24)), electric currents that produce, or at least modify, the magnetic field found at higher levels. (When $\mathscr{R}_m \gg 1$ lines of magnetic force move with the conducting fluid. Hence, the rotation period of sources of decameter and decimeter radiation, $9^h\, 55^m\, 29.7^s$ at the present time, might be a rough measure of the average zonal motion of Jupiter's lower atmosphere (see Hide (loc. cit)); this motion is westerly (eastward) relative to the surface underlying the atmosphere if the Great Red Spot, whose present rotation period is about 10 sec longer than the radio period, is rooted in that surface (Hide [25]).) The dynamical effects of ponderomotive forces associated with motional induction (the $\boldsymbol{j} \times \boldsymbol{B}$ term in (2.1)) on the fluid motions in the lower reaches of Jupiter's atmosphere, a measure of which involves the "magnetic Rossby number" $\mathscr{Y}_m$ (see Equations (2.15) and (2.20)), might be sufficiently important to warrant detailed investigation.

The bases of these suggestions and some of their consequences in regard to the momentum, angular momentum, and energy balance of Jupiter's interior have been discussed elsewhere. Suffice it to remark here (1) that $\mathscr{R}_m$ might attain values as high as 10 if $\eta < 10^{-3}\Omega$ m (the corresponding electrical conductivity η^{-1} being only 10^{-3} times that of the metallic hydrogen presumed to be the principal constituent of the main body of the planet (see Smoluchowski [63]), out to a distance from the center of about 0.8 times the radius of the planet (7×10^7 m)), and (2) that $\mathscr{Y}_m$ is $\sim 10^{-4}$ if for ΔB we take 5×10^{-3} Wb m^{-2} (50 gauss), its value at the visible surface, or about 10^{-3} or 10^{-2} if we accept certain arguments (Hide [31]) that ΔB might be as high as 10^{-1} or 1 Wb m^{-2} (10^3 or 10^4 gauss). Observe that when $\mathscr{Y}_m > \mathscr{Y}$ a quantity proportional to $\mathscr{Y}_m^2/\mathscr{Y}$ replaces $\mathscr{Y}$ as a measure of departures from geostrophy (see (4.3)).

[5] See pages 297–299 and 325–327.

A certain amount of work on the magnetohydrodynamics of rotating fluids has been carried out in connection with theories of solar phenomena and of fluid motions in the liquid core of the Earth, where the geomagnetic field is produced by magnetohydrodynamic processes (Bullard and Gellman [**4**], Elsasser [**13**]) but a great deal remains to be done. In the treatment of certain processes, e.g., inertial, internal,[6] and planetary waves (Lehnert [**44**], Chandrasekhar [7], Hide [**29**], [**33**]) and baroclinic instability (Gilman [**17**]), ν and λ have been set equal to zero (see (2.13), (2.10) and (2.18)). It is interesting to consider the dispersion relationship for planetary waves, which then takes the form (see Hide [**29**]):

$$\text{(4.1)}^{7} \qquad \omega \doteqdot -\beta k/(k^2 + l^2) + V^2(k\cos\theta + l\sin\theta)^2/\omega,$$

(cf. Equation (3.2)) when the direction of basic magnetic field, whose strength is proportional to the Alfvén speed V (cf. (2.16) and (A13)) makes an angle θ with latitude circles (and the fluid is incompressible). For planetary-scale disturbances, the larger root of (4.1) corresponds to a Rossby-Haurwitz wave (see Equation (3.2)), whose properties, as we have seen, are greatly influenced by compressibility effects (see Equation (3.1)). The other root,

$$\text{(4.2)} \qquad \omega \sim V^2(k^2 + l^2)(k\cos\theta + l\sin\theta)^2/(k\beta),$$

corresponds to a slow wave of a new type (and upon which it may be shown that effects due to compressibility should not be serious).

When ν and λ cannot be set equal to zero, further complications arise. The dimensionless parameter

$$\text{(4.3)} \qquad \alpha \equiv (\Delta B)^2/(\Omega\rho\eta) = (\Delta V)^2/(\Omega\lambda) = \mathscr{R}_m\mathscr{Y}_m^2/\mathscr{Y}$$

(cf. Equations (2.18), (2.15), (2.11) and (A21)) is then a measure of ageostrophic effects due to magnetohydrodynamic processes (see Hide [**24**]), $\mathscr{E}$, the Ekman number (Equation (2.13)), being the corresponding measure of the importance of viscous processes. α exceeds 10^{-2} in regions where $\mathscr{R}_m > 10$, $\mathscr{Y}_m > 10^{-3}$ and $\mathscr{Y} \sim 10^{-3}$ (see above). Future investigations should obviously include systematic theoretical studies of the effects of varying the parameter α, as well as attempts to determine the quantities (η, ΔB, etc.) upon which the numerical values of these parameters depend. The discussion of boundary layers presented in the following Appendix provides some insight into the physical significance of α [**77**].

---

[6] See pages 327–328.

[7] See pages 346–350.

Appendix: The Ekman–Hartmann boundary layer at the rigid bounding surface of an electrically-conducting rotating fluid in the presence of a magnetic field.

Summary. The effects of rotation and a magnetic field on the structure of the viscous boundary layer that arises near the rigid bounding surface Σ of an electrically-conducting fluid in slow hydrodynamical motion are analyzed theoretically. The relative importance of the magnetic field to rotation is measured in terms of the dimensionless parameter $\alpha \equiv V^2/f\lambda$, where V is the Alfvén speed based on the component of the magnetic field normal to Σ, $\frac{1}{2}f$ is the component of the basic rotation vector in the same direction, and $\lambda \equiv \eta/\mu$, μ being the magnetic permeability and η the electrical resistivity of the fluid. The boundary layer is of the Ekman type, thickness $(2\nu/f)^{1/2}$, when $\alpha \ll 1$ (where ν denotes kinematic viscosity), and of the Hartmann type, thickness $(\nu\lambda)^{1/2}/V$, when $\alpha \ll 1$.

The results of the analysis are used to establish the boundary conditions that must be satisfied at Σ by the fluid velocity and magnetic field in the treatment of theoretical models for which ν and λ are taken as zero in the main body of the fluid. If $(\nu/\lambda)^{1/2}H(\alpha) \ll 1$ then the tangential components of velocity can be discontinuous at Σ, but if $(\nu/\lambda)^{1/2}H(\alpha) \gg 1$ then the viscous "no-slip" condition applies, although an electric current sheet is then permitted, across which the tangential component of the magnetic field undergoes a jump in both magnitude and, because of rotation, in direction. The discontinuity required in the *normal* component of velocity in order to take boundary-layer suction into account is proportional to $(\nu/2f)^{1/2}G(\alpha)$ and the analogous discontinuity in the normal component of electric current density is proportional to $(\rho\nu/\eta)^{1/2}H(\alpha)$ (where ρ denotes density). Here $H(\alpha)$ and $G(\alpha)$ are explicit functions with the following properties: $H(0) = 0$, $H(\alpha) = (\alpha/2)^{1/2}$ when $\alpha \ll 1$; $H(\alpha) \doteq 1$ when $\alpha \gg 1$; $G(0) = 1$, $G(\alpha) = (2\alpha^3)^{-1/2}$ when $\alpha \gg 1$.

(a) INTRODUCTION. The boundary layer that arises near the rigid bounding surface of a rapidly rotating, incompressible fluid of low but nonzero viscosity in which slow steady relative motions are occurring is of the Ekman type (Ekman [**12**]; see also Prandtl [**53**], Greenspan [**21**]) with thickness $(\nu/|\mathbf{\Omega}\cdot\boldsymbol{n}|)^{1/2}$, where ν is the coefficient of kinematic viscosity, $\mathbf{\Omega}$ the angular velocity of basic (uniform) rotation, and $\boldsymbol{n}$ a unit vector normal to the surface and directed toward the fluid. The corresponding boundary layer at a rigid bounding surface of an electrically-conducting, nonrotating, incompressible fluid pervaded by a magnetic field is of the Hartmann type (Hartmann [**22**]; see also Cowling [**8**], Ferraro and Plumpton [**14**], Hide and Roberts [**38**], Alfvén and Fälthammar [**2**], Shercliff [**61**], Roberts [**56**]), with thickness $(\eta\nu\rho/B_0^2)^{1/2}$, where ρ is the density of the fluid, η its electrical resistivity, and B_0 the strength of the normal component of the magnetic field at the bounding surface (all in rationalized MKS units). Complications arise when the flow is not steady and combined rotational and magnetohydrodynamic effects occur (see Hide and Roberts [**37**]). Some of these complications are discussed further in what follows; related work has been reported by Backus [**3**] and Gilman and Benton [**18**].

According to Equations (2.1)–(2.10)

$$\begin{aligned} &\partial \boldsymbol{u}/\partial t + (\boldsymbol{u} \cdot \nabla)\boldsymbol{u} + 2\boldsymbol{\Omega} \times \boldsymbol{u} \\ &\quad = -(1/\rho)\nabla p + (1/\mu\rho)(\nabla \times \boldsymbol{B}) \times \boldsymbol{B} + \nu\nabla^2 \boldsymbol{u} + \boldsymbol{g}, \end{aligned} \tag{A1}$$

$$\nabla \cdot \boldsymbol{u} = 0, \tag{A2}$$

$$\partial \boldsymbol{B}/\partial t = \lambda \nabla^2 \boldsymbol{B} + \nabla \times (\boldsymbol{u} \times \boldsymbol{B}), \tag{A3}$$

(together, in general, with appropriate equations of state and heat transfer, see Equations (2.3) and (2.4)) govern the flow when the fluid is incompressible. These equations suffice, when combined with appropriate boundary conditions, to determine the four unknowns ρ, p, $\boldsymbol{u}$ and $\boldsymbol{B}$.

Assume that the conducting fluid is in contact over the whole or part of its boundaries with rigid, impermeable, solid surfaces that are fixed in the rotating frame. Consider an area Σ of the interface between the solid and the fluid; at that interface the boundary conditions

$$\boldsymbol{u} \cdot \boldsymbol{n} = 0, \qquad \nu \boldsymbol{u} \times \boldsymbol{n} = 0, \tag{A4}$$

must be satisfied.

(b) BOUNDARY LAYER STRUCTURE. When ν, though nonzero, is not too large, viscous forces will be weak except in a thin boundary layer of thickness Δ, where

$$\Delta \ll L, \tag{A5}$$

separating the interface Σ from the effectively inviscid "interior" region of the fluid, L being a length characteristic of the scale of the "interior" motions.

We denote by ϕ a dummy variable representing p, ρ or any component of $\boldsymbol{u}$ or $\boldsymbol{B}$, and separate ϕ into its "interior" part and its "boundary-layer" part by writing

$$\phi = \bar{\phi} + \hat{\phi}, \tag{A6}$$

where $\hat{\phi}$ is that part of ϕ which undergoes significant variations in the boundary layer, but only in directions parallel to $\boldsymbol{n}$ (see (A10)), and vanishes in the "interior," where $\phi = \bar{\phi}$. The gradients of p and $\boldsymbol{B} \cdot \boldsymbol{n}$ will be small within the boundary layer, and we assume here that the gradients of ρ are also negligible there, so that we can set[8]

(A7) $$\hat{\rho} = \hat{\boldsymbol{B}} \cdot \boldsymbol{n} = \hat{p} = 0.$$

On combining (A1), (A2), (A3), (A7) and (A8), making use of the usual boundary layer approximations, in which gradients in directions parallel to Σ are treated as negligible in comparison with gradients normal to Σ (see (A5)), and introducing the further assumption that the curvature of Σ is so small (i.e., $\ll L^{-1}$) that

(A8) $$|2\boldsymbol{\Omega} \times \boldsymbol{u}| \gg |(\boldsymbol{u} \cdot \nabla)\boldsymbol{u}|$$

everywhere within the boundary layer, we find that

(A9) $$\partial(\hat{u}, \hat{v})/\partial t + f(-\hat{v}, \hat{u}) = \nu D^2(\hat{u}, \hat{v}) + VD(\hat{B}_x, \hat{B}_y),$$

(A10) $$(\partial/\partial t)(\hat{B}_x, \hat{B}_y) = V(\mu\rho)^{1/2}D(\hat{u}, \hat{v}) + \lambda D^2(\hat{B}_x, \hat{B}_y).$$

Here

(A11) $$\hat{\boldsymbol{u}} = (\hat{u}, \hat{v}, \hat{w}), \qquad \hat{\boldsymbol{B}} = (\hat{B}_x, \hat{B}_y, \hat{B}_z),$$

(A12) $$f \equiv |2\boldsymbol{\Omega} \cdot \boldsymbol{n}|, \qquad D \equiv d/dz = \boldsymbol{n} \cdot \nabla,$$

(A13) $$V \equiv [\bar{\boldsymbol{B}}(z = 0) \cdot \boldsymbol{n}]/(\mu\rho)^{1/2},$$

where the modulus of V is the "Alfvén speed," (x, y, z) being the coordinates of a point fixed in the rotating frame, referred to a coordinate system whose x and y axes lie within the surface Σ, whose z axis is normal to Σ and directed into the fluid, and which is right-handed or left-handed according as $\boldsymbol{\Omega} \cdot \boldsymbol{n} \gtrless 0$; $\boldsymbol{n}$ is a unit vector along the positive z axis.

We eliminate any three of the four dependent variables $\hat{u}$, $\hat{v}$, $\hat{B}_x$, $\hat{B}_y$ from (A9) and (A10) and show that each of them satisfies

(A14) $$[(\partial/\partial t - \lambda D^2)(\partial/\partial t - \nu D^2) - V^2D^2]^2\hat{\phi} + f^2(\partial/\partial t - \lambda D^2)^2\hat{\phi} = 0$$

(cf. Equation (29) of Hide and Roberts [**37**]).

Neither the velocity nor the magnetic field can undergo discontinuous changes; hence

(A15a) $$\hat{\boldsymbol{u}} = -\bar{\boldsymbol{u}}, \quad \text{and } \boldsymbol{B} \text{ is continuous at } z = 0$$

---

[8] $\hat{\rho} = 0$ is satisfied trivially in the case of an incompressible barotropic fluid for which, by definition, $\nabla\rho = 0$ everywhere. However, setting $\hat{\rho} = 0$ in the case of a baroclinic fluid will be a valid assumption only when the field of temperature, etc., on which ρ depends does not undergo rapid variations in the boundary layer, i.e., when κ, the "density diffusion coefficient" (thermal diffusivity when baroclinicity is due solely to temperature variations) is large. The assumption should be valid when κ is not too small in comparison with ν and λ, as, for instance, in the case of the Earth's liquid core, for which $\kappa \sim 10^3$ m^2 sec^{-1} (see Bullard and Gellman [**4**]), $\lambda \sim 1$ m^2 sec^{-1}, and $\nu \ll 10^3$ m^2 sec^{-1}; the extension of the present work to cases when κ is so small that $\hat{\rho} \neq 0$ might be of general theoretical interest.

are boundary conditions under which (A13) must be solved. According to their definitions, $\hat{\boldsymbol{u}}$ and $\hat{\boldsymbol{B}}$ vanish in the interior region; that is to say,

$$\hat{\boldsymbol{u}} = \hat{\boldsymbol{B}} = 0, \quad \text{when } z \gg \Delta. \tag{A15b}$$

If time variations of the "interior" flow are harmonic, with positive angular frequency ω, and the z variation of $\hat{\phi}$ has the form $\exp(\gamma z)$, then

$$\hat{\phi} \propto \exp(i\omega t + \gamma z), \tag{A16}$$

and (A14) and (A15) are satisfied if

$$[(i\omega - \lambda\gamma^2)(i\omega - \nu\gamma^2) - V^2\gamma^2]^2 + f^2(i\omega - \lambda\gamma^2)^2 = 0, \tag{A17}$$

and we choose those roots for which

$$\mathrm{Re}(\gamma) < 0. \tag{A18}$$

If we interpret Δ, the boundary-layer thickness, as the distance from Σ in which $|\hat{\boldsymbol{u}}|$ decreases to $(1 - e^{-1})$ of its value at $z = 0$ (see (A20a)), then

$$\Delta^{-1} = -\mathrm{Re}(\gamma). \tag{A19}$$

Though in general quite complicated, in the absence of magnetohydrodynamic effects (i.e., when $V/\lambda = 0$), acceptable solutions of (A18) are

$$\gamma = ((\omega \pm f)/2\nu)^{1/2}(-1 - i), \quad \text{when } \omega > f,$$

$$\gamma = ((f + \omega)/2\nu)^{1/2}(-1 - i) \quad \text{and} \quad ((f - \omega)/2\nu)^{1/2}(-1 + i), \quad \text{when } \omega < f,$$

which reduce to the case first discussed by Ekman (see above) in connection with ocean currents when $\omega = 0$, and to the case first discussed by Stokes (see Rayleigh [**54**]) in an investigation of the attenuation of sound waves, when $f = 0$. When $f = 0$ but $V/\lambda \neq 0$, we have a variety of magnetohydrodynamic cases discussed in detail by a number of authors (see Hide and Roberts [**37**], Roberts [**56**]); when, in addition, $\omega = 0$ we have the Hartmann case for which

$$\gamma = -V/(\lambda\nu)^{1/2}$$

(see Stewartson [**67**]). In what follows we shall discuss in detail the case of steady flows subject to the combined influence of rotation and a magnetic field (i.e., $\omega = 0$, but $f \neq 0$, $V/\lambda \neq 0$).

(c) Steady Flows. When $\omega \ll f$, we can ignore time variations and set $\omega = 0$ in (A23), which then has acceptable solutions (see (A18)):

$$\gamma = -(f/\nu)^{1/2}(1 + \alpha^2)^{1/4}\{\cos\vartheta \mp i\sin\vartheta\}, \tag{A20a}$$

where

(A20b) $$\vartheta \equiv \tfrac{1}{2}\cot^{-1}\alpha,$$

(A21) $$\alpha \equiv V^2/(f\lambda) = \eta^{-1}B_0^2/(\rho|2\boldsymbol{\Omega}\cdot\boldsymbol{n}|),$$

α being (cf. Equation (4.3)) a dimensionless measure of the relative importance of magnetohydrodynamic to rotational effects (see Hide [**24**]); it is the ratio of the speed with which electromagnetic disturbances are transmitted by magnetohydrodynamic waves modified by rotation (see Lehnert [**44**], Chandrasekhar [7]), proportional to V^2/f, to that at which they are transmitted by diffusive processes, proportional to λ.

We denote by $\tilde{\boldsymbol{u}}$ and $\tilde{\boldsymbol{B}}$ the interior values of $\boldsymbol{u}$ and $\boldsymbol{B}$ evaluated at $z = 0$ (see (A6)), i.e., $\tilde{\boldsymbol{u}} \equiv \bar{\boldsymbol{u}}\ (z = 0)$, etc., and choose the orientation of the axes of coordinates so that $\tilde{v}$, the y component $\tilde{\boldsymbol{u}}$, is zero. By (A9), (A10), (A15) and (A19),

(A22a) $$\hat{u} + i\hat{v} = -\tilde{u}\exp - \{(1+\alpha^2)^{1/4}(\cos\vartheta + i\sin\vartheta)\zeta\},$$

(A22b)

$$\hat{B}_x + i\hat{B}_y = -\frac{\tilde{u}}{\lambda}\left(\frac{v}{f}\right)^{1/2}\frac{\tilde{\boldsymbol{B}}\cdot\boldsymbol{n}}{(1+\alpha^2)^{1/4}}\exp - \{(1+\alpha^2)^{1/4}(\cos\vartheta + i\sin\vartheta)\zeta + i\vartheta\},$$

where

(A23) $$\zeta \equiv z/(v/f)^{1/2},$$

a stretched coordinate. The dependence on α of the factor multiplying ζ in the exponential function is illustrated by Figure 1.

Equations (A22) reduce to well-known cases when $\alpha = 0$, corresponding to rotational effects only, and when $\alpha^{-1} = 0$, corresponding to magnetohydrodynamic effects only. In the former case we have the Ekman boundary layer, for which

(A24)

$$\hat{u} + i\hat{v} = \tilde{u}\exp(-\zeta/2^{1/2})[-\cos(\zeta/2^{1/2}) + i\sin(\zeta/2^{1/2})],\ \hat{B}_x + i\hat{B}_y = 0;$$

in the latter case, we have the Hartmann boundary layer, for which

(A25) $$\begin{aligned} \hat{u} &= -\tilde{u}\exp(-\alpha^{1/2}\zeta),\\ \hat{B}_x &= -(v\mu\rho/\lambda)^{1/2}\tilde{u}\exp(-\alpha^{1/2}\zeta),\\ \hat{v} &= \hat{B}_y = 0. \end{aligned}$$

Observe that $\zeta\alpha^{1/2} = zV/(v\lambda)^{1/2}$.

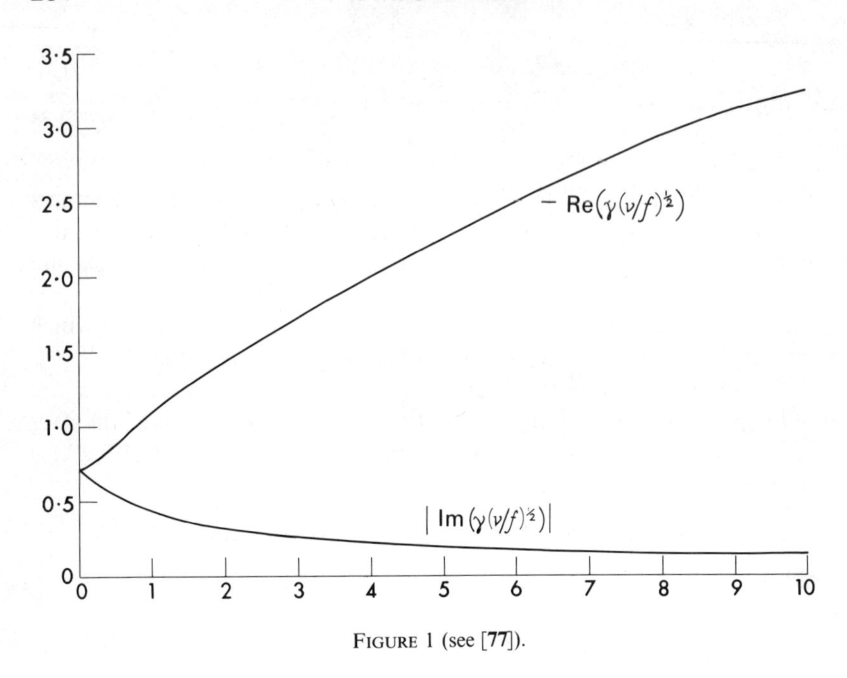

FIGURE 1 (see [77]).

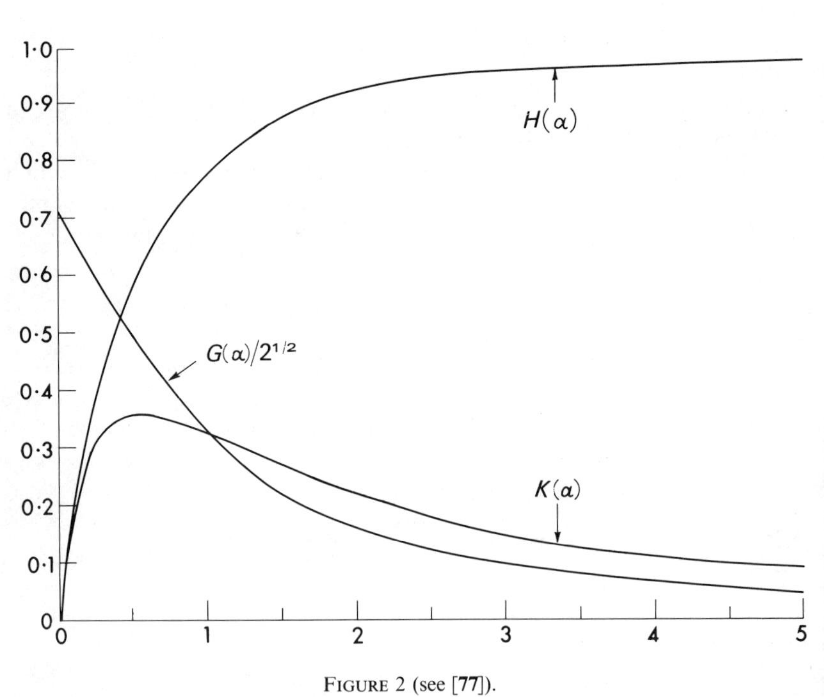

FIGURE 2 (see [77]).

(d) "JUMP CONDITIONS" AND BOUNDARY-LAYER SUCTION. We denote by $[\hat{\phi}]$ the change in $\hat{\phi}$ across the boundary layer, i.e.

$$[\hat{\phi}] \equiv \hat{\phi}(z \gg \Delta) - \hat{\phi}(z = 0). \tag{A26}$$

By (A6) and (A22),

$$[\hat{B}_x + i\hat{B}_y] = [\hat{u}](\mu\rho)^{1/2}[\operatorname{sgn}(\hat{\boldsymbol{B}} \cdot \boldsymbol{n})](\nu/\lambda)^{1/2}\{H(\alpha) - iK(\alpha)\}, \tag{A27a}$$

where

$$H(\alpha) - iK(\alpha) \equiv (\alpha^{1/2}/(1 + \alpha^2)^{1/2}) \exp - ((i/2) \cot^{-1} \alpha). \tag{A27b}$$

As shown in Figure 2, $H(\alpha) = 1$ and $K(\alpha) = 0$ when $\alpha^{-1} = 0$; rotational effects are then absent and the boundary layer is of the Hartmann type, with the velocity and magnetic field in the boundary layer parallel to the local interior flow. The expression for $[\hat{B}_x]$ then reduces to one found by Stewartson [**67**] in his discussion of the boundary conditions to be used at a rigid surface when dealing with problems in which the interior fluid is regarded as inviscid (i.e. $\nu = 0$) and perfectly conducting (i.e. $\lambda = 0$), namely,

$$[\hat{B}_x] = [\hat{u}] \operatorname{sgn}(\tilde{\boldsymbol{B}} \cdot \boldsymbol{n})(\mu\rho)^{1/2}(\nu/\lambda)^{1/2}. \tag{A28}$$

When $\nu \gg \lambda$ a current sheet can occur, across which $[\hat{B}_x] \neq 0$, but the "no-slip" condition must then be satisfied; at the other extreme, $\nu \ll \lambda$, a vortex sheet can occur, across which $[\hat{u}] \neq 0$, but there can then be no jump in $[\hat{B}_x]$. A current sheet and a vortex sheet cannot occur simultaneously, only one or the other.

When $\alpha \ll 1$ we have, in place of (A28),

$$\operatorname{sgn}(\tilde{\boldsymbol{B}} \cdot \boldsymbol{n})[\hat{B}_x + i\hat{B}_y] = [\hat{u}z](\mu\rho)^{1/2}(\nu/\lambda)^{1/2}(V^2/2f\lambda)^{1/2}(1 - i), \tag{A29}$$

which shows that $[\hat{B}_x]$ is very much less, by a factor $(\alpha/2)^{1/2}$, than in the Hartmann case, $\alpha^{-1} = 0$, and that when $f \neq 0$ there can be a jump in the direction as well as in the magnitude of the tangential component of the magnetic field.

The normal component of $\boldsymbol{u}$ may, as a result of "boundary-layer section," undergo a significant variation across the boundary layer. Because $w = 0$ when $z = 0$, the speed $\tilde{w}$ $[\equiv \bar{w}\ (z = 0)]$ with which fluid passes out of the boundary layer into the interior region is equal to $[\hat{w}]$ (A26), which by (A2), (A6), (A15a) and (A22), is given by

$$[\hat{w}] = -\frac{\partial}{\partial y}\int_0^\infty \hat{v}\, dz = -\left(\frac{\nu}{2f}\right)^{1/2} G(\alpha)\frac{\partial \tilde{u}}{\partial y}, \tag{A30a}$$

where

$$G(\alpha) \equiv \{2^{1/2} \sin(\tfrac{1}{2} \cot^{-1} \alpha)\}/(1 + \alpha^2)^{1/4}. \tag{A30b}$$

When $\alpha = 0$, $G(\alpha) = 1$ (see Figure 2) and (A26) then reduces to the well-known Ekman boundary-layer suction formula relating the normal component of the flow velocity at the edge of the interior region to $\boldsymbol{n} \cdot (\nabla \times \bar{\boldsymbol{u}})_{z=0}$, the component of relative vorticity of the interior flow normal to the bounding surfacc (Prandtl [**53**], Greenspan [**21**]). $G(\alpha)$ falls steadily to zero with increasing α, slowly at first, then more rapidly, and eventually, when $\alpha \to \infty$, as $1/(2^{1/2}\alpha^{3/2})$.

The effect of the boundary layer on the magnetic field in the interior region and in the region outside the fluid, which depends on the constraints imposed at large y on the electric currents induced by the flow, will not concern us here (see Hide and Roberts [**38**]). It is of interest, however, to evaluate $[\hat{j}_z]$, the jump in the normal component of the electric current density $\mu^{-1}\nabla \times \boldsymbol{B}$ (see Equation (2.5)), across the boundary layer. As

$$\text{(A31a)} \qquad \mu[\hat{j}_z] = -\mu \int_0^\infty \left(\frac{\partial \hat{j}_x}{\partial x} + \frac{\partial \hat{j}_y}{\partial y}\right) dz = \frac{\partial}{\partial x}[\hat{B}_y] - \frac{\partial}{\partial y}[\hat{B}_x],$$

by (A33) we have

$$\text{(A31b)} \qquad [\hat{j}_z] = -(\rho\nu/\eta)^{1/2} H(\alpha) \partial\tilde{u}/\partial y;$$

$[\hat{j}_z] \to (\rho\nu/\eta)^{1/2}(\partial\tilde{u}/\partial y)$ when $\alpha \to \infty$, the Hartmann limit, and vanishes when $\alpha = 0$, the Ekman limit (see Figure 2). Note the similarity in form of the expressions for $[\hat{w}]$ in the Ekman limit and $[\hat{j}_z]$ in the Hartmann limit, not a surprising result when we bear in mind the essential physical processes responsible for fluid flow in the boundary layer at right angles to $\tilde{\boldsymbol{u}}$ in the former case, and for electric current flow in the boundary layer in the latter case.

References, Lecture II

1. A. F. O'D. Alexander, *The planet Saturn*, Faber and Faber, London, 1962.

2. H. Alfvén and G. G. Fälthammar, *Cosmical electrodynamics*, 2nd ed. Clarendon Press, Oxford, 1963.

3. G. Backus, *Kinematics of the geomagnetic secular variation in a perfectly conducting core*, Philos. Trans Roy. Soc. London Ser. A **263** (1968), 239–266.

4. Sir E. Bullard and H. Gellman, *Homogeneous dynamos and terrestrial magnetism*, Philos. Trans. Roy. Soc. London Ser. A **247** (1954), 213–278. MR **17**, 327.

5. G. Campani, *On account of the improvement of optik glasses at Rome*, Philos. Trans. Roy. Soc. London Ser. A **1** (1665), no. 1, 2.

6. C. R. Chapman, *The discovery of Jupiter's red spot*, Sky and Telescope **35** (1968), 276–278.

7. S. Chandrasekhar, *Hydrodynamic and hydromagnetic stability*, Clarendon Press, Oxford, 1961. MR **23** #B1270.

8. T. G. Cowling, *Magnetohydrodynamics*, Interscience, New York, 1957. MR **20** #5013.

9. W. C. DeMarcus, *The constitution of Jupiter and Saturn*, Astronom. J. **63** (1958), 2–28.

10. W. C. DeMarcus and R. Wildt, *Jupiter's great red spot*, Nature, **209** (1966), 62.

11. J. Douglas and H. J. Smith, *Change in rotation period of Jupiter's decametric radio sources*, Nature **199** (1963), 1080–1081.

12. V. W. Ekman, *On the influence of the Earth's rotation on ocean currents*, Ark. Mat. Astronom. Fys. **2** (1905), 1–52.

13. W. M. Elsasser, *Hydromagnetism: A review*, Amer. J. Phys. **24** (1956), 85–110.

14. V. C. A. Ferraro and C. Plumpton, *An introduction to magneto-fluid mechanics*, Oxford Univ. Press, London, 1961. MR **23** #B2753.

15. J. H. Focas, *Preliminary results concerning the atmospheric activity of Jupiter and Saturn*, Mém. Soc. Roy. Sci. Liège (5) **7** (1962), 535–542.

16. P. J. Gierasch and P. H. Stone, *A mechanism for Jupiter's equatorial acceleration*, J. Atmospheric. Sci. **25** (1968), 1169–1170.

17. P. A. Gilman, *Stability of baroclinic flows in a zonal magnetic field.* III, J. Atmospheric. Sci. **24** (1967), 130–143.

18. P. A. Gilman and E. R. Benton, *The influence of a steady magnetic field on the steady, linear Ekman boundary layer*, Phys. Fluids **11** (1968), 2397–2401.

19. P. Goldreich and S. Soter, *Q in the solar system*, Icarus **5** (1966), 375–389.

20. G. S. Golitsyn and L. A. Dikii, *Oscillations of planetary atmospheres as a function of the rotational speed of the planet*, Izv. Atmos. Ocean. Phys. **2** (1966), 225–235; **3** (1966), 137–142.

21. H. P. Greenspan, *The theory of rotating fluids*, Cambridge Univ. Press, New York, 1968.

22. J. Hartmann, *Hg-dynamics* 1. *Theory of the laminar flow of an electrically conductive liquid in a homogeneous magnetic field*, Mat.-Fys. Medd. **15** (1937), no. 6, 1–26.

23. S. L. Hess, *The general atmospheric circulation of Jupiter*, Lowell Observatory Report on Planetary Atmospheres, 1952, pp. 47–57.

24. R. Hide, *Hydrodynamics of the earth's core*, Physics and Chemistry of the Earth, vol. 1, Pergamon Press, London, 1956, pp. 94–137.

25. ———, *Origin of Jupiter's great red spot*, Nature **190** (1961), 895–896.

26. ———, *On the hydrodynamics of Jupiter's atmosphere*, Mém. Soc. Roy. Sci. Liège (5) **7** (1962), 481–505.

27. ———, *On the circulation of the atmospheres of Jupiter and Saturn*, Planetary Space Sci. **14** (1966), 669–675.

28. ———, *Planetary magnetic fields*, Planetary Space Sci. **14** (1966), 579–586.

29. ———, *Free hydromagnetic oscillations of the earth's core and the theory of the geomagnetic secular variation*, Philos. Trans. Roy. Soc. London, Ser A **259** (1966), 615–647.

30. ———, *On the dynamics of rotating fluids and related topics in geophysical fluid dynamics*, Bull. Amer. Meteorol. Soc. **47** (1966), 873–885.

31. ———, "On the dynamics of Jupiter's interior and the origin of his magnetic field," in *Magnetism and the cosmos*, Oliver and Boyd, Edinburgh, 1965, pp. 378–393.

32. ———, *Theory of axisymmetric thermal convection in a rotating fluid annulus*, Phys. Fluids **10** (1967), 56–68.

33. ———, *On hydromagnetic waves in a rotating stratified fluid*, J. Fluid Mech. **39** (1969), 283–287.

34. ———, *Equatorial jets in planetary atmospheres*, Nature **225** (1970), 254–255.

35. R. Hide and A. Ibbetson, *An experimental study of Taylor columns*, Icarus **5** (1966), 279–290.

36. R. Hide, A. Ibbetson and M. J. Lighthill, *On slow transverse flow past obstacles in a rapidly rotating fluid*, J. Fluid Mech. **32** (1968), 251–272.

37. R. Hide and P. H. Roberts, *Hydromagnetic flow due to an oscillating plane*, Rev. Mod. Phys. **32** (1960), 799–806. MR **26** #5842.

38. ———, *Some elementary problems in magnetohydrodynamics*, Advances in Appl. Mech., vol. 7, Academic Press, New York, 1962, pp. 215–316. MR **27** #5461.

39. R. Hooke, *A spot in one of the belts of Jupiter*, Philos. Trans. Roy. Soc. London Ser. A **1** (1665), no. 1, 3.

40. W. B. Hubbard, *Thermal structure of Jupiter*, Astrophys. J. **162** (1968), 745–755.

41. A. P. Ingersoll, *Inertial Taylor columns and Jupiter's great red spot*, J. Atmospheric. Sci. **26** (1969), 744–752.

42. A. P. Ingersoll and J. N. Cuzzi, *Dynamics of Jupiter's cloud bands*, J. Atmospheric Sci. **26** (1969), 981–985.

43. H. Jeffreys, *On the dynamics of geostrophic winds*, Quart. J. Roy. Meteorol. Soc. **52** (1926), 85–104.

44. B. Lehnert, *Magnetohydrodynamic waves under the action of the Coriolis force*, Astrophys. J. **119** (1954), 647–654. MR **15**, 1008.

45. E. N. Lorenz, *The vertical extent of Jupiter's atmosphere*, Lowell Observatory Report on Planetary Atmospheres, 1952, pp. 136–145.

46. ———, *The nature and theory of the general circulation of the atmosphere*, World Meteor. Organ. Publ., #218, 1967.

47. *Final report on the study of planetary atmospheres*, Lowell Observatory, Contract A.F. 19(122)-162, 1952.

48. C. N. Olsen and A. G. Smith, *Apparent drift of the radio rotational period of Jupiter*, Astronom. J. **73** (1968), 530–535.

49. E. J. Öpik, *Jupiter: Chemical composition, structure and origin of a giant planet*, Icarus **1** (1962), 200–257.

50. P. J. Peebles, *The structure and composition of Jupiter and Saturn*, Astrophys. J. **140** (1964), 328–347.

51. B. M. Peek, *The planet Jupiter*, Faber and Faber, London, 1958.

52. N. A. Phillips, *Geostrophic motion*, Rev. Geophys. **1** (1963), 123–176.

53. L. Prandtl, *Essentials of fluid dynamics*, Blackie and Sons, London, 1952.

54. Lord Rayleigh, *The theory of sound*, 2nd ed., reprint, Dover, New York, 1945. MR **7**, 500.

55. E. J. Reese and B. A. Smith, *Evidence of vorticity in Jupiter's great red spot*, New Mexico State Univ. Report TN-701-69-23 (1969).

56. P. H. Roberts, *An introduction to magnetohydrodynamics*, American Elsevier, New York, 1967.

57. L. J. Robinson, *Observations of the rotation of Saturn*, Publ. Astronom. Soc. Pacific **73** (1961), 347–349.

58. S. K. Runcorn, *The rotation of the planets and their interiors*, Ist. Naz. Alta Mat. Sympos. Math. **3** (1970), 193–202.

59. Carl Sagan, *On the nature of the Jovian red spot*, Mém. Soc. Roy. Sci. Liège (5) **7** (1962), 506–515.

60. E. Schoenberg, *Die äquatoriale Beschleunigung bei Jupiter*, S.-B. Bayer. Akad. Wiss. Math.-Natur. K1. 1949.

61. J. A. Shercliff, *A textbook of magnetohydrodynamics*, Pergamon Press, Oxford, 1965. MR **32** #3421.

62. B. A. Smith and C. W. Tombaugh, *Observations of the red spot on Jupiter*, Publ. Astronom. Soc. Pacific **75** (1963), 436–440.

63. R. Smoluchowski, *Internal structure and energy emission of Jupiter*, Nature **215** (1967), 691–695.

64. H. G. Solberg, *Jupiter's red spot in* 1966–1967, Icarus **9** (1968), 212–216.
65. ———, *A three-month oscillation in the longitude of Jupiter's great red spot*, New Mexico State University Report TN-701-69-27, Contract NGR 32-003-027, 1969.
66. V. P. Starr, *The physics of negative viscosity phenomena*, McGraw-Hill, New York, 1968.
67. K. Stewartson, *On the motion of a non-conducting body through a perfectly conducting fluid*, J. Fluid Mech. **8** (1960), 82–96. MR **22** #10506.
68. P. H. Stone, *On non-geostrophic baroclinic stability*, J. Atmospheric. Sci. **23** (1966), 390–400.
69. ———, *An application of baroclinic stability theory to the dynamics of the Jovian atmosphere*, J. Atmospheric. Sci. **24** (1967), 642–652.
70. P. H. Stone and D. J. Baker, *Concerning the existence of Taylor columns in atmospheres*, Quart. J. Roy. Meteorol. Soc. 94 (1968), 576–580.
71. C. W. Titman, "Experiments on Taylor columns over holes," in *Magnetism and the cosmos*, Oliver and Boyd, Edinburgh, 1965, pp. 348–351.
72. J. W. Warwick, *Radiophysics of Jupiter*, Space Sci. Rev. **6** (1967), 841–891.
73. J. Wasiutyński, *Studies in hydrodynamics and structure of stars and planets*, Astrophys. Norv. **4** (1946), 1–497.
74. R. Wildey, B. C. Murray and J. Westphal, *Thermal infrared emission of the Jovian disk*, J. Geophys. Res. **70** (1965), 3711–3719.
75. R. Wildt, "Planetary interiors," in *Planets and satellites*, Univ. of Chicago Press, Chicago, Ill., 1961, pp. 150–212.
76. R. Hide, *Dynamics of the atmospheres of the major planets*, J. Atmospheric Sci. **26** (1969), 841–847.
77. ———, *The viscous boundary layer at the rigid bounding surface of an electrically-conducting rotating fluid in the presence of a magnetic field*, J. Atmospheric Sci. **26** (1969), 847–853.

III. Jupiter's Rotation, Magnetism, Internal Dynamics and Structure[1]

Summary. Observations of the appearance and motion of markings on Jupiter's visible surface and of various characteristics of Jovian decimetre and decametre radio emission provide information about the internal structure of the planet. Horizontal and vertical transfer of angular momentum is implied by these observations. The variable period of rotation of the Great Red Spot and radio sources have been interpreted as evidence of a gross torsional oscillation of Jupiter's internal layers involving a toroidal magnetic field of over 1000 gauss within the planet. The electric currents responsible for this magnetic field and for the comparatively weak (50 gauss at the visible surface) magnetic field in the neighbourhood of Jupiter that has been inferred from radio-observations are probably produced by hydromagnetic dynamo action, maintained by convection driven by gravitational energy release within the planet.

[1] Supplementary notes for Lecture III, based on an article by the lecturer [**42**].

1. **Introduction.** Present knowledge of the internal structure of Jupiter is so meagre that it may seem surprising that anyone should consider the discussion of the dynamics of the interior of the planet worthwhile. Nevertheless, as has been shown elsewhere in connection with the "Taylor column" theory of the Great Red Spot (G.R.S.) [**16**], [**17**], observations of markings on Jupiter's visible surface, especially when combined with radio-astronomical observations, lead directly to certain information on Jupiter's interior that may be obtainable in no other way. The reconciliation of the variable rotation speed of the G.R.S. with that of the pattern of radio sources presents a fascinating dynamical problem [**16**], [**17**], [**38**], [**41**]. As one leading worker has remarked "... this is the first instance in astronomy when the distribution of angular momentum within a rotating cosmic body (perhaps even including the Earth) manifests itself in observational effects measurable within a short time scale."

According to the above-mentioned theory of the G.R.S., Jupiter rotates so rapidly that planetary-scale hydrodynamical disturbances near the bottom of the fluid layer in which the clouds that make up Jupiter's visible disk are suspended (region A, the "lower atmosphere"; see below) should penetrate upward throughout the whole vertical extent of the layer (see [**16**]) in a manner reminiscent of the Taylor columns that—according to both theory and laboratory experiments [**16**], [**20**], [**34**][2]—are associated with interactions of slow hydrodynamical motion of a rapidly rotating fluid with "topographical features" of the fluid container. Variations in the rotation period of the G.R.S. imply corresponding variations in angular momentum of the region underlying the lower atmosphere (region B; see below). In putting forward the Taylor column theory it was necessary to discuss possible mechanisms to account for these angular momentum variations [**16**], [**17**].

For the total angular momentum of the planet to remain constant, variations in the angular momentum of region B must be associated with equal and opposite variations in angular momentum of the remainder of the planet. The (obvious) suggestions made [**16**], [**17**] were that these opposite variations arise in the lower atmosphere, region A, and possibly in a central fluid core, region C (see below). It was shown that, under plausible assumptions concerning the thickness of the fluid regions of Jupiter and the speed of the fluid motions involved, it is possible to account for these angular momentum variations in terms of angular momentum transfer between different regions within the planet.

It was also pointed out that the concomitant variations in kinetic energy may be difficult to explain if this kinetic energy changes irreversibly into

[2] See pages 305–307.

heat. One way out of this difficulty is to postulate that the kinetic energy is transformed reversibly into magnetic energy [**17**]. This postulate, if correct, implies that changes in rotation period of the G.R.S. and of radio-sources on Jupiter are manifestations of a gross internal hydromagnetic torsional oscillation of the planet. The electrical conductivity of Jupiter, the strength of his internal magnetic field, and possible locations within the planet of the fluid layer in which the planet's magnetic field is produced (which may, in fact, be the lower reaches of the atmosphere, region A (see below)) are not inconsistent with the (admittedly limited) traditional theoretical models of the planet's internal constitution. The discussion of the dynamics and the electrical properties of Jupiter's interior should lead eventually to refinements in these theoretical models.

In what follows we shall first summarize observations of Jupiter's visible surface (§2) and of radio-waves from the planet (§3). These observations lead to information on Jupiter's internal structure, discussed briefly in §4, and on the origin of the magnetic field of the planet, discussed in §5. In §6, a general discussion of the hydrodynamics of the fluid regions of Jupiter is presented and the rotation of Jupiter's interior is considered in §7.

2. **Observations of Jupiter's visible surface.** The mean radius, R_A, of the visible surface of Jupiter is 69,300 km. The reflection of sunlight by this surface is due largely to the presence of opaque clouds suspended in an atmosphere of hydrogen, helium and methane. These clouds, which may be composed of ammonia crystals, effectively shroud the underlying planet from view, so that the depth of the atmosphere below the visible cloud layer cannot be found by direct visual observation.

The most prominent markings on Jupiter's surface (see [**27**] and [**12**] are the bright cloud zones, of which there are usually about seven or eight. These zones run parallel to the equator and are separated by darker belts. The zones and belts are not entirely regular: dark patches often appear on the bright regions and bright patches on the dark regions, and the boundaries between belts and zones often take on a serrated shape. The most striking marking of all, the Great Red Spot, is elliptical in shape, having its long axis along zeno-centric latitude 22°S, and occupies about 30 degrees of longitude and 10 degrees of latitude.

The belts and zones are permanent features only in the sense that the planet always exhibits a banded appearance; the bands themselves are subject to perceptible variations in latitude. The G.R.S., on the other hand, has appeared in the same latitude ever since the first telescopic observations, by Robert Hooke, revealed its presence in 1664 (see [**16**]).

The motion of irregular markings across the visible disk yields information on the rotation of the planet. Rotation periods thus measured depend on both time and postion on the disk; they suggest a complicated variation with latitude which is unsymmetrical about the equator. The rotation period of the equatorial zone, which lies between approximately $\pm 7°$ latitude (see Figure 1), is roughly five minutes *less* than that found at higher latitudes; hence, astronomers have found it convenient to use two separate reference systems for measuring the longitude of features of Jupiter's surface: System I with a rotation period of $9^h\ 50^m\ 30^s.003$, and System II with a period of $9^h\ 55^m\ 40^s.632$.

Within the equatorial zone the rotation is most rapid roughly on the equator itself. Temporal fluctuations in period, with amplitudes of up to 30 sec, are indicated by the observations. At higher latitudes, spatial variations of the mean (with respect to time) period range from $9^h\ 55^m\ 5^s$ to $9^h\ 55^m\ 54^s$. Temporal variations there are erratic and range from 10 sec to 20 sec, depending upon position.

During the past hundred years, the period of rotation of the G.R.S. has varied between $9^h\ 55^m\ 31^s$ and $9^h\ 55^m\ 44^s$, the average value being $9^h\ 55^m\ 38^s$. The average (with respect to position) period of ephemeral markings of the South Tropical Zone, within which the G.R.S. is situated, has varied between $9^h\ 55^m\ 27^s$ and $9^h\ 55^m\ 36^s$. The corresponding average linear velocity of the South Tropical Zone with respect to the

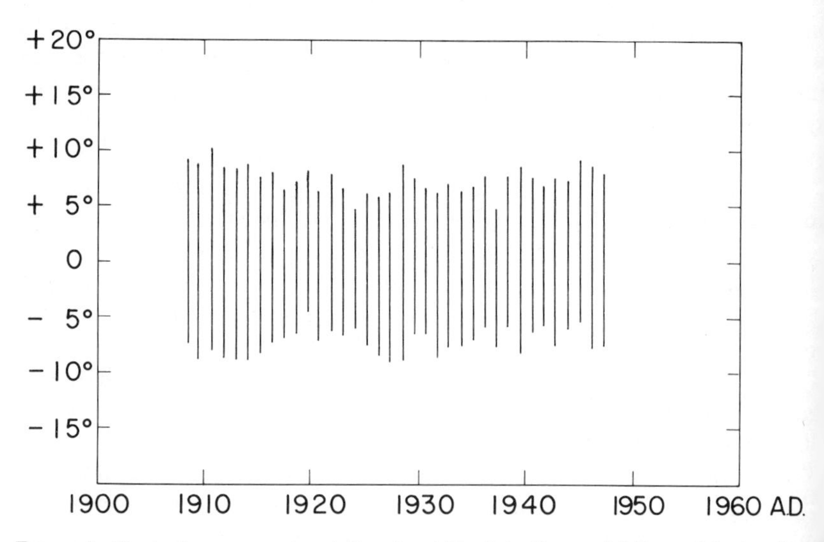

FIGURE 1. Illustrating apparent variations in width of the Equatorial Zone of Jupiter from 1908 to 1947 [**42**]. Each line indicates the range of latitude occupied by the Zone. (Based on a table given by Peek [**27**].)

G.R.S. is about +2m/sec, the positive sign indicating that this motion is eastward referred to the planet (see Figure 2). For the sake of comparison, note that points on the equator of Jupiter rotate about the axis of the planet at 12 km/sec, and that the linear velocity of the

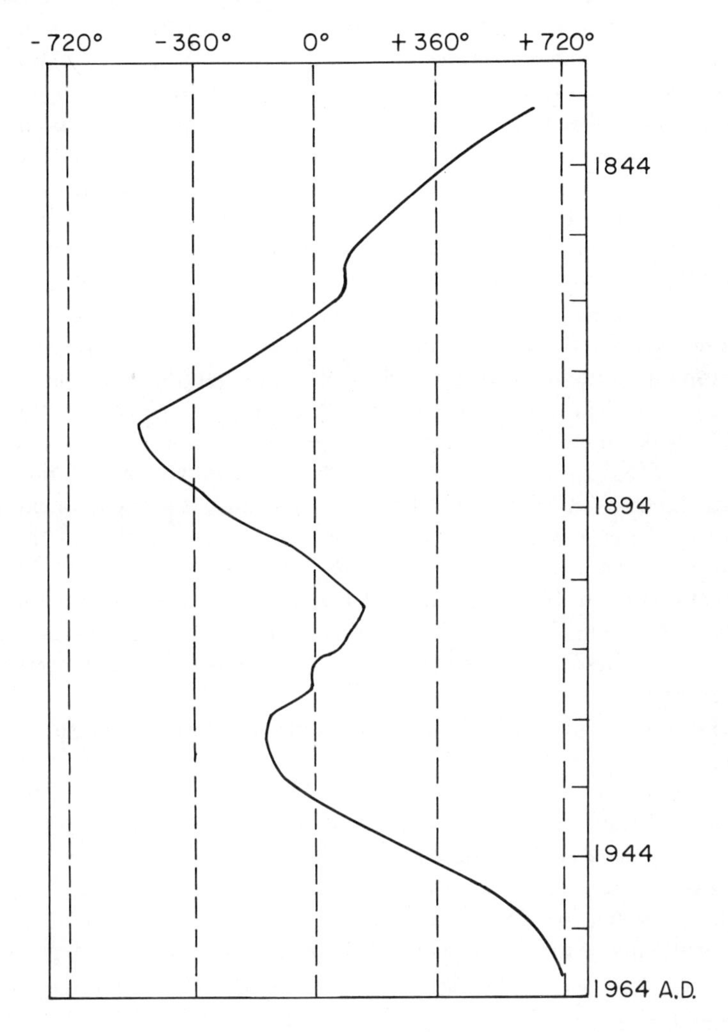

FIGURE 2. The wanderings of the Great Red Spot in longitude, 1831 to 1960. The abscissa is $\lambda_2 - 264°.3 + 28°.62t$, where λ_2 is longitude in System II and t is time measured in units of 398.88 days, the mean intervals between oppositions of Jupiter. The ordinate is time in years. (Based on a diagram by Peek [**27**] and reproduced by kind permission of the author and of Faber and Faber Ltd.) According to [**16**]–[**19**] and [**41**], these wanderings may be a manifestation of a gross hydromagnetic torsional oscillation of Jupiter's interior.

equatorial zone relative to points on the equator of a hypothetical surface rotating with angular velocity of higher-latitude regions is about +100 m/sec.

Spinrad [**32**] and Spinrad and Trafton [**33**], from an examination of the Doppler shift of ammonia lines in the spectrum of sunlight reflected from the equatorial regions of Jupiter, concluded that the material responsible for the absorption lines, which cannot lie too far above the level of the visible disk, does not always rotate at the speed of the visible planet (see also Owen and Staley [**24**]). Moreover, the maximum departures in speed, if the spectra have been interpreted correctly, are considerable; they correspond to "winds" of several kilometers per second. Whether or not this interpretation of the spectra is the only acceptable one has not yet been established; the magnitude of the "Spinrad effect" [**17**] happens to be zero at the present time!

Typical lifetimes of individual ephemeral markings on Jupiter range from days for the smallest ones to decades for markings of dimensions comparable in size with the radius of the planet. An example of one of these large features is the so-called "South Tropical Disturbance" (see [**27**]) which made its appearance in 1901 as a short dark streak in the South Tropical Zone at some distance "downstream" of the G.R.S. For the next four decades it grew in length until it stretched nearly two-thirds of the way around the planet. The rotation period of the South Tropical Disturbance was somewhat less than that of the G.R.S. Conjunctions of these two features occurred on nine occasions before 1939, when the South Tropical Disturbance vanished, giving way, apparently, to three white spots in the belt just south of the G.R.S., which have existed ever since.

During a typical conjunction the South Tropical Disturbance skirted around the edge of the G.R.S. *at a speed up to ten times the speed at which it approached and receded from the G.R.S.* Tentative explanations of this bizarre behavior and of the long lifetime of the South Tropical Disturbance have been proposed (see §6 below).

There is direct spectroscopic evidence for the presence of methane and ammonia in Jupiter's atmosphere. The presumption that the main constituents of the atmosphere are hydrogen and helium is based on the low mean density of the planet, and on a measurement of the light variation of a star during an occultation, leading to an estimate of the mean molecular weight above the cloud level [**1**].

The mean temperature of Jupiter's disk is about 160°K (see [**23**]). Temperature variations across the disk have also been estimated from infrared measurements [**22**]. It has been suggested that colour variation across Jupiter's visible disk may be due to traces of alkali metals or t certain free radicals (see [**25**] and [**28**]) and that these colour variations may reflect temperature variations at that level.

3. **Radio waves from Jupiter.** Nonthermal electromagnetic radiation from Jupiter is so strong on decimetre and decametre wavelengths that the planet is one of the brightest radio sources in the sky. Since this fact was first discovered ten years ago, by Burke and Franklin, numerous papers, both observational and theoretical, have appeared on the subject (for reviews see [**3**], [**14**], [**29**], [**13**], [**38**], [**39**], [**31**]; also [**11**], [**36**], [**9**], [**30**]).

The radio-waves are emitted by charged particles trapped in van Allen-type radiation belts surrounding Jupiter. Spectrum and polarization studies indicate that electrons in these radiation belts may be 10^3 times more energetic than those in the Earth's radiation belts, and that Jupiter's magnetic moment is about 8×10^{30} e.m.u. (corresponding to a polar field of about 50 gauss), 10^5 times that of the Earth, and inclined about 11° from the rotation axis. Asymmetries in the total radiation, together with dynamic spectra of radiation bursts on decametre wavelengths, suggest that the magnetic field may be much more complicated than that of a centered dipole [**37**], [**38**]. A dipole displaced a distance $0.7\ R_A$ from the centre of the planet evidently fits the observations fairly well, although this result should be treated with caution until nondipolar contributions to the field have been assessed.

Radio-astronomical observations indicate that the magnetic field has a characteristic rotation period of its own. This has led to the introduction of yet another system, System III, with a rotation period of $9^h\ 55^m\ 29^s.37$, for use in measuring the longitude of radio sources. The radio period when first determined was apparently constant and close to that of System III. Recently, however, fluctuations of the order of a second in amplitude have been detected [**14**], [**7**]. Unfortunately, it will take several decades to gain knowledge of the radio period that is comparable in extent with our present knowledge of the motion of the G.R.S. The acquisition of this knowledge will be of the greatest importance; of especial theoretical interest will be the accurate determination of the radio period at different wavelengths (see §7 below).

4. **Jupiter's internal structure.** The mean density, ρ, of Jupiter is 1.334×10^3 kg/m^3. The moment of inertia of the planet about its axis of rotation is 2.3×10^{42} kg m^2. As this value is only 0.6 of that of a homogeneous sphere of density $\bar{\rho}$ and radius R_A, the density, ρ, of Jupiter must be a generally decreasing function of r, the distance from the centre of the planet.

The low mean density of Jupiter is usually taken as evidence that the planet is composed mainly of hydrogen, with a small admixture of helium and heavier elements (see §2 above).

For the purpose of our discussion we shall follow II and assume that Jupiter consists of four regions: region C, a central core of mean density ρ_C and mean radius R_C, an intermediate region, B, of mean density ρ_B occupying $R_C < r < R_B$, a "lower atmosphere," region A, of mean density ρ_A, occupying $R_B < r < R_A$, and region E of mean density ρ_E (presumably much less than ρ_A, ρ_B and ρ_C) occupying $r > R_A$, which thus includes the upper atmosphere, van Allen radiation belts, etc. Tentative estimates of R_B are discussed in §6 below. At the present time it is difficult to hazard even a guess at R_C; it may not even be necessary to invoke the existence of region C (see §7 below), in which case $R_C = 0$.

Theoretical models of Jupiter's internal structure have been constructed by several workers (see [**6**], [**40**], [**23**], [**26**]). These models give density, ρ, and pressure, p, of the main body of the planet (i.e. the regions below $r = R_A$) as a function of r with fair accuracy; but they do not lead to precise information on temperature, thermal conductivity, electrical conductivity (see §5 below), viscosity, mechanical strength etc., and cannot predict whether the material is gaseous, liquid or solid. Hence, whilst the distinction between region A (the fluid lower atmosphere) and region E (mainly gas and plasma above the cloud layers) is a fairly obvious one, the other regions, as we shall see, have not been clearly identified.

Region B is defined here as being that region at whose interface with region A occurs the "topographical feature" responsible (on the Taylor column theory) for the G.R.S. As the G.R.S. is at least 300 years old—it may be much older, possibly by many orders of magnitude [**16**], [**17**]—the distinguishing property of region B is that typical time scales of mechanical processes occurring within it must be far longer than the corresponding time scales for region A. Region B could be solid, plastic, or some very viscous fluid.

The characterization of region C is even more imprecise. It is that fluid region which, together with region A, applies to region B the torques required to account for the observed variations in rotation period in the G.R.S. If these torques can be adequately accounted for in terms of processes occurring in region A, there is no need to postulate the existence of region C.

Motions in the lower atmosphere (region A) of Jupiter may derive their kinetic energy not only from solar heating but also from internal sources. Let H_s be the rate at which energy of solar radiation is intercepted by the planet, and H_i the rate at which energy is released by internal energy heat sources. $H_s \sim 8 \times 10^{17}$ watts; Öpik [**23**] has suggested that in order to account for the measured surface temperature (see §2 above) internal heat sources must supply energy to the atmosphere at a rate comparable with H_s. It is not inconceivable that this internal heating is due to the release of gravitational energy [**18**]. The gravitational energy of the planet is 3.5×10^{36} joule, and only about 3% of this energy would, if released gradually over 4×10^9 years, result in a value of H_i equal to H_s.

5. **Jupiter's magnetic field.** In §3 we have seen that the magnetic field within region E of Jupiter (see §4) has been represented roughly by a dipole of 8×10^{30} e.m.u. (corresponding to a field strength of 50 gauss near $r = R_A$), inclined to the rotation axis at an angle of about 11° and displaced by a considerable distance, about 0.7 R_A, from the centre of the planet. The properties of this *poloidal* field within the planet (i.e. within regions A, B and C), cannot, however, be determined *uniquely* from the field in region E without introducing further physical hypotheses. Moreover, lines of force of any *toroidal* magnetic field that may exist within the planet cannot get through to the surface, since by definition toroidal fields have no radial component.

If the magnetic field of Jupiter is due to a homogeneous dynamo process of the type discussed by Bullard [**2**] and Elsasser [**10**] in connection with the Earth's magnetism (see [**17**]) in which hydromagnetic (magnetohydrodynamic) flow amplifies and maintains the energy of the magnetic field against dissipation due to ohmic heating, the field must originate in one or both of the fluid regions of the planet (A and C). The fluid motions involved will be quite complicated and characterized by a very low degree of symmetry. The interaction of horizontal motions with the poloidal magnetic field produces a toroidal magnetic field. Interaction of vertical motions with both types of field, poloidal and toroidal, then leads to the regeneration of the original poloidal field.

These hydromagnetic interactions can occur only when the electrical conductivity, σ, of the fluid is so high that a magnetic Reynolds number,

$$G \equiv \mu\sigma LU, \tag{5.1}$$

exceeds a certain critical value, G_d, where μ denotes the magnetic permeability of the fluid and L is a length scale associated with the fluid motion, in which typical velocity shears are of order U/L. G measures the degree to which magnetic lines of force more with the fluid particles. When $G > G_d$, the coupling between the fluid motion and the magnetic lines of force is strong and the fluid flow stretches the magnetic lines of force and thus increases the magnetic energy of the system until magnetic energy production due to "motional induction" is offset by ohmic dissipation. When $G < G_d$, ohmic losses are so strong that the magnetic lines of force are only slightly distorted by the fluid motions.

G_d depends upon the particular system under consideration. For the homogeneous dynamo discussed by Bullard, $G_d \sim 10$. If we take this value and assume that $\mu = 4\pi \times 10^{-7}$ henry/m and $L = 10^7$ m (about $0.14\, R_A$) it follows that $\sigma U > 1\,\Omega^{-1}\,s^{-1}$ (where σ is measured in reciprocal ohms per meter and U in meters/sec) in the region of Jupiter where the dynamo operates.

Although the electrical conductivity, σ, will be negligibly small at $r = R_A$, it should be quite high at great depths within the planet where, owing to the high prevailing pressure, the material is probably metallic. Hence, σ should undergo a general increase with depth below the visible surface.

The magnetic field originates in those parts of region A or region C for which $\sigma U > 1\,\Omega^{-1}\,s^{-1}$. For region A, if we can take $U = 1$ m/sec (see §§2 and 7) then $\sigma U > 1\,\Omega^{-1}\,s^{-1}$ when $\sigma > 1\Omega^{-1}m^{-1}$. Typical values of σ for semiconductors exceed $10\,\Omega^{-1}\,m^{-1}$, and for metals σ is even higher, by five or more orders of magnitude. Whether or not the possibility that Jupiter's magnetic field originates within his lower atmosphere, region A (evidence in favour of which suggestion being the high eccentricity of the dipole field) should be considered seriously at the present time may hinge on the validity of the supposition that σ exceeds $1\,\Omega^{-1}\,m^{-1}$ at a level lying somewhere within $R_B < r < R_A$ [**19**]. Present knowledge of Jupiter's interior seems inadequate to settle this point.

If σ nowhere exceeds $1\,\Omega^{-1}\,m^{-1}$ within region A, it is necessary to postulate that $U\sigma > 1\,\Omega^{-1}\,s^{-1}$ within the hypothetical region C. Although U for region C is unknown, if region C exists at all σ might attain such high values there (see above) that extremely small values of U would suffice for dynamo action to occur. It is important, however, to recognize that very high values of σ do not necessarily favor the efficient production of a poloidal magnetic field by the dynamo process. Lines of magnetic force produced in a perfectly conducting fluid would remain confined to the fluid.

When the quantity U appearing in the definition of G (see Equation (5.1)) is associated with zonal shearing motions, as one might expect would be the case for a rapidly rotating planet, the average strength, B_T, of the toroidal magnetic field is related to that, B_P, of the poloidal magnetic field by the equation

$$B_T/B_P \doteqdot \bar{G} \tag{5.2}$$

where $\bar{G}$ is an average value of G for the whole planet. The corresponding ratio of M_T, the energy of the toroidal magnetic field, to M_P, that of the poloidal field, is as follows:

$$M_T/M_P \doteqdot \bar{G}^2. \tag{5.3}$$

If $B_P = 100$ oersted (10^{-2} weber/m^2), then $M_P = 10^{25}$ joule. To fix ideas, take $G = 10^2$, whence, by Equations (5.2) and (5.3), $B_T = 10^4$ oersted and $M_T = 10^{29}$ joule.

H_M, the rate of magnetic energy dissipation due to ohmic heating, is given by

$$H_M = (M_T + M_P)/\tau_M, \tag{5.4}$$

where τ_M is the time constant associated with the free decay of the magnetic field that would occur if the generating mechanism suddenly ceased to function. If $\bar{\sigma}$ is an average value of σ for the whole planet, then

$$\tau_M \doteqdot \mu\bar{\sigma}R_A^2. \tag{5.5}$$

$\bar{\sigma}$ is not known. If we assume (arbitrarily) that $\sigma = 3 \times 10^4\ \Omega^{-1}\ \mathrm{m}^{-1}$ (that of a poor metallic conductor), then by the last equation $\tau_M = 3 \times 10^{13}$ sec $= 10^6$ years. As the corresponding value of H_M, namely 3×10^{15} watts, is over two orders of magnitude less than H_s and H_i (see §4 above) the values of $\bar{\sigma}$ and B_T etc., assumed in the foregoing calculation, may not be too wide of the mark.

6. **Hydrodynamics of Jupiter's internal fluid layers.** The variable appearance and rotation periods of surface markings on Jupiter (see §2) are clear indications of motions within the fluid region A, the lower atmosphere. Whether or not the existence of a distinct time-varying radio period can be accounted for only by invoking the occurrence of hydrodynamical motions in region C (see §4) cannot be readily assessed (see §§5 and 7), owing largely to ignorance of the electrical conductivity of the planet.

The parameters required to characterize the dynamics of a fluid layer of a rapidly rotating planet ([17], [18]) are a Rossby number

$$Q \equiv U/2L\Omega \tag{6.1}$$

(where Ω is the angular velocity of basic rotation), a rotational Mach number,

$$S \equiv L\Omega/c \tag{6.2}$$

(where c is the speed of sound in the fluid), a parameter measuring the vertical gradient, Γ, of potential density, namely

$$K \equiv \omega^2/4\Omega^2 \tag{6.3}$$

where ω, the "Brunt-Väisälä frequency," depends on Γ as follows:

$$\omega^2 \equiv -g\Gamma/\rho \tag{6.4}$$

(if g is the acceleration of gravity), an Ekman number, N, measuring the effective (eddy) viscosity ν_E, namely

$$N \equiv \nu_E/L^2\Omega, \tag{6.5}$$

and an "aspect ratio,"

$$D \equiv 2d/R_A \tag{6.6}$$

where d is a length scale characteristic of the vertical structure of the fluid layer. When hydromagnetic phenomena have to be taken into account, to Q, S, K, N and D should be added a magnetic Reynolds number, G (see Equation (5.1)), and a further parameter that measures the ratio of magnetic energy to kinetic energy per unit volume, namely

$$F \equiv (B_T^2 + B_P^2)/\mu\rho U^2. \tag{6.7}$$

For the sake of definiteness in what follows we shall consider planetary scale features only, i.e. those for which $L \sim R_A$.

For the Earth's atmosphere, a planetary fluid layer about which a great deal is known, $Q \sim 10^{-1}$, $S \sim 1$ and $D \sim 10^{-3}$. The vertical lapse rate is subadiabatic, so that ω is real and K is positive. The pressure forces that give rise to atmospheric motions result entirely from the horizontal north-south temperature gradient. In the theory of these motions KD^2 is a key parameter, and it is significant not only that KD^2 is positive for the Earth's atmosphere but also that it is comparable with, though less than, unity; only in these circumstances can the nonaxisymmetric flow associated with the energy-producing eddies—the cyclones—occur [**4**], [**8**].

For the values of U suggested by the observations outlined in §2 above, Q for region A of Jupiter ranges from 10^{-2} in equatorial regions to 10^{-4} at higher latitudes. At such low values of Q hydrodynamical effects of those "topographical features" of the interface between region A and region B whose vertical dimensions exceed $Q(R_A - R_B)$ will extend upward throughout the depth of the atmosphere, a simple example of such a phenomenon being a "Taylor column" [**16**], [**20**]. This result was used in [**16**] in an attempt to explain the G.R.S. (see §1 above). If K (see Equation (6.3)) is fairly small, the Taylor column theory of the G.R.S. indicates that the thickness $(R_A - R_B)$ of region A is less than 2800 km.

The physical nature of the "topographical feature" is not yet known, and it should be emphasized that the use of the words "mountains" and "craters" as descriptive terms may be misleading. It is not inconceivable that the "topographical feature" of Hide's Taylor column model of the G.R.S. [**16**], [**17**] is an electric current loop of hydromagnetic origin or a thermal anomaly [**18**].

The existence of rapid equatorial currents in the fluid layers of rapidly rotating planets seems fairly general [**15**], [**17**]. Jupiter and Saturn exhibit such currents, and on earth there are the Cromwell current in the ocean [**21**] and the Berson westerlies in the stratosphere. No satisfactory theory of these currents has yet been proposed [**5**].

The angular latitudinal width of the current should be of the order of $Q^{1/2}$ radians, and it has been argued that if the currents represent sinks of kinetic energy and angular momentum originating at higher latitudes, they might build up until $Q \sim D$ (see Equations (6.1) and (6.6)) (III). Observations (see, for example, Figure 1) are at least consistent with the first of these results, and if the second result is correct, $d \sim 250$ km, which, according to its definition, should be comparable with or less than $(R_A - R_B)$.

Typical values of S for the atmosphere of Jupiter (and of Saturn) are of the order of 10, and in having values of S significantly greater than unity the major planets may differ in a fundamental way from the rotating inner planets (Earth and Mars) [**17**], [**35**]. The long lifetime of the South Tropical Disturbance and the dramatic acceleration it undergoes when it encounters the G.R.S. (see §2) have been interpreted in [**17**] as observational evidence favoring this suggestion.

The sign of K (see Equation (6.3)), let alone its numerical value, is unknown for any part of Jupiter, although the Taylor column hypothesis of the Great Red Spot, if correct, implies that K is a good deal less than unity. If internal energy sources produce atmospheric heating comparable with that due to solar radiation (see §4), K would be small or even negative in region A; if K is negative, in contradistinction to the Earth's atmosphere, vertical overturning associated with a superadiabatic lapse rate could constitute the principal mode of hydrodynamical flow.

If hydromagnetic effects are absent in region A, energy dissipation will be due to viscous friction, as measured by the parameters N (see Equation (6.5)). According to the arguments given in §5, however, it is possible that energy dissipation in Jupiter's atmosphere is due mainly to ohmic heating, as measured by the parameters G and F (see Equations (5.1) and (6.7)) [**18**]. If $B_T = 1$ weber/m^2 (10^4 oersted), $\rho = 10^2$ kg/m^3 and $U = 1$ m/sec (see §5 above), then F, the ratio of magnetic energy to kinetic energy of hydrodynamical motion, is large, about 10^4.

7. **Rotation of Jupiter's interior.** Denote by I_A, I_B and I_C the respective moments of inertia of regions A, B and C, and let the corresponding mean angular speeds of rotation about the axis of the planet be $\bar{\Omega}_A$, $\bar{\Omega}_B$ and $\bar{\Omega}_C$.

If $\bar{I}$ is the moment of inertia of the whole planet and ρ_E, the mean density of region E, is so much less than ρ_A, ρ_B and ρ_C that the contribution of region E to $\bar{I}$ can be neglected, then

$$I_A + I_B + I_C = \bar{I} = 2.3 \times 10^{42} \text{ kg m}^2. \tag{7.1}$$

Assume that no external torques act on the planet so that P, its total angular momentum, remains constant. P is related as follows to the other quantities just defined:

$$I_A \bar{\Omega}_A + I_B \bar{\Omega}_B + I_C \bar{\Omega}_C = P = 4.1 \times 10^{38} \text{ kg m}^2 \text{ sec}^{-1}. \tag{7.2}$$

Now introduce for each region a measure $\mathscr{E}$ of the total rotational energy at a given instant of time, and define a quantity $\Delta\Omega$ for each region such that

$$\mathscr{E} = \tfrac{1}{2} I \bar{\Omega}^2 (1 + 2\Delta\Omega/\bar{\Omega}), \tag{7.3}$$

where $\Omega = \Omega_A, \Omega_B$ or Ω_C as the case may be. $\Delta\Omega$ is a measure of departures from solid body rotation; when $\Delta\Omega/\bar{\Omega} \ll 1$, $\Delta\Omega$ is the amplitude of spatial variations in rotation speed.

The total rotational energy of the planet, $\bar{\mathscr{E}}$, is given by

$$\begin{aligned} \bar{\mathscr{E}} &= \mathscr{E}_A + \mathscr{E}_B + \mathscr{E}_C \\ &= \tfrac{1}{2} I_A \bar{\Omega}_A^2 (1 + 2\Delta\Omega_A/\bar{\Omega}_A) + \tfrac{1}{2} I_B \bar{\Omega}_B^2 (1 + 2\Delta\Omega_B/\bar{\Omega}_B) \\ &\quad + \tfrac{1}{2} I_C \bar{\Omega}_C^2 (1 + 2\Delta\Omega_C/\bar{\Omega}_C) \doteqdot 3.63 \times 10^{34} \text{ joule}. \end{aligned} \tag{7.4}$$

If, as a result of angular momentum transfer between regions A, B and C, $\bar{\Omega}_A$ changes by an amount $\delta\bar{\Omega}_A$, $\bar{\Omega}_B$ by $\delta\bar{\Omega}_B$ and $\bar{\Omega}_C$ by $\delta\bar{\Omega}_C$ while the moment of inertia of each of these regions remains unaltered, then, according to Equation (7.2)

$$I_A \delta\bar{\Omega}_A + I_B \delta\bar{\Omega}_B + I_C \delta\bar{\Omega}_C = 0. \tag{7.5}$$

Associated with this transfer of angular momentum will be changes $\delta\mathscr{E}_A$ in $\mathscr{E}_A$, $\delta\mathscr{E}_B$ in $\mathscr{E}_B$, $\delta\mathscr{E}_C$ in $\mathscr{E}_C$, $\delta\bar{\mathscr{E}}$ in $\bar{\mathscr{E}}$, $\delta(\Delta\Omega_A)$ in $\Delta\Omega_A$, $\delta(\Delta\Omega_B)$ in $\Delta\Omega_B$ and $\delta(\Delta\Omega_C)$ in $\Delta\Omega_C$. By Equation (7.3),

$$\begin{aligned} \delta\mathscr{E}_A &= I_A \bar{\Omega}_A \delta\bar{\Omega}_A (1 + \delta(\Delta\Omega_A)/\delta\bar{\Omega}_A + \Delta\Omega_A/\bar{\Omega}_A), \\ \delta\mathscr{E}_B &= I_B \bar{\Omega}_B \delta\bar{\Omega}_B (1 + \delta(\Delta\Omega_B)/\delta\bar{\Omega}_B + \Delta\Omega_B/\bar{\Omega}_B), \\ \delta\mathscr{E}_C &= I_C \bar{\Omega}_C \delta\bar{\Omega}_C (1 + \delta(\Delta\Omega_C)/\delta\bar{\Omega}_C + \Delta\Omega_C/\bar{\Omega}_C), \end{aligned} \tag{7.6}$$

and, by Equation (7.4),

$$\delta\bar{\mathscr{E}} = \delta\mathscr{E}_A + \delta\mathscr{E}_B + \delta\mathscr{E}_C \neq 0. \quad (7.7)$$

Now consider the numerical values of the foregoing quantities.

As R_C is completely unknown and R_B is very uncertain (see §6 above), the individual values of I_A, I_B and I_C cannot be estimated.

According to [**16**], $\bar{\Omega}_B$ and $\delta\bar{\Omega}_B$ are given with some precision by the motion of the G.R.S., and $\Delta\Omega_B$ and $\delta(\Delta\Omega_B)$ can be taken as zero on time scales over which $\bar{\Omega}_B$ changes by $\delta\bar{\Omega}_B$. Observations of the G.R.S. (see §2 above, especially Figure 2) suggest that

$$\delta\bar{\Omega}_B/\bar{\Omega}_B = 3 \times 10^{-4} \quad \text{where } \bar{\Omega}_B = 1.74 \times 10^{-4} \text{ rad/sec.} \quad (7.8)$$

If vertical variations in angular velocity are not greater than the horizontal variations across the visible surface (see §2), then

$$(\bar{\Omega}_C - \bar{\Omega}_A) < 10^{-2}\bar{\Omega}_A \quad \text{and} \quad (\bar{\Omega}_B - \bar{\Omega}_A) < 10^{-2}\bar{\Omega}_A. \quad (7.9)$$

Owing to the presence of a strong equatorial current at the visible surface of Jupiter (see §§2 and 6 above), it is not clear how to estimate $\Delta\Omega_A$ from the observations of motion of the surface markings. If the equatorial current penetrates within region A to a depth comparable with $(R_A - R_B)$, then

$$\Delta\Omega_A/\bar{\Omega}_A \sim 10^{-3} \quad (7.10)$$

and, as $\delta\bar{\Omega}_A$ should be comparable with $\Delta\Omega_A$,

$$\delta\bar{\Omega}_A/\bar{\Omega}_A \sim 10^{-3}. \quad (7.11)$$

The interpretation of the radio period of Jupiter is of some interest. The plasma responsible for the radio emission is tied to magnetic lines of force in region E. The radio waves that reach the Earth come from a certain limited region of this plasma. Extrapolate the lines of magnetic force that pass through this limited region back into the main body of the planet. At any level within the planet, these lines of force will define a limited area of the whole spherical surface. At those levels for which the magnetic Reynolds number, G, is less than G_d (see Equation (5.1)) (where G_d can be taken as 10 without fear of serious error), the lines of force are unaffected by the material motion; but at that level $r = R_M$ (say), at which G first attains the value G_d, the lines of force will remain tied to the material. Measurements of the radio period, therefore, yield information on the motion of that material at the level $r = R_M$ which is connected by lines of magnetic force to the emitting plasma in region E. To relate the radio period to the motion of the remainder of the spherical surface

$r = R_M$ requires further information or, in the absence of this information, additional physical hypotheses.

As the radio period differs from that of the G.R.S. (see §§ 2 and 3), to postulate that the surface $r = R_M$ lies within region B, for which $\Delta\Omega_B \doteqdot 0$, is inconsistent with the suggestion that the rotation of the G.R.S. is that of region B. It is more likely that the level $r = R_M$ lies either within region A or within region C, for which $\Delta\Omega$ does not vanish. If this is so, as these regions are fluid and it is unlikely that the positions of the sources of decametric wavelength radiation are identical with those of the decimetre sources, the decametre and decimetre radio periods will not be identical. They would differ by up to a small fraction, possibly one-tenth (depending on the degree of asymmetry about the axis of rotation that Jupiter's magnetic field exhibits) of $(2\pi/\bar{\Omega}_A)(\Delta\Omega_A/\bar{\Omega}_A)$, a second or so (see Equation (7.11)). A careful scrutiny of the radio period data should reveal whether or not a significant difference exists. I have not been able to find sufficiently reliable information on this matter in the literature.

Attempts have been made [**16**], [**17**] to account for the magnitude of $\delta\bar{\Omega}_B$ (see Equation (7.8)) in terms of angular momentum transfer within the planet. In one of the models considered, it was assumed that $I_C\delta\bar{\Omega}_C \ll I_A\delta\bar{\Omega}$ and that $\delta\bar{\Omega}_A$ has the value given by Equation (7.11). It follows from Equation (7.5) that this model requires that $I_A/I_B = 0.3$ and from Equation (7.7) that, associated with the angular momentum transfer between region A and region B, the change $\delta\bar{\mathscr{E}}$ in the kinetic energy of rotation of the planet would amount to about $10^{-7}(I_B/\bar{I})\bar{\mathscr{E}}$ or $4 \times 10^{27}(I_B/\bar{I})$ joule.

Now consider the fate of this energy. If regions A and B are coupled together by friction, then the energy will be dissipated into heat. If, on the other hand, frictional forces are ineffective and the mechanical coupling is due to the presence of lines of magnetic force linking the two regions [**17**] then the rotational energy will be transformed reversibly into magnetic energy by the process of amplification discussed in §5. In the latter case, the motion of the G.R.S. is to be regarded as a manifestation of a hydromagnetic torsional oscillation of Jupiter's interior [**17**].

If the energy is transformed into heat, the amount of heat released is unlikely to exceed H (see §4 above) multiplied by the time taken for the changes $\delta\bar{\Omega}_B$ in $\bar{\Omega}_B$ to occur. The first of these quantities is less than 10^{18} watts (see §4) and the second is less than 10^8 sec (see §2, especially Figure 2), so that the product is probably much less than 10^{26} joule. For this product to equal $4 \times 10^{27}(I_B/\bar{I})$ joule, $I_B/\bar{I}$ must be much less than 0.025. If, on the other hand, $\delta\bar{\mathscr{E}}$ is transformed into magnetic energy, then the strength B_T of the toroidal magnetic field will have to exceed about 1000 oersted, a value which, according to §5 above, may not be unreasonably high. The implied value of $I_B/\bar{I}$ is then about 0.25.

Appendix: On geostrophic motion of a nonhomogeneous fluid [45]. Certain general results concerning the properties of strictly geostrophic motion of a rotating nonhomogeneous fluid are evidently less widely known and understood than might be expected, having regard for their usefulness in the investigation of flows in many natural and laboratory systems (see §1 above).

Indefinitely slow, perfectly steady hydrodynamical flow of an inviscid fluid of density ρ which otherwise rotates uniformly with angular velocity $\mathbf{\Omega}$ relative to an inertial frame satisfies the equation for strictly geostrophic motion, characterized by an exact balance between Coriolis acceleration and the nonhydrostatic component of the pressure gradient; thus

$$2\rho\mathbf{\Omega} \times \boldsymbol{u} = -\nabla p + \rho \boldsymbol{g} \tag{A1}$$

where $\boldsymbol{u}$ is the Eulerian flow velocity relative to the rotating system, p denotes pressure, and $\boldsymbol{g}$ is the acceleration due to gravity and centrifugal effects. Take the *curl* of this equation, making use of the equation of continuity for steady flow

$$\nabla \cdot (\rho \boldsymbol{u}) = 0 \tag{A2}$$

and the fact that $\boldsymbol{g}$ is irrotational, and show that

$$2\Omega\partial(\rho \boldsymbol{u})/\partial s = \boldsymbol{g} \times \nabla\rho \tag{A3}$$

(cf. Equation (2.22) of Lecture II), where $\partial/\partial s$ denotes differentiation with respect to distance *measured parallel to* $\mathbf{\Omega}$. If we define local coordinate axes (x, y, z) such that z is the downward vertical, i.e. $\boldsymbol{g} = (0, 0, g)$, then by Equation (A3)

$$\partial(\rho w)/\partial s = 0 \tag{A4}$$

and

$$\left[\frac{\partial}{\partial s}(\rho u), \frac{\partial}{\partial s}(\rho v)\right] = \frac{g}{2\Omega}\left[-\frac{\partial \rho}{\partial y}, \frac{\partial \rho}{\partial x}\right] \tag{A5}$$

(essentially the meteorologist's thermal-wind equation), if $\boldsymbol{u} = (u, v, w)$. When $\boldsymbol{g} \times \nabla\rho = 0$ we have, by Equation (3), $\partial(\rho\boldsymbol{u})/\partial s = 0$, which reduces when $\nabla\rho = 0$ to $\partial\boldsymbol{u}/\partial s = 0$, a celebrated result due to Proudman [**44**] and Taylor [**34**] (see Greenspan [**43**]).

Write

$$u + iv = Ue^{i\psi} \tag{A6}$$

where $U = (u^2 + v^2)^{1/2}$ and $\psi = \tan^{-1}(v/u)$, respectively the magnitude of and the angle made to the x axis by the horizontal component of $\boldsymbol{u}$. By Equations (A5) and (A6),

$$\text{(A7)} \quad \left[\frac{\partial \psi}{\partial s}, \frac{\partial}{\partial s} \ln(\rho U)\right] = \frac{g}{2\Omega\rho U^2}\left[\left(\frac{u\,\partial\rho}{\partial x} + \frac{v\,\partial\rho}{\partial y}\right), \left(\frac{v\,\partial\rho}{\partial x} - \frac{u\,\partial\rho}{\partial y}\right)\right].$$

The case when $(u\,\partial\rho/\partial x + v\,\partial\rho/\partial y)$, a measure of horizontal density advection, is very much less in magnitude than each of the individual terms $u\,\partial\rho/\partial x$, $v\,\partial\rho/\partial y$, $u\,\partial\rho/\partial y$, and $u\,\partial\rho/\partial x$ is of some practical significance. When $u\,\partial\rho/\partial x + v\,\partial\rho/\partial y = 0$, by Equation (A7)

$$\text{(A8)} \qquad \frac{\partial \psi}{\partial s} = 0, \quad \frac{\partial}{\partial s}\ln(\rho U) = -\frac{g}{2\Omega u}\frac{\partial}{\partial y}\ln \rho$$

(where $u = U$ if, without loss of generality, we choose the local x direction so that $v = 0$); the *pattern* of horizontal flow is then independent of s, notwithstanding the possibility of strong variations with s of the *magnitude* of the horizontal flow.

If the entropy and chemical composition of each individual fluid element remain unchanged during the motion, then, by the first law of thermodynamics and Equations (A1) and (A2),

$$\text{(A9)} \qquad u\,\partial\rho/\partial x + v\,\partial\rho/\partial y + w(\partial\rho/\partial z - \Gamma) = -\partial\rho/\partial t = 0$$

where Γ is the adiabatic density gradient, which vanishes for an incompressible fluid, and t denotes time. Combine this equation with the first component of Equation (A7) and show that for *strictly geostrophic isentropic flow*:

$$\text{(A10)} \qquad \partial\psi/\partial s = -gw(\partial\rho/\partial z - \Gamma)/2\Omega\rho U^2.$$

$\partial\psi/\partial s$ vanishes when there is no vertical advection of potential density.

Equations (A1)–(A7) for strictly geostrophic motion are valid in the multiple limit when in magnitude $\partial\rho/\partial t$ and each component of $\{\rho(\partial \boldsymbol{u}/\partial t + (\boldsymbol{u}\cdot\nabla)\boldsymbol{u}) + \nabla \times (\mu\nabla \times \boldsymbol{u})\}$ (respectively the terms neglected in Equations (A2) and (A1), where μ is the coefficient of viscosity) divided respectively by $\nabla\cdot(\rho\boldsymbol{u})$ and the corresponding component of $2\rho\boldsymbol{\Omega} \times \boldsymbol{u}$ (the terms retained in those equations) tend to zero. The set of dimensionless parameters formed by these ratios (which include the familiar Rossby and Ekman numbers) constitutes a measure of average departures from strict geostrophy. Likewise, Equations (A9) and (A10) for strictly isentrophic and geostrophic motion are valid when we also require that the magnitude of the term representing irreversible thermodynamic processes that was neglected in writing down Equation (A9) divided by that of a typical term retained (a dimensionless quantity equal to the reciprocal of a Péclét number) tends to zero.

Equations (A1) to (A10) are mathematically degenerate; they do not suffice when combined with appropriate boundary conditions or initial conditions to determine the field of flow. Nevertheless, they express with good accuracy important properties that slow hydrodynamical motions in rapidly rotating systems must possess *nearly* everywhere and can, when judiciously applied, lead to the discovery of the location and nature of the highly ageostrophic phenomena (e.g. viscous or inertial boundary layers or detached shear layers, inertial oscillations) that are typical concomitants of highly geostrophic motion.

REFERENCES. LECTURE III

1. W. A. Baum and A. D. Code, *A photometric observation of the occultation of σ-Arietis by Jupiter*, Astronom. J. **58** (1953), 108–112.

2. E. C. Bullard and H. Gellman, *Homogeneous dynamos and terrestrial magnetism*, Philos. Trans. Roy. Soc. Ser. A **247** (1954), 213–278. MR **17**, 327.

3. B. F. Burke, "Radio observations of Jupiter," in I, *Planets and satellites*, G. P. Kuiper and B. M. Middlehurst, (editors), 1961, pp. 473–499.

4. J. G. Charney, *The dynamics of long waves in a baroclinic westerly current*, J. Meteorol. **4** (1947), 135–162. MR **9**, 163.

5. Deep-sea Research **6** (1960), no. 4. (A series of papers devoted to the Cromwell Current).

6. W. C. DeMarcus, *The constitution of Jupiter and Saturn*, Astronom. J. **63** (1958), 2–28.

7. J. N. Douglas and H. J. Smith, *Changes in rotation period of Jupiter's decameter radio sources*, Nature **199** (1963), 1080–1081.

8. E. T. Eady, *Long waves and cyclone waves*, Tellus **1** (1949), no. 3, 33–52. MR **13**, 86.

9. G. R. A. Ellis and P. M. McCulloch, *The decametric radio emissions of Jupiter*, Austral. J. Phys. **16** (1963), 380–397.

10. W. M. Elsasser, *Hydromagnetism: a review*, Amer. J. Phys. **23** (1955), 590–609; **24** (1956), 85–110.

11. G. B. Field, *The source of radiation from Jupiter at decimeter wavelengths*, J. Geophys. Res. **65** (1960), 1661–1671.

12. J. F. Focas, *Preliminary results concerning the atmospheric activity of Jupiter and Saturn*, Les Congrès et Colloq. Univ. Liège 24, La Physique des Planètes, also: Mém. Soc. Roy. Sci. Liège (5) **7** (1962), 535–540.

13. K. L. Franklin, *Radio-waves from Jupiter*, Sci. Amer. (July), (1964), 35–42.

14. R. M. Gallet, "Radio observations of Jupiter. II," in *Planets and satellites*, G. P. Kuiper and B. M. Middlehurst, (editors), Univ. of Chicago Press, Chicago, Ill., pp. 500–533.

15. R. Hide, *On the effect of rotation on fluid flow, with application to geophysics, solar physics and Jupiter's atmosphere* (unpublished ms.), see also: 1962. *Some thoughts on rotating fluids*, Scientific Report HRF/SR2, Geophysical Fluid Dynamics Laboratory, Department of Geology and Geophysics, M.I.T., Cambridge, Mass., 1960.

16. ———, *Origin of Jupiter's great red spot*, Nature **190** (1961), 895–896.

17. ———, *On the hydrodynamics of Jupiter's atmosphere*, Les Congrès Colloq. Univ. Liège 24, La Physique des Planètes, also: Mém. Soc. Roy. Sci. Liège (5) **7** (1962), 481–505.

18. ———. *On the circulation of the atmospheres of Jupiter and Saturn*, Planetary and Space Science **14** (1966), 669–675.

19. ———, *Planetary magnetic fields*, Planetary and Space Science **14** (1966), 579–586.

20. R. Hide and A. Ibbetson, *An experimental study of Taylor columns*, Icarus **5** (1966), 279–290.

21. J. A. Knauss, "Equatorial current systems", in *The sea*, Vol. 2, M. N. Hill (editor), Interscience, New York, 1963, p. 235.

22. B. C. Murray, R. L. Wildey and J. A. Westphal, *Observations of Jupiter and the Galilean satellites at* 10 *microns*, Astrophys. J. **139** (1964), 986–993.

23. E. J. Öpik, *Jupiter: chemical composition, structure and origin of a giant planet*, Icarus **1** (1962), 200–257.

24. T. C. Owen and D. O. Staley, *A possible Jovian analogy to the terrestrial stratospheric wind reversal*, J. Atmospheric Sci. **20** (1963), 347–350.

25. C. Payne-Gaposchkin, *Introduction to astronomy*, Eyre and Spottiswoode, London; 1956.

26. P. J. E. Peebles, *The structure and composition of Jupiter and Saturn*, Astrophys. J. **140** (1964), 328–347.

27. B. M. Peek, *The planet Jupiter*, Faber & Faber, London, 1958.

28. R. O. Rice, *The chemistry of Jupiter*, Sci. Amer. (June), (1956).

29. J. A. Roberts, *Radio emission from the planets*, Planetary and Space Science **11** (1963), 221–259.

30. J. A. Roberts and M. M. Komesaroff, *Observations of Jupiter's radio spectrum and polarization in the range from* 6 cm *to* 100 cm, Icarus **4** (1965), 127–156.

31. A. G. Smith and T. D. Carr, *Radio exploration of the planetary system*, Van Nostrand, Princeton, N.J., 1964.

32. H. Spinrad, *The anomalous inclination of the Jovian ammonia lines*, Astrophys. J. **136** (1962), 311.

33. H. Spinrad and L. M. Trafton, *High dispersion spectra of the outer planets*. I: *Jupiter in the visual and infrared*, Icarus **2** (1963), 19–28.

34. G. I. Taylor, *Experiments on the motion of solid bodies in rotating fluids*, Proc. Roy. Soc. Ser. A **104** (1923), 213–218.

35. A. Toomre, *Effect of rotation on two-dimensional flow of a slightly compressible fluid around a long cylinder*, 1961 (unpublished manuscript).

36. J. W. Warwick, *Dynamic spectra of Jupiter's decametric emission* 1961, Astrophys. J. **137** (1963), 41–59.

37. ———, *The position and sign of Jupiter's magnetic moment*, Astrophys. J. **137** (1963), 1317–1318.

38. ———, *Radio emission from Jupiter*, Ann. Rev. Astronom. Astrophys. **2** (1964), 1–22.

39. ———, *Jupiter's magnetic field as inferred from decametric and decimetric radio-observations*, (this volume).

40. R. Wildt, "Planetary interiors," in *Planets and satellites*, G. P. Kuiper and B. M. Middlehurst, (editors), Univ. of Chicago Press, Chicago, Ill., 1961.

41. P. Goldreich and S. J. Peale, *The dynamics of planetary rotations*, Ann. Rev. Astronom. and Astrophys. **6** (1968), 287–320.

42. R. Hide, "On the dynamics of Jupiter's interior and the origin of his magnetic field," in *Magnetism and the cosmos*, Oliver and Boyd, Edinburgh, 1965, pp. 378–393.

43. H. P. Greenspan, *The theory of rotating fluids*, Cambridge Univ. Press, New York, 1968.

44. J. Proudman, *On the motion of solids in liquids possessing vorticity*, Proc. Roy. Soc. London, Ser. A **92** (1916), 408–424.

45. R. Hide, *On geostrophic motion of a nonhomogeneous fluid*, J. Fluid Mech. (1971) (to appear).

IV. Planetary Magnetic Fields[1]

Summary. Present knowledge and modern theories of the Earth's main magnetic field are outlined and the state of knowledge of the magnetic fields of the other planets is sketched.

1. **Introduction.** The Earth and Jupiter are the two objects in the planetary system that are known to possess magnetic fields of their own, the evidence in the case of Jupiter being indirect, stemming from radio-astronomical observations. Space probe magnetometer measurements in the vicinity of the Moon ("Lunik 2") (Dolginov et al. [**30**]), Venus ("Mariner 2" and "Mariner 5") (Smith et al. [**85**], [**86**]) and Mars ("Mariner 4") (Dryer and Heckman [**32**], Smith and Carr [**83**], Smith et al. [**84**], Davis and Williams [**27**]) indicate that if these objects are magnetic, they are only very weakly so. The determination of the magnetic fields of the remaining planets will doubtless constitute a major part of the growing programme of planetary exploration made possible by recent advances in space technology.

Most geophysicists accept that the Earth's magnetic field is due to electric currents within the Earth, that these currents flow mainly in the liquid core, and that they are generated by a magnetohydrodynamic "dynamo" process involving the interaction of hydrodynamical motions in the core with the magnetic field present there. The rotation of the Earth probably enhances the efficiency of the dynamo process through its influence on the pattern of core motions.

It has been suggested that Jupiter's magnetic field might be produced by magnetohydrodynamic processes in the lower atmosphere of the planet, where the electrical conductivity might be high enough for fluid motions to produce magnetic energy without being so high that the magnetic lines force cannot escape from their region of origin [**53**]. Alternatively, Jupiter's magnetic field might originate largely in a fluid core, if Jupiter has one, as in the case of the Earth. It will be surprising if any large, rapidly rotating and partially fluid planet, such as Saturn and the other major planets, is found to be nonmagnetic.

It is obviously impossible to discuss planetary magnetic fields without giving the Earth's magnetism pride of place, and in what follows three sections are devoted to it. Jupiter's magnetic field is discussed in §5.

2. **Description and analysis of the main geomagnetic field.**

Definition of Main Geomagnetic Field. The magnetic field at the surface of the Earth undergoes complicated changes with time (see Table I).

---

[1] Supplementary notes for Lecture IV, based on a recent article by the lecturer [**10**].

The most rapid changes, occurring on time scales ranging from fractions of a second (subacoustic oscillations) to several days (magnetic storms) have amplitudes less than 1% of the total field and are revealed only by sensitive instruments. They are due to varying electric currents flowing well above the Earth's surface, in the ionosphere and beyond. When, by taking annual mean values, these rapid variations have been eliminated from the magnetic record, the "main geomagnetic field" remains. The main field originates within the Earth and undergoes variations on time scales of decades and upwards.

MAGNETIC ELEMENTS. The geomagnetic field, a vector quantity $\boldsymbol{B}$, has at the Earth's surface the components (X, Y, Z) directed northward, eastward and vertically downward respectively. The traditional magnetic elements, H (horizontal force), D (declination) and I (inclination of dip) are related to (X, Y, Z) as follows:

$$(1) \qquad H = (X^2 + Y^2)^{1/2}, \qquad Y = X \tan D, \qquad Z = H \tan I.$$

Direct measurements of these elements, which go back only a century or so, have been made at both permanent and temporary observatories, located mainly in the populated regions of the Earth. Consequently, secure and detailed knowledge of the behavior of the geomagnetic elements is available for only a limited region of space over a short (geologically speaking) period of time. Thanks, however, to recent studies of the magnetization of human artifacts, such as hearths, pottery and kilns, and of sedimentary and igneous rocks, we are not completely ignorant of the history of the geomagnetic field.

Fairly thorough accounts of observatory data have been given and summarized in a number of recent publications. Vestine and his collaborators of the Carnegie Institution of Washington and others (Finch and Leaton [**41**], Vestine et al. [**98**], [**99**]) have made intensive studies of all the available data for the period 1900–1945, and Gaibar-Puertas [**45**] of the Observatorio del Ebro has covered a somewhat longer period, extending back to the middle of the last century, but in a little less detail. Detailed reviews of paleomagnetic work have been given by a number of writers (Blackett [**7**], Cox and Doell [**23**], Irving [**60**], Nagata [**69**], Runcorn [**82**]).

The principle of the compass is that one end of a magnetic needle freely suspended about a vertical axis points near the true north, the angle between these two directions, D (declination), being small except near the geomagnetic poles. Direct measurements of D and of I ("inclination" or "angle of dip") go back nearly 500 years at some observatories. The discovery of the geomagnetic secular variation (g.s.v.) is usually ascribed to Gellibrand who, in 1635, found that D varies with time; the discovery that all the elements undergo secular changes came much later (see Chapman and Bartels [**19**], Fleming [**42**], Runcorn [**81**]).

Temporal changes in the direction of $\boldsymbol{B}$ at a given station may be represented by the motion of a point on a sphere cut by a unit vector drawn parallel to the $\boldsymbol{B}$, the motion on a sphere being transferred, for convenience, on to a plane, by projection. Such curves representing the change in the direction of $\boldsymbol{B}$ have been prepared for a large number of observatories (Gaibar-Puertas [**45**]). The longest series of direct observations come from London and Paris and cover four centuries; archeomagnetic studies (Cook and Belshé [**20**]) of fired clays have yielded additional information (Figure 1). The shape of this curve suggests that it could be explained in terms of the geomagnetic pole precessing around the geographic pole at a rate corresponding to a period of about 500 years. However, the correlation expected on this picture between similar observations at different observatories all over the world is not, in fact, present, although there is one feature common to most stations, namely, the sense of motion along the curve (clockwise viewed along the vector). This lack of correlation is due to the regional, rather than planetary, nature of the g.s.v.

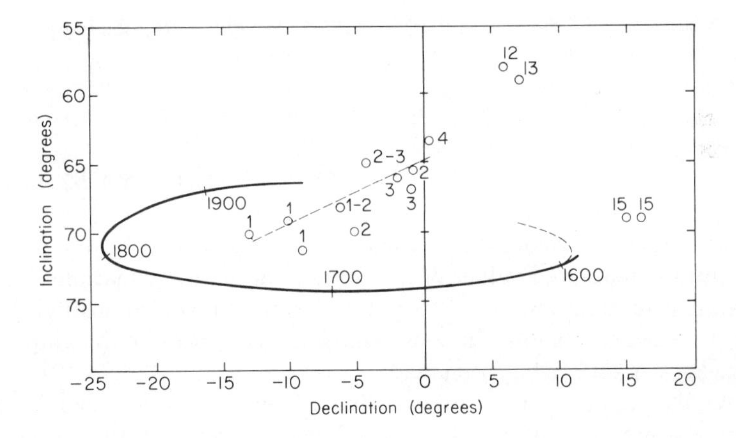

FIGURE 1. Changes in the mean direction of the Earth's magnetic field since the first century A.D. The continuous black line represents observations at London since A.D.1578 and the extension in short dashes represents Bauer's extrapolations from observations at Rome. The circles show the average readings, corrected for Cambridge, of 14 samples of fired clay, and the line of longer dashes the trend for the Roman period (after Cook and Belshé [**20**]).

Spherical Harmonic Analysis. The main geomagnetic field was first subjected to spherical harmonic analysis by Gauss (Chapman and Bartels, [**19**], Vestine et al. [**98**], [**99**]). If no electric currents are present in the neighbourhood of the Earth's surface, $\boldsymbol{B}$ may be written as $-\nabla V$, the gradient of a scalar potential which satisfies Laplace's equation $\nabla^2 V = 0$. V can therefore be expressed as a series of spherical harmonics

$$V = R_s \sum_{n=1}^{\infty} \sum_{m=0}^{n} n^{-1} P_n^m(\cos\theta)$$

$$\cdot \left[\left\{ C_n^m \left(\frac{r}{R_s}\right)^n + (1 - C_n^m)\left(\frac{R_s}{r}\right)^{n+1} \right\} A_n^m \cos m\phi \right. \tag{2}$$

$$\left. + \left\{ S_n^m \left(\frac{r}{R_s}\right)^n + (1 - S_n^m)\left(\frac{R_s}{r}\right)^{n+1} \right\} B_n^m \sin m\phi \right]$$

where R_s is the radius of the Earth, (r, θ, ϕ) are spherical coordinates, ϕ being the *geographical* colatitude, C_n^m, S_n^m are the portions of the harmonic terms of *external* origin, and A_n^m, B_n^m are the coefficients of the spherical harmonic series. By differentiating Equation (2) it may be shown that at the surface of the Earth

$$X = \sum_{n=1}^{\infty} \sum_{m=0}^{n} n^{-1}(dP_n^m(\cos\theta)/d\theta)[A_n^m \cos m\phi + B_n^m \sin m\phi],$$

$$Y = \sum_{n=1}^{\infty} \sum_{m=0}^{n} mn^{-1} \sin\theta \; P_n^m(\cos\theta)[A_n^m \sin m\phi - B_n^m \cos m\phi], \tag{3}$$

$$Z = \sum_{n=1}^{\infty} \sum_{m=0}^{n} n^{-1} P_n^m(\cos\theta)[\{nC_n^m - (n+1)(1 - C_n^m)\}A_n^m \cos m\phi$$

$$+ \{nS_n^m - (n+1)(1 - S_n^m)\}B_n^m \sin m\phi].$$

Two values of each of the coefficients A_n^m, B_n^m, C_n^m, S_n^m may be obtained by substituting measured values of X, Y and Z in the last equations. These separate determinations agree to within the errors involved, justifying the use of a scalar potential. In some analyses allowance for the ellipticity of the Earth's surface has been made.

The first result of the analysis is that C_n^m and S_n^m are not significantly different from zero, within the errors involved, implying that the main field originates entirely within the Earth. Setting C_n^m and S_n^m equal to zero and following the convention of denoting A_n^m and B_n^m by g_n^m and h_n^m respectively, the field may be regarded as comprising the separate fields of hypothetical multipoles, situated at the geocentre, of magnitude given by the g_n^m and h_n^m coefficients. g_1^0 corresponds to the axial dipole and g_1^1 and h_1^1 to the two equatorial dipole components. There are five quadrupoles, g_2^0, g_2^1, g_2^2, h_2^1, h_2^2, and so on. The results of the principal spherical harmonic analyses based on an earlier compilation by Runcorn [**82**] are reproduced in Table I.

TABLE I. The Spherical Harmonic Analysis of the Main Field Units 10^{-4} gauss (after [**81**]).

Source	Epoch	g_1^0	g_1^1	h_1^1	g_2^0	g_2^1	h_2^1	g_2^2	h_2^2
Gauss	1835	−3235	−311	+625	+ 51	+292	+ 12	− 2	+157
Erman–Petersen	1839	−3201	−284	+601	− 8	+257	− 4	− 14	+146
Adams	1845	−3219	−278	+578	+ 9	+284	− 10	+ 4	+135
Adams	1880	−3168	−243	+603	− 49	+297	− 75	+ 61	+149
Fritsche	1885	−3164	−241	+591	− 35	+286	− 75	+ 68	+142
Schmidt	1885	−3168	−222	+595	− 50	+278	− 71	+ 65	+149
Neumayer–Petersen	1885	−3157	−248	+603	− 53	+288	− 75	+ 65	+146
Dyson–Furner	1922	−3095	−226	+592	− 89	+299	−124	+144	+ 84
Afana-sieva	1945	−3032	−229	+590	−125	+288	−146	+150	+ 48
Vestine–Lange	1945	−3057	−211	+581	−127	+296	−166	+164	+ 54

Third-order Coefficients

Source	Epoch	g_3^0	g_3^1	h_3^1	g_2^3	h_3^2	g_3^3	h_3^3
Neumayer–Petersen	1885	+ 98	−129	− 24	+144	+ 2	+ 41	+ 70
Vestine–Lange	1945	+115	−173	− 52	+121	+ 18	+ 88	+ 3

The predominance of the centered axial dipole, g_0^1, is a conspicuous feature of these results, the equatorial dipole amounting to about 0.15 of the axial dipole. The higher harmonics correspond to the "nondipole" field. The r.m.s. strength of the nondipole field amounts to only 0.05 of the dipole field, but because of the more rapid variation with depth of the former than the latter, in the upper reaches of the Earth's core (see below), where the geomagnetic field originates, these fields are probably comparable in strength.

Maps showing all the geomagnetic elements at different times have been prepared by several workers (Finch and Leaton [**41**], Gaibar-Puertas [**45**], Vestine et al. [**98**], [**99**]) (see, for example, Figures 2a and 2b).

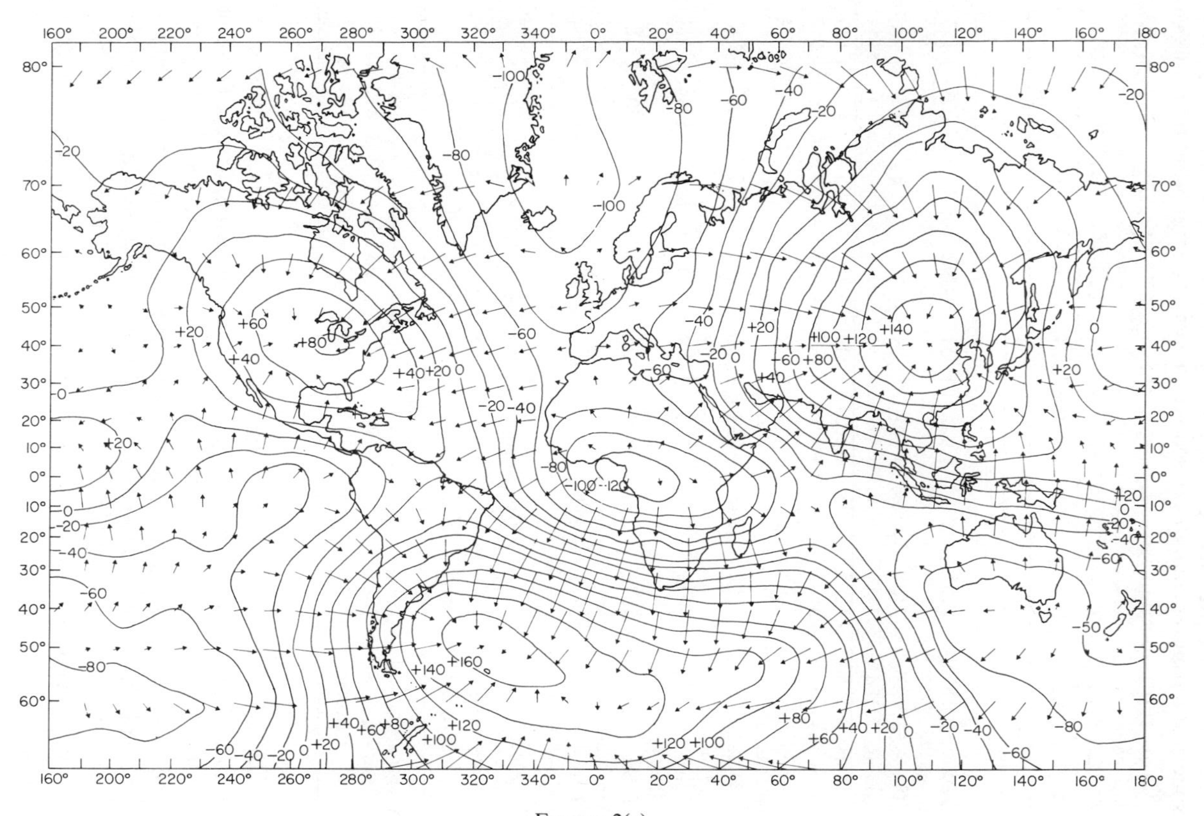

160° 200° 220° 240° 260° 280° 300° 320° 340° 0° 20° 40° 60° 80° 100° 120° 140° 160° 180°
80° 70° 60° 50° 40° 30° 20° 10° 0° 10° 20° 30° 40° 50° 60°

FIGURE 2(a)

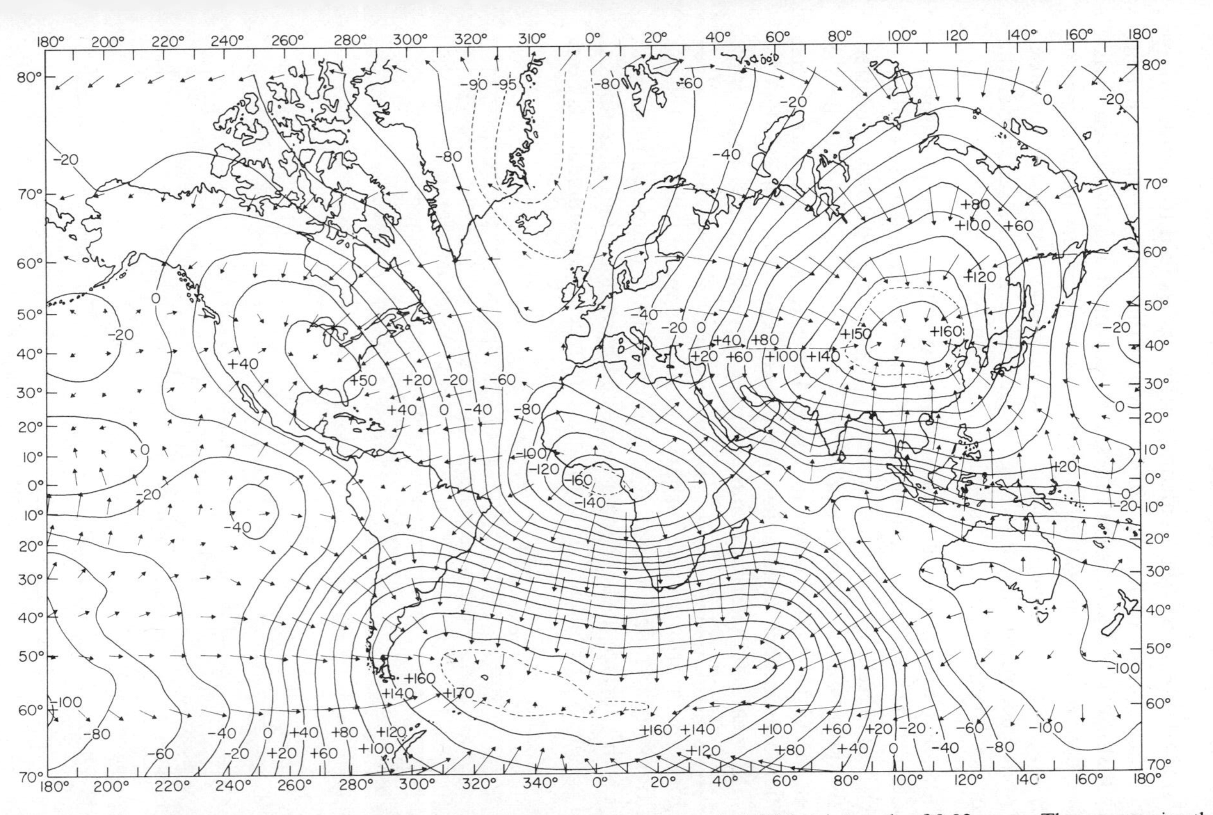

FIGURE 2(b). Nondipole fields for (a) 1907.5 and (b) 1945. The contours give the vertical field at intervals of 0.02 gauss. The arrows give the horizontal components. An arrow equal in length to the distance between the lines of longitude drawn in the diagram represents 0.068 gauss (after Bullard et al. [**15**]).

The characteristic features of the g.s.v. are much more readily ascertained from such maps than from the spherical harmonic coefficients.

PALEOMAGNETIC STUDIES. Paleomagnetic studies lead to useful, if only approximate, information on the *direction* of the main geomagnetic field over historical and geological times, but they have not, as yet, given accurate data on field *intensity*. The most striking discovery made by paleomagnetic workers is that the dipole field undergoes "reversals" in sense, no preference being shown over geological time for one sense over the other (see Irving [**60**] for a comprehensive review).

PRINCIPAL PROPERTIES OF THE MAIN GEOMAGNETIC FIELD. The principal properties of the main geomagnetic field may be summarized as follows:

(1) The present field is predominantly that of an hypothetical centred axial dipole, with a moment of 8×10^{25} e.m.u., giving a surface field of about 0.5 G (5×10^{-5} weber/m^2) (see Table I). The assumption that the field was approximately that of centered axial dipole during the past 5×10^7 years (Quaternary and Tertiary periods) is consistent with paleomagnetic data (see Figure 3). (The inconsistencies present in paleomagnetic data prior to the Tertiary are more apparent than real if the positions of the poles of rotation of the Earth have wandered relative to the Earth's surface in the manner suggested by various lines of geological and geophysical evidence. A very important recent development has been

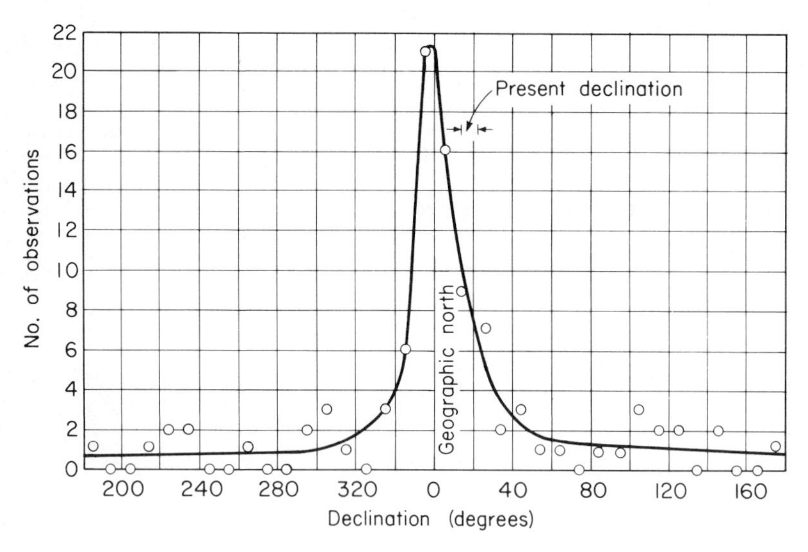

FIGURE 3. Frequency distribution of declination measurements on Mesozoic rock samples (after Torreson et al. [**92**]).

the interpretation of paleomagnetic pole positions prior to the Tertiary in terms of polar wandering and continental drift.)

(2) The polarity of the (assumed) dipole field may have reversed several hundred times since the Pre-Cambrian, at irregular intervals. (No reversals occurred, for example, during the Permian, which lasted from 280 My B.P. (before the present) to 230 My B.P.) The most accurate studies of "reversals" are those by Cox et al. [**26**] (see Figure 4).

(3) The small but significant nondipole field undergoes changes that are so rapid in comparison with those of the dipole field that the r.m.s. amplitudes of their respective contributions to the geomagnetic secular variation, 5×10^{-4} G/year and 3.5×10^{-4} G/year respectively, are comparable with one another. (The corresponding *maximum* amplitude is 1.5×10^{-3} G/year.) On typical magnetic maps, lines of equal change of any element (isopors) form a series of sets of oval curves surrounding points at which the changes are most rapid (see Figure 5). At any epoch, the sets of isopors cover areas of continental size and are separated by areas over which changes are small. Whilst the whole pattern of nondipole field and the geomagnetic secular variation shows no striking correlation with topographical features of the Earth's surface it may be noteworthy that the g.s.v. is systematically higher in the Antarctic and

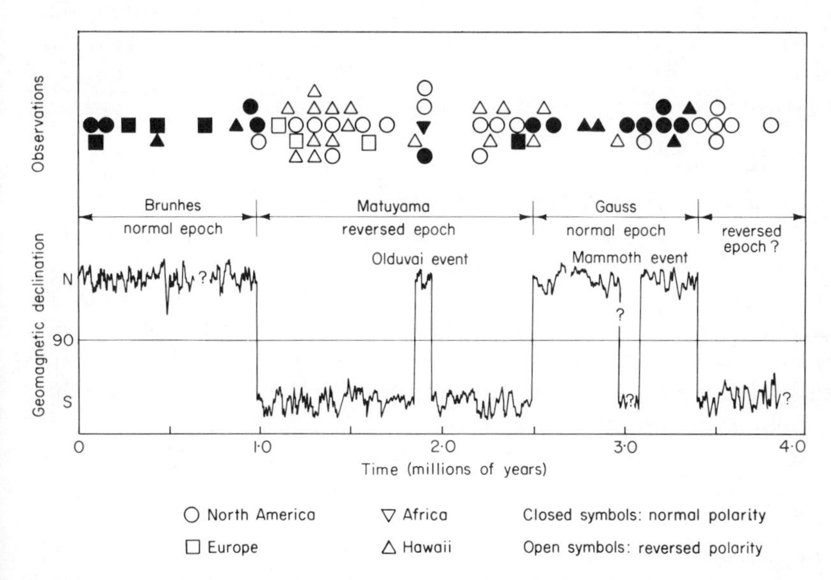

FIGURE 4. Magnetic polarities of 64 volcanic rocks and their potassium–argon ages. Geomagnetic declination for moderate latitudes is indicated schematically (after Cox et al. [**26**]).

lower in the Pacific hemisphere than the world-wide average. Paleomagnetic work indicates that the properties of the g.s.v. revealed by observatory data have persisted for the past few thousand years at least (Blackett [7], Cox and Doell [**23**], Irving [**60**], Nagata [**69**], Runcorn [**81**]), and that the low value of the g.s.v. over the Pacific may have persisted for the past million years (Doell and Cox [**29**]).

(4) Isoporic foci move across the Earth's surface in a generally westward direction at a fraction of a degree of longitude per year (Bullard et al. [**15**], Lowes [**65**], Nagata [**70**], Yukutake [**106**]) as does the pattern of the nondipole field (see Figures 5 and 2); indeed a significant fraction of the geomagnetic secular variation—about one half—can be represented by the changes due to the uniform westward drift at 0.2° long/yr of the instantaneous nondipole field. The drift of the dipole component of the field is much slower than that of the nondipole field. Aitken and Weaver [**1**] have adduced evidence from a study of the magnetic properties of ancient pottery kilns in Britain that the motion of the equatorial dipole may have been eastward (relative to the mantle) during the period A.D. 900 to A.D. 1350. Brynjolfsson [**10**], from measurements of Icelandic lava flows, concludes that "the higher harmonics (periods of the order of 300 to 500 yr) drift westward while lower harmonics (periods of the order of 4000 to 5000 yr) drift eastward." Related Japanese work (Kawai [**62**]) is in general agreement with that of Aitken and Weaver and of Brynjolfsson.

An acceptable theory of the origin of the main geomagnetic field must account for the foregoing properties of the phenomenon in terms of physical processes capable of operating within the Earth.

3. **The Earth's interior.**

PRINCIPAL REGIONS. The Earth is an almost spherical body of radius (R_s) 6500 km. Seismology shows that beneath a thin surface crust of variable thickness (~ 30 km), the internal structure of the Earth is nearly spherically symmetric. In the region lying between the radii of (R_c) 3500 km and R_s, the Earth can transmit shear waves, showing that it is a solid there so far as short period disturbances are concerned. This region is called the mantle, mean density 4.4 g/cm^3. Below the mantle is a fluid region, as evinced by its inability to transmit shear waves. The density (ρ) of this region—the core—is about 10 g/cm^3. Below the fluid region is a small solid central body, some 1300 or 1400 km in radius.

The mantle is thought to consist mainly of magnesium and iron-rich silicates. The temperature and, of course, the pressure in it increases with depth and these produce variations in its electrical properties. At a depth of 500 km, the electrical conductivity is at least $10^{-1}\,\Omega^{-1}\,m^{-1}$, while at the base of the mantle it probably reaches $10^2\,\Omega^{-1}\,m^{-1}$ (Tozer [**94**]).

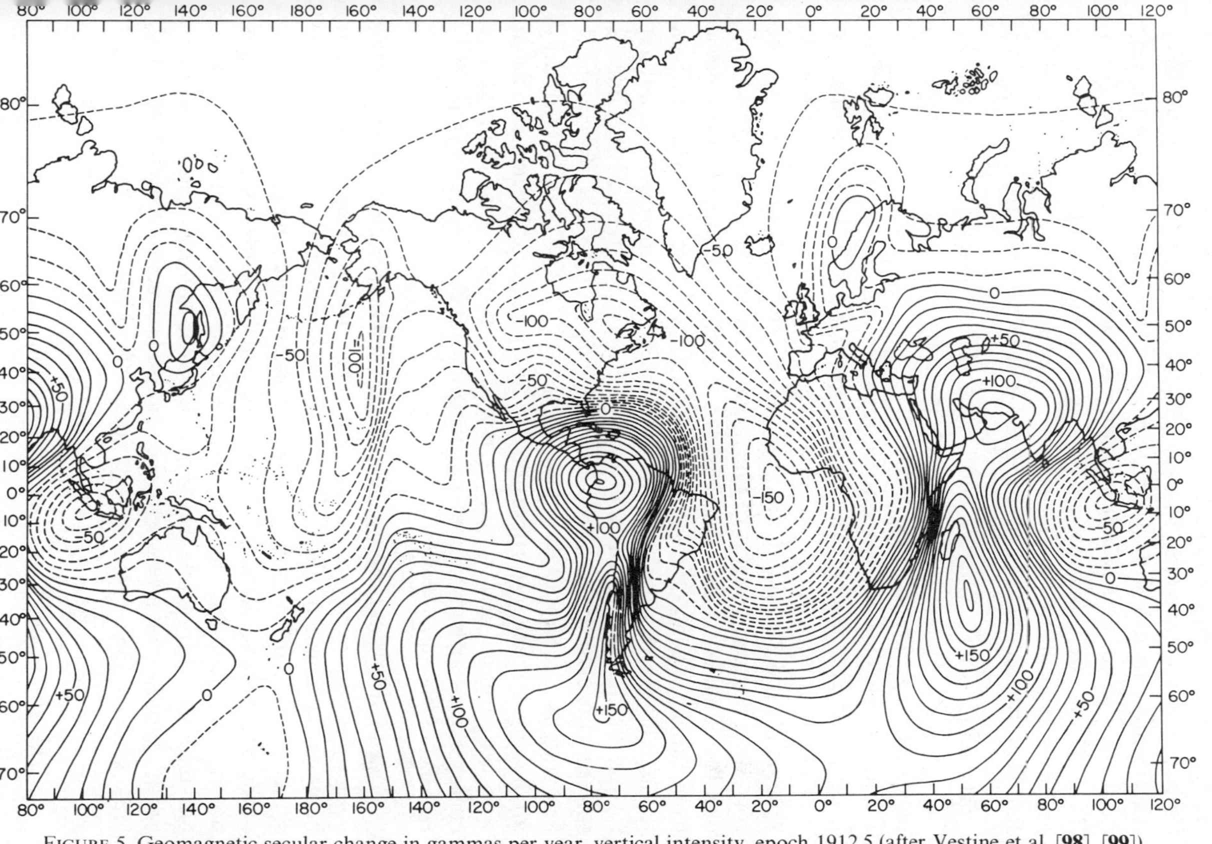

FIGURE 5. Geomagnetic secular change in gammas per year, vertical intensity, epoch 1912.5 (after Vestine et al. [**98**], [**99**]).

The core is thought to consist mainly of iron, with a little nickel and other elements. Its electrical conductivity (σ) is probably about $7 \times 10^5 \Omega^{-1} m^{-1}$ (Bullard and Gellman [**16**], Elsasser [**37**]; see also Hide [**57**], Stacey [**88**]). Its (kinematical) viscosity (ν) is extremely uncertain, and estimates varying from 10^{-7} to $10^3 m^2 sec^{-1}$ have been offered (see Bullard and Gellman [**16**], Jeffreys [**61**]). Because of the high temperature of the core, it is generally considered that its magnetic permeability (μ) does not differ significantly from that of free space (see, however, Weiss [**105**]).

MAGNETIC FIELD WITHIN THE EARTH. The magnetic field within the Earth cannot be determined uniquely from the field at the surface without making further physical hypotheses.

The electric currents responsible for the main geomagnetic field flow mainly in the liquid core. The magnetic field $\boldsymbol{B}$ due to a general distribution of currents, density $\boldsymbol{j}$, in such a region can be decomposed into its "poloidal" and "toroidal" components, $\boldsymbol{B}_P$ and $\boldsymbol{B}_T$ respectively; let the corresponding components of $\boldsymbol{j}$ be $\boldsymbol{j}_P$ and $\boldsymbol{j}_T$. As $\boldsymbol{B}$ (unlike $\boldsymbol{B}_T$ whose lines of force lie on horizontal surfaces) possesses lines of force which emerge from the core, penetrate the upper reaches of the solid Earth and pass through the Earth's surface into space, the gross properties of $\boldsymbol{j}_P$ (but not $\boldsymbol{j}_T$) can be determined directly from the magnetic field at the Earth's surface. However, modern theories of the origin of the main geomagnetic field and its secular variation indicate that $|\boldsymbol{B}_T/\boldsymbol{B}_P| \sim 10$ to 10^2, virtually all the magnetic energy (proportional to $\boldsymbol{B} \cdot \boldsymbol{B}$) of the Earth being associated with the toroidal field, so that it is entirely possible that the magnetic energy of the Earth undergoes only insignificant changes during a polarity reversal of the dipole field.

CORE MOTIONS. It is now generally accepted that electric currents in the core are produced by "motional induction" there, due to the interaction between the magnetic fields present there and hydrodynamical flow, at rather less than 0.1 cm/sec, and the next section is devoted to a discussion of the processes involved. The agency responsible for stirring the core has not yet been identified with certainty; radioactive heating within the Earth, gravitational energy released if the Earth is still condensing, and the Earth's precessional motion are the principal suggestions to date (Bullard [**13**], Elsasser [**38**], Frenkel [**44**], Hide [**50**], Malkus [**67**], Roberts and Stewartson [**78**], Taylor [**90**], Toomre [**91**], Urey [**95**], Verhoogen [**96**]).

4. Theories of the main geomagnetic field.

MOTIONAL INDUCTION IN THE EARTH'S CORE. Modern theoretical work on the main geomagnetic field and its secular variations stems from a suggestion by Larmor [**63**] that the Sun's magnetic field is due to electric currents resulting from motional induction involving the circulation of matter within the body of the Sun. Cowling [**22**] subsequently proved that the axisymmetric system invoked by Larmor is incapable of acting as a "homogeneous dynamo" mechanism, but further work (see Backus [**3**], Backus and Chandrasekhar [**5**], Braginskiĭ [**8**], [**9**], Bullard and Gellman [**16**], Herzenberg [**48**], Inglis [**59**], Lowes and Wilkinson [**66**], Parker [**71**], Rikitake [**74**], Roberts [**77**], Lilley [**80**], Tough and Roberts [**93**], and others) pioneered independently by Bullard [**11**], [**14**]) and Elsasser [**34**], [**35**], [**36**], [**37**], [**39**], has demonstrated not only that homogeneous dynamos can occur in principle, but that such a mechanism operating in the Earth's core is the most likely cause of the main geomagnetic field.

Within the core $\boldsymbol{B}$ and $\boldsymbol{u}$, the Eulerian velocity vector, satisfy

$$\partial \boldsymbol{B}/\partial t = \lambda \nabla^2 \boldsymbol{B} + \nabla \times (\boldsymbol{u} \times \boldsymbol{B}), \qquad \nabla \cdot \boldsymbol{B} = 0, \tag{4a, b}$$

the former equation being the result of eliminating current density $\boldsymbol{j}$, and the electric field vector between the first and second of Maxwell's equations (Ampère's and Faraday's laws) and Ohm's law. The parameter

$$\lambda \equiv (\mu\sigma)^{-1} \tag{5}$$

is assumed independent of position in Equation (4a); according to §3 above, $\lambda \sim 1\ \text{m}^2/\text{sec}$.

The steady, or quasi-steady, energy balance in the core is such that the rate at which the magnetic field is built up by motional induction, represented by the term $\nabla \times (\boldsymbol{u} \times \boldsymbol{B})$ in Equation (4), is equal to the rate at which the field decays through Ohmic dissipation, represented by $\lambda \nabla^2 \boldsymbol{B}$. Therefore, when dealing with the long-period behaviour of the field both terms on the right-hand side of Equation (4) have to be taken into account. However, when dealing with the comparatively rapid geomagnetic secular variation, in magnitude the term $\lambda \nabla^2 \boldsymbol{B}$ is probably only a small fraction of the other terms, so that Equation (4) is well approximated by

$$\partial \boldsymbol{B}/\partial t = \nabla \times (\boldsymbol{u} \times \boldsymbol{B}). \tag{6}$$

The last equation is the differential form of a famous theorem due to Alfvén, which states that in a perfect conductor, lines of magnetic force move with the fluid particles. Thus, it might be possible to interpret the secular variation as a process of continual rearrangement—as opposed to creation and destruction—of lines of magnetic forces of the Earth's poloidal magnetic field (see Backus [**4**] for general discussion and comprehensive list of references).

THEORIES OF THE GEOMAGNETIC SECULAR VARIATION. The geomagnetic secular variation is due to fluctuations in the magnetohydrodynamic flow within the core. A successful theory of the phenomenon will involve the determination of simultaneous solutions to Equation (4) and the equations of hydrodynamical motion

$$\partial \boldsymbol{u}/\partial t + (\boldsymbol{u} \cdot \nabla)\boldsymbol{u} + 2\boldsymbol{\Omega} \times \boldsymbol{u} = -(1/\rho)\nabla p + \nu\nabla^2\boldsymbol{u} + \nabla\Phi + (1/\mu\rho)(\nabla \times \boldsymbol{B}) \times \boldsymbol{B}, \tag{7a}$$

$$\nabla \cdot \boldsymbol{u} = 0, \tag{7b}$$

(where $\boldsymbol{\Omega}$ is the angular velocity of the Earth's rotation if $\boldsymbol{u}$ is measured relative to the rotating Earth, ρ denotes density, p pressure, ν kinematic viscosity (assumed uniform), and Φ the potential due to gravitational and centripetal forces) under appropriate boundary conditions.

The observed displacements and alterations of the pattern of the geomagnetic field at the Earth's surface suggest that the magnitude U, of the horizontal component of core motions is about 3×10^{-4} m/sec. In the earliest theoretical work on the geomagnetic secular variation (Allan and Bullard [**2**], Bullard [**12**], Coulomb [**21**], Herzenberg and Lowes [**49**]) which is based on models of core motions that exclude effects of magnetohydrodynamic waves, U is interpreted as a lower limit to U_H, the speed of horizontal motion of material particles in the core. For example, Bullard (Bullard [**12**], Bullard et al. [**15**]) has proposed the following explanation of the westward drift of the geomagnetic field (see §3 above). Assuming that during their motion, individual fluid particles tend to conserve angular momentum, he argued that if meridional flow occurs in the core, the concomitant inward advective transfer of angular momentum would cause the inner parts of the core to rotate more rapidly than the outer parts. For such an angular velocity distribution to remain steady on the average, advective angular momentum transfer inward must be balanced by an outward transfer due to friction. Bullard showed that although viscosity would be inadequate to supply this frictional transfer, electromagnetic forces might suffice. Further, at the outer extremity of the core, where viscosity demands that there be no motion relative to the mantle, electromagnetic coupling between the core and mantle would allow the outer part of the core to move westward relative to the mantle. Hence, if secular variation sources are located mainly in the upper reaches of the core they would appear to drift westward relative to a fixed observer at the Earth's surface.

An alternative model of the westward drift and an explanation of the general time-scale of the geomagnetic secular variation have been proposed by Hide [**55**], who showed that if the strength of the toroidal magnetic field in the core is about 100 G (10^{-2} weber/m^2) then it might be possible to account for these properties of the geomagnetic field in terms of the interaction of free magnetohydrodynamic oscillations of the core with the Earth's poloidal magnetic field. On the "free oscillations" model (the underlying theory of which is a tentative solution to Equation (6) and the equation to which Equation (7) simplifies when dissipation is ignored and departures from an assumed equilibrium state are treated as small), U is the average speed of propagation of magnetohydrodynamic waves in the core and may exceed U_H. The tendency for the nondipole field to drift westward more rapidly than the dipole field (see §2 above) is interpreted as a manifestation of the dispersive properties, due to the Earth's rotation, of the waves. Although eastward-propagating waves are possible in principle, according to Hide's arguments they are not expected to occur frequently in the core because the conditions for their excitation, mainly that the vertical density gradient is stable and exceeds a certain critical value, are unlikely to be met. (For further discussion see Appendix.)[2]

THEORIES OF THE MAIN FIELD. Fluid motions in the core at the speeds suggested by the properties of the geomagnetic secular variation could, by motional induction, interact with the magnetic fields there (which are probably mainly toroidal (see above)) to produce electric currents of sufficient strength to sustain these magnetic fields. This is the rough quantitative basis of the "homogeneous dynamo theory" (see above).

If a self-maintaining dynamo is to occur in the core, then the system of fluid motions there must be quite complicated and highly asymmetric (Cowling [**22**], Bullard [**13**], [**14**], Elsasser [**34**], [**35**]). The interaction of horizontal motions with the poloidal magnetic field leads to a toroidal magnetic field. Interaction of vertical motions with these toroidal and poloidal fields then leads to the regeneration of the original poloidal field.

With such a system, a dipole field of either sign would be maintained. For a given toroidal field the sign of the dipole depends on the details of the velocity field (Lilley [**80**]), and reversal of the external dipole may involve comparatively minor changes in the magnetic field in the core and in the pattern of fluid motions there.

It has recently been suggested (Hide [**55**]) that the effect on core motions of horizontal variations in the properties (topography (see Hide and Horai [**58**]) temperature) of the core-mantle interface that would escape detection by modern seismological techniques might be sufficient to produce measurable geomagnetic effects (see also Lecture V). Thus,

---

[2] See pages 327–328 and 346–349.

the nondipole field might reflect in part the properties of the core-mantle interface. Moreover, motions in the mantle might occasionally change the properties of the core-mantle interface sufficiently to bring about reversals of the dipole field.

A quantitative requirement for dynamo action is that a "magnetic Reynolds number"

$$R \equiv \tilde{U}L/\lambda, \tag{8}$$

the ratio of the order of magnitude of the second term on the right-hand side of Equation (4a) to that of the first term (where $\tilde{U}$ is a typical value of $\boldsymbol{u}$ and L is a typical length scale) should satisfy

$$R_1 < R < R_2, \tag{9}$$

where $R_1 \sim 10$ or 100 and $R_2 \gg R_1$. The lower limit corresponds to the requirement that motional induction be sufficient to overcome Ohmic dissipation and thus give rise to an increase in the magnetic energy. The upper limit corresponds to the requirement that σ, though large, is not so great that magnetic lines of force are unable to leak out of the fluid.

Most theoretical work on the dynamo mechanism concerns itself with finding solutions to Equations (4) and (7b) by postulating $\boldsymbol{u}$ a priori. In the most recent developments, however, (see Parker [**71**], Braginskiĭ [**8**], [**9**]) use is also made of simplified versions of the equation of motion, Equation (7a). Finding simultaneous solutions of Equations (4) and (7) under geophysically realistic conditions presents formidable mathematical problems, most of which have not yet been solved.

Simple order of magnitude arguments applied to Equation (7) show, as several workers have pointed out, that the Coriolis acceleration dominates the left-hand side and that the viscous term on the right-hand side is probably quite negligible. In these circumstances Coriolis forces align core eddies, thus giving the geomagnetic field its approximate alignment with the Earth's rotation axis. Less obvious, but probably very important, is the role played by Coriolis forces in ensuring that (irrespective of the size of the kinetic energy producing eddies in the core) the largest eddies are comparable in size with the dimensions of the core. Thus may be achieved the low degree of symmetry and the large magnetic Reynolds number necessary for the dynamo mechanism to work efficiently. These large eddies gain kinetic energy by nonlinear interactions with smaller eddies, a process which, although impossible in isotropic homogeneous turbulence (where energy cascades in the opposite direction, from large to small eddies), should be quite common in large-scale natural systems, where anisotropy, due to the Earth's rotation in the case of the core, is the rule rather than the exception. This type of mechanism may underlie the observational result that magnetic astronomical bodies (stars, planets) usually rotate.

5. **Jupiter's magnetic field.**

INTRODUCTION. Jupiter is the largest of the nine planets that revolve around the Sun. Its radius is 70,000 km, eleven times that of the Earth, and its mass is 2.5 times that of all the other planets put together. It lies fifth in order of distance from the Sun, which it circles once every 11.9 years in an orbit lying between those of Mars and Saturn.

NONTHERMAL RADIATION. Electromagnetic radiation from Jupiter on wavelengths ranging from decimetres to decametres has many fascinating and quite unexpected properties (see Warwick [**104**] for the most recent review). On these wavelengths Jupiter is one of the brightest radio sources in the sky, emitting much more energy, by several powers of 10, than the thermal radiation expected on the basis of observations in the infra-red part of the spectrum. The decametre radiation is emitted in intermittent bursts from relatively localized sources. The frequency of occurrence of the bursts has been shown to depend on the position in its orbit of Io, the innermost of the Galilean satellites, which revolves around the planet once every 1.8 days at a distance of six times the planetary radius (Bigg [**6**], Burns [**18**], Lebo et al. [**64**], Piddington and Drake [**73**], Warwick [**103**]).

Since Jupiter's nonthermal radiation was discovered, by accident as recently as 1955, intensive studies of its intensity, degree of polarization, spectral characteristics and other properties have been made by several groups of radio-astronomers in different parts of the world (Burke [**17**], Douglas and Smith [**31**], Ellis and McCullough [**33**], Field [**40**], Franklin [**43**], Gallet [**46**], Gulkis and Carr [**47**], Roberts, J. A. [**75**], Roberts and Komesaroff [**76**], Rose [**79**], Warwick [**100**]–[**104**]). No completely satisfactory theory of the radiation has yet been proposed and the Io effect in particularly mysterious (see [**108**] and [**109**]), but there seems to be no doubt that most of the radiation originates in belts of electrically charged particles surrounding Jupiter. The electrons in these belts are evidently on the average about 10^3 as energetic as those trapped in the van Allen belts surrounding the Earth. The magnetic field required to keep these charged particles trapped in the vicinity of Jupiter originates, presumably, within the planet; the field strength at the visible surface is probably about 50 gauss, two orders of magnitude greater than the field at the Earth's surface.

Radio-astronomical observations have also led to tentative information about the form and direction of Jupiter's magnetic field, the careful study of which will almost certainly have high priority in the first space-probe investigations of Jupiter. The field is approximately that of a dipole and roughly parallel to Jupiter's rotation axis (the Earth's present dipole axis being roughly antiparallel to the rotation axis).

JUPITER'S INTERIOR. The mean density of Jupiter is 1.334 gm/cm, 0.25 times that of the Earth, indicating that the light elements, such as hydrogen, are probably the most important chemical constituents of the planet. Its moment of inertia is 0.6 times that of a homogeneous sphere of the same mass and radius, showing that the density ρ must be a general decreasing function of radial distance r from the centre.

Theoretical models of Jupiter's internal structure have been constructed by several workers (DeMarcus [**28**], Peebles [**72**]). These models give $\rho(r)$ and $p(r)$ fairly precisely, where p denotes pressure, indicate that 80% of Jupiter is hydrogen, 18% helium and 2% heavier elements, that the planet possesses a deep, well-stirred atmosphere, and that from the centre to $r = 0.8$ times the radius of the planet, owing to the high prevailing pressure, the electrical conductivity should be quite high in the metallic range. But because these models fail to predict temperature to anything like the degree of accuracy with which ρ and p are found, we do not know how much of Jupiter's deep interior is in the fluid state, and how much is solid. The "Taylor column" theory of the origin of Jupiter's Great Red Spot [**51**] indicates that the atmosphere may be several thousand kilometres thick and that the region immediately below the atmosphere may be solid or very viscous. The variations in the rotation period of this solid region that are thus implied by the observed motion of the Red Spot can be accounted for only if (a) the fluid regions of Jupiter—the atmosphere and possibly a liquid core underlying the solid region—are sufficiently massive and well agitated [**51**], [**52**], and (b) Jupiter's internal magnetic field is mainly toroidal and over 1000 gauss in strength [**56**].

ORIGIN OF JUPITER'S MAGNETIC FIELD. Two lines of evidence rule out the possibility that the *Earth's* magnetic field is a remnant of a primordial field, namely, the occurrence of reversals in sign of the geomagnetic dipole and the fact that the time constant of free decay of electric currents in the Earth's core, proportional to σ and to the square of the core radius, is four or five orders of magnitude less than the accepted age of the Earth (about 4×10^9 years). As the electrical conductivity of Jupiter's deep interior could exceed that of the Earth's core by one or two orders of magnitude (Douglas-Hamilton (unpublished), Smolukowsi [**87**]) and Jupiter i much bigger than Earth, the possibility that Jupiter's poloidal magnetic field is primordial, being the remnant of the magnetic field in the material that condensed to form the planet in the first instance, is not ruled out by the evidence at present available. The condensation process could amplify the magnetic field strength to well above its ambient value in the interplanetary medium.

As the toroidal magnetic field of Jupiter (if it exists) would not arise through the condensation process, we must also consider the possibility that the poloidal field is *not* primordial. If both fields are due to a "dynamo" mechanism (see §4), then they must arise in one or more of the fluid regions of the planet. The magnetic Reynolds number (see Equation 8) associated with fluid motions of order 10^{-2} ms^{-1} on a length scale comparable with Jupiter's radius would satisfy Equation (9) if the electrical conductivity is of order $10^2\ \Omega^{-1}\ \mathrm{m}^{-1}$, a very low value, comparable with that of sea water, and much less than the metallic values expected in Jupiter's deep interior. This calculation, when combined with the demonstration that the concomitant Ohmic dissipation of energy would not be unreasonably high, is the basis of the suggestion that Jupiter's external magnetic field originates in the lower atmosphere, where it would not be unreasonable to expect such values of σ and $\tilde{U}$ [**53**], [**54**], [**56**]. Thus, it may not be necessary to invoke dynamo action in an hypothetical fluid core.

Because Jupiter rotates very rapidly—once in 10 hours (approximately)—atmospheric motions at the level of the visible surface of dense clouds should be highly correlated with motions in the lower reaches of the atmosphere [**51**]. Therefore, the hypothesis that the latter motions cause Jupiter's external magnetic field can be tested, in principle, by determining the extent to which horizontal variations in Jupiter's magnetic field are correlated with the appearance of the visible surface and with temperature variations at that level.

Appendix: Free hydromagnetic oscillations of the Earth's core. Each eigenmode of hydromagnetic oscillation of a rotating fluid occupying a cavity bounded by concentric spherical walls of radii $R \pm H$ may be classed as "magnetic" or "inertial" according as the absolute value of the corresponding eigenfrequency is less than or greater than the corresponding quantity in the case of no rotation, and as "spherical" or "non-spherical" according as displacements of fluid elements are or are not largely confined to spherical surfaces concentric with the cavity walls.

The elucidation of these eigenmodes is a very difficult problem, but a little progress has been made in the past few years. The case when the basic magnetic field is parallel to latitude circles has been analysed by Malkus [**68**], who gave a formal solution for both inertial and magnetic eigenmodes in a sphere ($H = R$) and investigated a few modes in detail, and by Stewartson [**89**], who considered magnetic modes only in a thin spherical shell ($H \ll R$). An approximate treatment of both "thin body" and "fat body" cases (namely $H \ll R$ and $H \sim R$ respectively) when the angle between the magnetic field and latitude circles is not necessarily equal to zero has been given by Hide [**55**] (see Equation (4.1) of Lecture II above).

Both spherical and nonspherical modes can occur when $H \not\ll R$, but only spherical modes are possible when $H \ll R$. Call the magnetic eigenfrequency ω_m. Apparently $\omega_m > 0$ for typical spherical modes and there is no evidence that the eigenmodes of a "fat body" have ω_m predominantly of one sign. It is plausible, therefore, (and consistent with the physical model proposed in [**55**]) to suppose that $\omega_m < 0$ for typical nonspherical modes of a "fat body."

The Earth's liquid metallic core (where the (poloidal) magnetic field of about 0.5 gauss at the surface of the Earth has its origin) is a "fat body," with $H/R \sim 0.5$. Arguing, effectively, that it might be hard to excite spherical modes in the core because the density stratification there is probably much too weak to constrain fluid elements to move on horizontal surfaces, Hide [**55**] (see pages 321–323 above) suggested that the preferred modes in the core should be nonspherical and that the preferred sign of ω_m would therefore be negative, the corresponding sense of phase and group propagation associated with the corresponding transverse wave motions being mainly westward (see also [**33**] of Lecture II above). He showed that if B_T, the (unknown) strength of the toroidal magnetic field in the core, is about 100 gauss, then these waves would propagate with speeds and dispersive characteristics reminiscent of the secular changes in the pattern of the magnetic field at the Earth's surface; he made on this basis the proposal that certain features at least of the geomagnetic secular variation might be a manifestation of the interaction of free hydromagnetic oscillations of the Earth's core with poloidal magnetic field there; and he discussed some of the geophysical consequences of this proposal (see Lecture V).

In their criticisms of Hide's proposal, Stewartson [**69**] assumed that spherical modes are more likely to be excited in the core than nonspherical modes, while Malkus [**68**] emphasized the need to consider "selection mechanisms" (see also Busse, F., *J. Fluid Mech.* **33** (1968), 739), but did not comment on the mechanism outlined above [**55**]. Further work is clearly needed in order to clarify and settle these questions and to extend the important theoretical work of Stewartson and Malkus.

References. Lecture IV

1. M. J. Aitken and G. H. Weaver, *Recent archaeomagnetic results in England*, J. Geomag. Geoelect. Kyoto **17** (1965), 391–394.

2. D. W. Allan and E. C. Bullard, *Distortion of a toroidal field by convection*, Rev. Mod. Phys. **30** (1958), 1087–1088.

3. G. Backus, *A class of self-sustaining dissipative spherical dynamos*, Ann. Phys. **4** (1958), 372–447. MR **20** #1512.

4. ———, *Kinematics of the geomagnetic secular variation in a perfectly conducting core*, Philos. Trans. Roy. Soc. A **263** (1968), 239–266.

5. G. E. Backus and S. Chandrasekhar, *On Cowling's theorem on the impossibility of self-maintained axisymmetric homogeneous dynamos*, Proc. Nat. Acad. Sci. U.S.A. **42** (1956), 105–109.

6. E. K. Bigg, *Influence of the satellite Io on Jupiter's decametric emission*, Nature London **203** (1964), 1008–1010.

7. P. M. S. Blackett, In: *Lectures on rock magnetism*, Weizmann Science Press of Israel, Jerusalem, 1956.

8. S. I. Braginskiĭ, *Kinematic models of the Earth's hydromagnetic dynamo*, Geomagnetism i Aeronomy. **4** (1964), 572–583.

9. ———, *Magnetohydrodynamics of the Earth's nucleus*, Geomagnetism i Aeronomy. **4** (1964), 698–712.

10. A. Brynjolfsson, Philos. Mag. **23** (1957), suppl. 6, 247–254.

11. Sir Edward Bullard, *The stability of a homopolar dynamo*, Proc. Cambridge Philos. Soc. Math. Phys. Sci. **51** (1955), 744–760.

12. ———, *The secular change in the Earth's magnetic field*, Mon. Not. Roy. Astronom. Soc. Geophys **5** (1948), suppl. 5, 248–257.

13. ———, Proc. Roy. Soc. London Ser. A **197** (1949), 433–453.

14. ———, *Electromagnetic induction in a rotating sphere*, Proc. Roy. Soc. London Ser. A **199** (1949), 413–443.

15. E. C. Bullard, G. Greedman, H. Gellman and J. Nixon, *The westward drift of the Earth's magnetic field*, Trans. Roy. Soc. London Ser. A **243** (1950), 67–92.

16. Sir Edward Bullard and H. Gellman, *Homogeneous dynamos and terrestrial magnetism*, Philos. Trans. Roy. Soc. London Ser. A **247** (1954), 213–278. MR **17**, 327.

17. B. F. Burke, "Radio observations of Jupiter," In: *Planets and satellites*, G. P. Kuiper and B. M. Middlehurst (editors), Univ. of Chicago Press, Chicago, Ill., 1961, pp. 473–499.

18. J. A. Burns, *Jupiter's decametric radio emission and the radiation belts of its Galilean satellites*, Science New York **159** (1968), 971–972.

19. S. Chapman and J. Bartels, In: *Geomagnetism*. Vol. 1, Clarendon Press, Oxford, 1940.

20. R. M. Cook and J. C. Belshe, Antiquar. Survival **32** (1958), 167–178.

21. J. Coulomb, *Variation séculaire par convergence ou divergence à la surface du noyau*, Ann. Geophys. **11** (1955), 80–82.

22. T. G. Cowling, *The magnetic field of sunspots*, Mon. Not. Roy. Astronom. Soc. **94** (1934), 39–48.

23. A. Cox and R. R. Doell, *Review of paleomagnetism*, Bull. Geolog. Soc. Amer. **71** (1960), 645–768.

24. ———, *Long period variations of the geomagnetic field*, Bull. Seismol. Soc. Amer. **54** (1964), 2243–2270.

25. A. Cox, R. R. Doell and G. B. Dalrymple, *Geomagnetic polarity epochs*, Science New York **143** (1964), 351–352.

26. ———, *Reversals of the Earth's magnetic field*, Science New York **144** (1964), 1537–1543.

27. R. D. Davies and D. Williams, In: *Magnetism and the cosmos*, Hindmarsh, Lowes, Roberts and Runcorn (editors), Oliver and Boyd, Edinburgh, 1966, pp. 288–291.

28. W. C. DeMarcus, *The constitution of Jupiter and Saturn*, Astronom. J. **63** (1958), 2–28.

29. R. R. Doell and A. Cox, *Paleomagnetism of Hawaiian lava flows*, J. Geophys. Res. **70** (1965), 3377–3405.

30. Š. Š. Dolginov, E. G. Erošenko, L. N. Žuzgov and N. V. Puškov, *Studies of the Moon's magnetic field*, Geomagnetism i Aeronomy. **1** (1961), 18–25.

31. J. N. Douglas and H. J. Smith, *Changes in rotation period of Jupiter's decameter radio sources*, Nature London **199** (1963), 1080–1081.

32. M. Dryer and G. R. Heckman, *Application of the hypersonicanalog to Mars' magnetic field*, Trans. Amer. Geophys. Un. **47** (1966), 155–156.

33. G. R. A. Ellis and P. M. McCulloch, *The decametric radio emissions of Jupiter*, Austral. J. Phys. **16**, (1963), 380–397.

34. W. M. Elsasser, *Induction effects in terrestrial magnetism*. I. *Theory*, Phys. Rev. (2) **69** (1946), 106–116. MR **7**, 401.

35. ———, *Induction effects in terrestrial magnetism*. II. *The secular variation*, Phys. Rev. (2) **70** (1946), 202–212. MR **8**, 186.

36. ———, *Induction effects in terrestrial magnetism*. III. *Electric nodes*, Phys. Rev. (2) **72** (1947), 821–833. MR **9**, 258.

37. ———, *Earth's interior and geomagnetism*, Rev. Mod. Phys. **22** (1950), 1–35.

38. ———, *Causes of motions in the Earth's core*, Trans. Amer. Geophys. Un. **31** (1950), 454–462.

39. ———, *Hydromagnetism: A review*, Amer. J. Phys. **24** (1956), 85–110.

40. G. B. Field, *The source of radiation from Jupiter at decimeter wavelengths*, J. Geophys. Res. **65** (1960), 1661–1671.

41. H. F. Finch and B. R. Leaton, *The Earth's main magnetic field—epoch* 1955.0, Mon. Not. Roy. Astronom. Soc. Geophys. **7** (1957), suppl. 7, 314–317.

42. J. A. Fleming, In: *Physics of the Earth*. VIII. *Terrestial magnetism and electricity*, McGraw-Hill, New York, 1939.

43. K. L. Franklin, *Radio-waves from Jupiter*, Sci. Amer. **214** (1964), 35–42.

44. J. Frenkel, C. R. (Dokl.) Sci. USSR **49** (1945), 98–101.

45. C. Gaibar-Puertas, Observatorio del Ebro Memo No. 11 (1953).

46. R. M. Gallet, In: *Planets and satellites*, G. P. Kuiper and B. M. Middlehurst (editors), Univ. of Chicago Press, Chicago, Ill., 1961, pp. 500–533.

47. S. Gulkis and T. D. Carr, *Radio rotation period of Jupiter*, Science New York **154** (1966), 257–259.

48. A. Herzenberg, *Geomagnetic dynamos*, Philos. Trans. Roy. Soc. London Ser. A **250** (1958), 543–585.

49. A. Herzenberg and F. J. Lowes, *Electromagnetic induction in rotating conductors*, Philos. Trans. Roy. Soc. London Ser. A **249** (1957), 507–584. MR **19**, 807.

50. R. Hide, "Hydrodynamics of the Earth's core," In: *Physics and chemistry of the Earth*. Vol. 1, L. Ahrens, K. Rankama and S. K. Runcorn (editors), Pergamon Press, London, 1956, pp. 94–137.

51. ———, *Origin of Jupiter's great red spot*, Nature London **190** (1961), 895–896.

52. ———, *On the hydrodynamics of Jupiter's atmosphere*, Mém. Soc. Roy. Sci. Liège (5) 7 (1963), 481–505.

53. ———, *Planetary magnetic fields*, Planet. Space Sci. **14** (1966), 579–586.

54. ———, *On the circulation of the atmospheres of Jupiter and Saturn*, Planet. Space Sci. **14** (1966), 669–675.

55. ———, *Free hydromagnetic oscillations of the Earth's core and the theory of geomagnetic secular variation*, Philos. Trans. Roy. Soc. London Ser. A **259** (1966), 615–647.

56. ———, In: *Magnetism and the cosmos*, Hindmarsh, Lowes, Roberts and Runcorn (editors), Oliver and Boyd, Edinburgh, 1967, pp. 141–147.

57. ———, In: *International dictionary of geophysics*. Vol. 1, S. K. Runcorn (editor), Pergamon Press, Oxford, 1967, p. 358.

58. R. Hide and K-I. Horai, *On the topography of the core-mantle interface*, Phy. Earth Planet. Interiors **1** (1968), 305–308.

59. D. R. Inglis, *Theories of the Earth's magnetism*, Rev. Mod. Phys. **27** (1955), 212–248.

60. E. Irving, *Paleomagnetism*, Wiley, New York, 1964.

61. H. Jeffreys, *The Earth*, 4th ed., Cambridge Univ. Press, London, 1959.

62. N. Kawai, *Geomagnetic secular variation revealed in the baked earths in West Japan*, Paper presented at a meeting of the International Association of Geomagnetism and Aeronomy, Pittsburgh, Pa., November 1964.

63. J. Larmor, *How could a rotating body such as the Sun become a magnet*?, Rep. Br. Assoc. Advancement Sci. **1919,** 159–160.

64. G. R. Lebo, A. G. Smith and T. D. Carr, *Jupiter's decametric emission correlated with the longitudes of the first three Galilean satellites*, Science New York **148** (1965), 1724–1725.

65. F. J. Lowes, *Secular variation and the non-dipole field*, Ann. Geophys. **11** (1955), 91–94.

66. F. J. Lowes and I. Wilkinson, *Geomagnetic dynamo: a laboratory model*, Nature London **198** (1963), 1158–1160.

67. W. V. R. Malkus, *Precessional torques as the cause of geomagnetism*, J. Geophys. Res. **68** (1963), 2871–2886.

68. ———, *Hydromagnetic planetary waves*, J. Fluid Mech. **28** (1967), 793–802. MR **36** #4869.

69. T. Nagata, *Rock magnetism*, Maruzen, Tokyo, 1953.

70. ———, (editor), Proceedings of the Benedum Earth-magnetism symposium, Univ. of Pittsburgh Press, Pittsburgh, Pa., 1962, pp. 39–55.

71. E. N. Parker, *Hydromagnetic dynamo models*, Astrophys. J. **122** (1955), 293–314. MR **18**, 92.

72. P. J. E. Peebles, *The structure and composition of Jupiter and Saturn*, Astrophys. J. **140** (1964), 328–347.

73. J. H. Piddington and J. F. Drake, *Electrodynamic effects of Jupiter's satellite Io*, Nature London **217** (1968), 935–937.

74. T. Rikitake, In: *Electromagnetism and the Earth's interior*, Elsevier, Amsterdam and New York, 1968, p. 308.

75. J. A. Roberts, *Radio emission from the planets*, Planet. Space Sci. **11** (1963), 221–259.

76. J. A. Roberts and M. M. Komesaroff, *Observations of Jupiter's radio spectrum and polarization in the range from* 6 cm *to* 100 cm, Icarus **4** (1965), 127–156.

77. P. H. Roberts, *An introduction to magnetohydrodynamics*, American Elsevier, New York, 1967.

78. P. H. Roberts and K. Stewartson, *On the motion of a liquid in a spheroidal cavity of a processing rigid body*, J. Fluid Mech. **17** (1963), 1–20.

79. W. K. Rose, In: *Magnetism and the cosmos*, Hindmarsh, Lowes, Roberts and Runcorn (editors), Oliver and Boyd, Edinburgh, 1966, pp. 292–294.

80. F. E. M. Lilley, *On kinematic dynamos*, Proc. Roy. Soc. London Ser. A **316** (1970), 153–167.

81. S. K. Runcorn, *Magnetization of rocks*, Handbuch Phys., XLVII, Springer-Verlag, Berlin, 1956, pp. 470–497.

82. ———, *The magnetism of the Earth's body*, Handbuch Phys., XLVII, Springer-Verlag, Berlin, 1956, pp. 498–533.

83. A. G. Smith and T. D. Carr, *Radio exploration of the planetary system*, Van Nostrand, Princeton, N.J., 1964, p. 148.

84. E. J. Smith, L. Davies, P. J. Coleman and D. E. Jones, *Magnetic field measurements near Mars*, Science New York **149** (1965), 1241–1242.

85. E. J. Smith, L. Davies, P. J. Coleman and C. P. Sonett, *Magnetic field* (in Mariner II preliminary reports), Science New York **139** (1963), 909–910.

86. ———, *Magnetic measurements near Venus*, J. Geophys. Res. **70** (1965), 1571–1586.

87. R. Smolukowski, *Internal structure and energy emission*, Nature London **215** (1967), 691–695.

88. F. D. Stacey, *Electrical resistivity of the Earth's core*, Earth Planet. Sci. Lett. **3** (1967), 204.

89. K. Stewartson, *Slow oscillations of fluid in a rotating cavity in the presence of a toroidal magnetic field*, Proc. Roy. Soc. London Ser. A **299** (1967), 173–187.

90. J. B. Taylor, *The magnetohydrodynamics of a rotating fluid and the Earth's dynamo problem*, Proc. Roy. Soc. London Ser. A **274** (1963), 274–283.

91. A. Toomre, In: *The Earth-Moon system*, B. J. Marsden and A. G. W. Cameron (editors), Plenum Press, New York, 1966, pp. 33–43.

92. O. W. Torreson, T. Murphy and J. W. Graham, *Magnetic polarization of sedimentary rocks and the Earth's magnetic history*, J. Geophys. Res. **54** (1949), 111–129.

93. J. G. Tough and P. H. Roberts, *Nearly symmetric hydromagnetic dynamos*, Phys. Earth Planet Interiors **1** (1968), 288–296.

94. D. C. Tozer, In: *Physics and chemistry of the Earth*. Vol. 3, L. H. Ahrens, F. Press, K. Rankama and S. K. Runcorn (editors), Pergamon Press, London, 1959, pp. 414–437.

95. H. C. Urey, *The planets* (*their origin and development*), New Haven Univ. Press, Conn., 1952, p. 245.

96. J. Verhoogen, *Heat balance of the Earth's core*, Geophys. J. Roy. Astronom. Soc. **4** (1961), 276–281.

97. E. H. Vestine and Anne B. Kahle, *The small amplitude of magnetic secular change in the Pacific area*, J. Geophys. Res. **71** (1966), 527.

98. E. H. Vestine, L. Laporte, C. Cooper, I. Lange and W. C. Hendrix, *Description of the Earth's main magnetic field and its secular change* 1905–1945, Publ. #578, Carnegie Inst., Washington, 1947.

99. E. H. Vestine, L. Laporte, I. Lange and W. E. Scott, *The geomagnetic field, its description and analysis*, Carnegie Inst., Washington, 1947.

100. J. W. Warwick, *Dynamic spectra of Jupiter's decametric emission* 1961, Astrophys. J. **137** (1963), 41–59.

101. ———, *The position and sign of Jupiter's magnetic moment*, Astrophys. J. **137** (1963), 1317–1318.

102. ———, *Radio emission from Jupiter*, An. Rev. Astronom. Astrophys. **2** (1964), 1–22.

103. ———, In: *Magnetism and the cosmos*, Hindmarsh, Lowes, Roberts, and Runcorn (editors), Oliver and Boyd, Edinburgh, 1965.

104. ———, *Radiophysics of Jupiter*, Space Sci. Rev. **6** (1967), 841–891.

105. R. J. Weiss, *Origin of the Earth's magnetic field*, Nature London **197** (1963), 1289–1290.

106. T. Yukutake, *The westward drift of the magnetic field of the Earth*, Bull. Earthquake Res. Inst. Tokyo Univ. **40** (1962), 1–65.

107. R. Hide, "Planetary magnetic fields," In: *Surfaces and interiors of the planets and satellites*, A. Dollfus (editor), Academic Press, London and New York, 1970, pp. 511–534.

108. P. Goldreich and D. Lynden-Bell, *Io: a Jovian unipolar inductor*, Astrophys. J. **156** (1969), 59–78.

109. S. F. Dermott, *Modulation of Jupiter's decametric radio emission by Io*, Mon. Not. Roy. Astronom. Soc. **149** (1970), 35–44.

V. The Core-Mantle Interface and the Earth's Magnetism, Gravitation and Rotation[1]

Summary. Recent work on the magnetohydrodynamics of the liquid core of the Earth and on the interaction between core motions and the overlying solid mantle has yielded several novel results. New types of magnetohydrodynamic waves have been discovered through an attempt to explain the geomagnetic secular variation; the nature of the horizontal stresses at the core-mantle interface that are implied by the "decade" fluctuations in the length of the day has been elucidated; and evidence pointing to a previously-unsuspected correlation between the Earth's gravitational and magnetic fields has been found. Global-scale undulations in level of the core-mantle interface that are so shallow (a kilometre or so) as to be unresolvable by contemporary seismological techniques are strongly implied by some of these results.

1. **Introduction:[2] Some preliminaries.** The surface of the Earth nowhere departs by more than several kilometres from a sphere of radius R = 6371.17 km centered on the Earth's centre of mass. A typically spherically-symmetric model of the Earth's interior, beneath its thin crust of mean

[1] Supplementary notes for Lecture V, based on a recent article by the lecturer [**71**].

[2] Original references and important review material can be found in various recent papers (Backus [**4**], Bullard [**9**], Goldreich and Peale [**22**], Kaula [**40**], [**41**], Lilley [**43**], Malkus [**46**], Moffatt [**48**], Namikawa and Matsuchita [**52**], Parker [**54**], Suffolk [**64**] and Wilson [**67**]), monographs (Bullen [**12**], Irving [**35**], Jeffreys [**37**], Kaula [**39**], Munk and MacDonald [**49**], Rikitake [**55**], Roberts [**56**] and Stacey [**62**]) and conference reports (Blackett et al. [**6**], Bullen et al. [**13**], Gaskell [**20**], Hindmarsh et al. [**33**], Hurley [**34**], Marsden and Cameron [**47**], Nagata [**51**], Runcorn [**59**], [**60**] and Tozer [**66**]).

density 2.8 gm cm^{-3} and variable thickness (ca. 30 km), that (a) satisfies data on the transmission of shear and compressional waves generated by earthquakes and man-made explosions, (b) gives a moment of inertia equal to 0.83 times that of a homogeneous sphere, and (c) is so lacking in perfect rigidity as to account for the 430 days (rather than 305 days) period of the free Chandlerian wobble of the Earth's figure axis relative to its rotation axis, comprises three regions: a solid inner core of mean density 12.3 gm cm^{-3} extending from $r = 0$ to $r = 1251$ km (where r denotes distance from the Earth's centre of mass), a liquid outer core of mean density 10.9 gm cm^{-3} extending from $r = 1251$ km to $r = 3473$ km, and a solid mantle of mean density 4.5 gm cm^{-3} extending from $r = 3473$ km to $r = 6352$ km (see Bullen [**12**], Jeffreys [**37**], Kanamori and Press [**38**], Kaula [**39**]).

It is generally supposed that the temperature within the Earth rises gradually with depth and attains about 4000 K in the core. However, owing to ignorance of the circumstances under which the Earth came into being some 4.5×10^9 years ago and its subsequent evolution, and of the details of long-term creep processes in the mantle that are implied by the recent revolutionary discoveries concerning continental drift and sea-floor spreading, the radial variation of temperature within the Earth cannot be determined accurately from measurements of the thermal properties of surface rocks and the feeble heat flow out of the Earth's surface (see articles in Blackett et al. [**6**], Gaskell [**20**], Hurley [**34**], Runcorn [**59**], Tozer [**66**]).

The dominant mineral of the mantel is olivine $(MgFe)_2SiO_4$ which is a solid solution of forsterite Mg_2SiO_4 and fayalite Fe_2SiO_4 and a typical electrical semiconductor; metallic iron is the dominant material in the core. Most geophysicists accept that the main geomagnetic field is due to electric currents within the Earth and that these currents flow mainly in the metallic core, where they experience least electrical resistance. They are generated by as yet imperfectly understood magnetohydrodynamic (hydromagnetic) processes in the liquid outer core (see Bullard [**9**], Hindmarsh et al. [**33**], Kaula [**39**], Rikitake [**55**], Roberts [**56**], Runcorn [**60**]). That the Earth has possessed a substantial liquid core for the past 2.7×10^9 years is evinced by the existence of rocks of that age with primary remanent magnetization (McElhinny and Evans [**50**]).

Many geophysical, especially seismological, data pertaining to the Earth's deep interior can be reconciled by means of theoretical models characterized by spherical or axial symmetry, but there now exists a substantial and growing body of evidence—to which the use of artificial satellites in geophysical research is making a leading contribution—that renders symmetric models increasingly inadequate and demands refinements. These refinements are required not merely to "gild the lily"; they must reflect physical and dynamical processes within the Earth and therefore will provide a key to the Earth's past and future evolution.

I shall be concerned in this talk with properties that the Earth reveals about its deep interior through its inability to keep time accurately, attract falling bodies exactly towards its centre of mass and align the compass needle along a meridian!

2. **Decade variations in the length of the day and core-mantle coupling.** Observations of lunar eclipses and transits of the planet Mercury over the face of the Sun since the seventeenth century, and accurate meridian observations of the Sun since 1770 and Venus since 1840 led to the discovery of tiny but complicated variations in the length of the day (i.e. the time taken for the Earth to rotate once on its axis); further variations came to light after the introduction of quartz clocks calibrated against atomic frequency standards. The variations comprise three distinct components: seasonal fluctuations of about 10^{-3} s and irregular "decade" fluctuations of up to 5×10^{-3} s, on time scales of a few yards upwards, superimposed on a steady increase in the length of the day by 10^{-3} s per century.

Dynamical processes in various parts of the Earth are responsible for length of day variations, which are therefore of considerable interest to geophysicists, though a nuisance to astronomers. The steady increase is associated with angular momentum transfer from the Earth to the Moon, caused by the action of gravitational torques associated with the tidal bulge whose orientation relative to the Earth-Moon line is determined by dissipative processes in the ocean and other parts of the Earth ("tidal friction"). The seasonal fluctuations are due largely to torques on the mantle produced by the combined effects of atmospheric winds and ocean currents, which cause the Earth's rotation to accelerate during the northern summer and decelerate during the northern winter. However, the amplitude of the "decade" fluctuations is too large to be accounted for in terms of interactions of the atmosphere and oceans with the mantel, and geophysicists generally agree that, in the absence of any reasonable alternative, these fluctuations must therefore be due to angular momentum transfer between the mantle and the liquid core (Munk and MacDonald [**49**], Rikitake [**55**] and various articles in Marsden and Cameron [**47**], Nagata [**51**], Runcorn [**60**]).

The distortion and displacement of the geomagnetic field pattern at the Earth's surface, including the well-known westward drift at some 0.03 cm s^{-1} (Bullard et al. [**10**]) is a direct manifestation of core motions, but the accurate determination of even the broadest features of these motions from geomagnetic observations is an unsolved theoretical problem (see Backus [**4**]). Nevertheless the quantitative requirement that the time scale and the r.m.s. value of fluctuations in zonal speed of core motions be generally compatible with the amplitude of the decade fluctuations in the length of the day is not particularly restrictive. The principal quantitative difficulties arise when the nature of the horizontal stresses that couple the core to the mantle, across the core-mantle interface, is considered.

These stresses must suffice both quantitatively and qualitatively to account for the fluctuating torques at the core-mantle interface that are implied by the foregoing interpretation of the decade fluctuations in the length of the day. Following Hide [**28**] we write

$$F = F_V + F_E + F_T = 0.4 \text{ dyn cm}^{-2} \tag{2.1}$$

where F is the average magnitude of these horizontal stresses. F_V is the contribution to F associated with molecular viscosity and, if the boundary layer at the surface of the core is turbulent, with small-scale eddy viscosity; if δ is the thickness of that boundary layer then

$$F_V = C_V \Omega \rho U \delta \tag{2.2}$$

where Ω is the angular speed of the Earth's rotation, ρ is the density of the core, U is a typical relative horizontal speed of flow in the core and C_V is a numerical coefficient whose exact value, though on general grounds less than or equal to about unity, cannot be determined without detailed theoretical considerations of specific models of the coupling process. The electric currents responsible for the main geomagnetic field leak out of the metallic core into the weakly-conducting lower mantle and thus give rise to electromagnetic coupling, fluctuations in which are represented by the term F_E in Equation (2.1). If B_h is the fluctuating horizontal component of the magnetic field at the core-mantle interface and B_v is the vertical component, and μ denotes magnetic permeability (not significantly different from unity, that of "free space") then

$$F_E = C_E B_h B_v / 4\pi\mu. \tag{2.3}$$

C_E is a numerical coefficient bearing the same relationship to F_E as does C_V to F_V (see Equation (2.2)). If shallow irregular topographic features are present on the core-mantle interface and h, a typical value of the vertical dimensions of these features, exceeds the boundary layer thickness δ, then "topographic coupling" also has to be considered; the corresponding contribution to F is given by

$$F_T = C_T \Omega \rho U(h - \delta) \tag{2.4}$$

where C_T is a numerical coefficient to be determined by detailed theoretical considerations.

In proposing a theory of the westward drift of the geomagnetic field, Bullard (see [**10**] and [**72**]), arguing that viscous coupling between core and mantle will be insignificant, introduced the idea of electromagnetic coupling and suggested a specific model of the interaction mechanism. Fears that there were quantitative difficulties in accounting for the biggest decade fluctuations in the length of the day in terms of fluctuating electromagnetic torques on the mantle were confirmed by the detailed calculations of Rochester [**57**] and Roden [**58**], who showed that even under the most optimistic assumptions about the radial distribution of electrical conductivity in the lower mantle, F_E falls short by a factor of 10 of the required value (see Equation (2.1)). However, Rochester and Roden implicitly neglect eddy currents in the lower mantle due to possible magnetic fluctuations in the core on time-scales so very short—less than a few years—that the magnetic record at the surface of the Earth is unaffected, and it is possible, therefore, that a view expressed by Runcorn [**61**]—that "some quantitative difficulties in the theory of (electromagnetic) coupling remain but do not seem sufficient to doubt the essential correctness of the explanation of this complex question"—may eventually be upheld. Indeed Hide [**25**], in proposing an explanation of the westward drift and general time-scale of the geomagnetic secular variation in terms of free hydromagnetic oscillations of the core, has argued that the spectrum of magnetic fluctuations in the core could contain very short-period components and suggested that they might enhance the effectiveness of electromagnetic coupling. However, no one has demonstrated quantitatively that the discrepancy revealed by the work of Rochester and Roden can be removed in this way.

The idea of topographic coupling was put forward by Hide [**28**] in an attempt to resolve the foregoing difficulty in the theory of the decade variations in the length of the day (and a related difficulty in the theory of the Chandler wobble). F_T can be evaluated if we can estimate each of the terms on the right-hand side of Equation (2.4). We take $U = 10^{-2}$ cm s^{-1} [**28**]—slower by a factor of three than the westward drift of the nondipole field, thus recognizing that, in addition to material motions in the core, magnetohydrodynamic wave motions there might contribute to the distortion and displacement of the geomagnetic field pattern at the Earth's surface [**25**], [**27**]—and for Ω and ρ we take 10^{-4} rad s^{-1} and 10 gm cm^{-3} respectively.

The coefficient of kinematical viscosity, ν, of the core is unknown and the structure of the boundary layer at the surface of the core is complicated by magnetohydrodynamic effects. It may be shown (see [**30**]) that

$$\delta = (\nu/\Omega)^{1/2} f(\sigma B_v^2/2\rho\Omega) \tag{2.5}$$

where σ is the electrical conductivity of the core and

$$f(x) = 2^{1/2} \sin(\tfrac{1}{2}\cot^{-1} x)/(1 + x^2)^{1/4}.$$

For the core $\sigma \doteq 5 \times 10^{-6}$ e.m.u. and $B_v \doteq 5$ gauss, so that $f \doteq 0.1$ and $\delta \doteq 10^2 \nu^{1/2}$ cm (Hide [**28**]). If we follow Bullard and take $\nu \doteq 10^{-3}\,\text{cm}^2\,\text{s}^{-1}$ then $\delta \doteq 3$ cm [**10**], [**72**]; if we accept Jeffreys' value of $10^7\,\text{cm}^2\,\text{s}^{-1}$ as an extreme upper limit to ν then $\delta \ll 3$ km. In any event, δ is very small indeed compared with the radius of the core.

The best value to take for C_T is not known, but laboratory experiments by Collier and Hide [**15**] support the assumption that taking $C_T > 0.25$ would not lead to serious error in the present calculation. It might be thought on the basis of certain mathematical work on "spin-up" (see Greenspan [**23**]) that this value of C_T is much too high, but boundary layer separation, clearly visible in the experiments, has hitherto been neglected in mathematical models, which are evidently not directly relevant here. With $C_T = 0.25$, $U = 10^{-2}\,\text{cm}\,\text{s}^{-1}$, $\Omega = 10^{-4}\,\text{s}^{-1}$, $\rho = 10$ gm cm^{-3} and $\delta \ll 3$ km we find by Equation (2.4) that $h \doteq 1$ km.

The nonhydrostatic (i.e. baroclinic) stresses in the lower mantle due to the presence of topographic features of this height (assuming a density jump at the core-mantle interface of 4 gm cm^{-3}) should not, at 2×10^8 dyn cm^{-2}, be geophysically unacceptable (see Hide and Horai [**31**]).

3. **The Earth's regional gravitational field and nondipole magnetic field.** The absence in seismological data, such as travel times of compressional waves reflected at the core-mantle interface (PcP), of clearcut evidence of topography indicates that h must be less than a few tens of kilometres, the present limit of resolution of seismological techniques (see Bullen [**12**], Buchbinder [**8**]). The exploitation of the increased precision expected as a result of the recent introduction of elaborate seismometer arrays and sophisticated methods of data analysis for the detection and identification of earthquakes and underground nuclear explosions might eventually yield information pertinent to the problem in hand, and the results of investigations along these lines will be awaited with interest. Meanwhile, however, it is more promising to seek a test for the hypothesis that the core-mantle interface has topography with h not less than about a kilometre by considering the fine details of the Earth's gravitational and magnetic fields.

The potential Ψ of the gravitational field of the Earth at a point above its surface with spherical polar co-ordinates (r, ϕ, λ) can be expressed as a spherical harmonic series, each term of which represents the contribution to Ψ due to a hypothetical gravitational multipole source placed at the Earth's centre of mass. Thus

$$\Psi = \frac{MG}{R}\left\{\frac{R}{r} + \sum_{n=1}^{\infty}\sum_{m=0}^{n} S(\alpha_n, \alpha_n^m, \beta_n^m)\right\}, \tag{3.1}$$

where r denotes distance from the Earth's centre of mass, ϕ co-latitude, λ east longitude, R the mean radius of the Earth (6371.17 km) and M its mass (5.977×10^{27} gm), G the universal gravitational constant and $S(\alpha_n, \alpha_n^m, \alpha_n^m)$, a typical term in the series, satisfies

$$\begin{aligned} \sum_{n=1}^{\infty}\sum_{m=0}^{n} & S(\alpha_n, \alpha_n^m, \beta_n^m) \\ & \equiv \sum_{n=1}^{\infty} (R/r)^{n+1}\left\{\alpha_n P_n(\cos\phi) + \sum_{m=1}^{n} P_n^m(\cos\phi)\right. \\ & \qquad \left. \times [\alpha_n^m \cos m\lambda + \beta_n^m \sin m\lambda]\right\}, \end{aligned} \tag{3.2}$$

where m denotes the order of the harmonic and n its degree, m and n being integers, and P_n (cos ϕ) and P_n^m (cos ϕ) are certain polynomials (see Jeffreys [**37**], Kaula [**39**]) whose exact form need not concern us here.

Values of α_n up to $n = 13$ and of α_n^m and β_n^m up to $n = 8$ and $m = 8$ have been determined from perturbations of the orbits of artificial satellites. Because the co-ordinate system for gravitational measurements has its origin at the Earth's centre of mass and the Earth rotates about its axis of maximum moment of inertia (a state of minimum kinetic energy), the terms α_1, α_1^1, β_1^1, α_2^1 and β_2^1 are all zero, as would all the other coefficients be if the distribution of matter within the Earth were spherically symmetric. The coefficient α_2 is a measure of the Earth's equatorial bulge and is of order 10^{-3}. The other coefficients, which describe what Jeffreys has termed the Earth's "regional gravitational field," are typically 10^{-6} or less.

The Earth's gravitational field is completely determined by the distribution of matter, density ρ (r, ϕ, λ), within the Earth, but even perfect knowledge of Ψ $(r \geqq R)$ would not suffice to solve the inverse problem of determining $\rho(r, \phi, \lambda)$ from gravitational measurements alone; further geophysical data must be brought to bear on the problem. At about 10^9 dyn cm^{-2} the baroclinic stresses within the Earth that are implied by observations of the regional gravitational field are much too large, by several powers of ten, to be associated with magnetohydrodynamic motions in the liquid core; therefore the field must arise in the solid part of the Earth. However, the relationship between $\Psi (r \simeq R)$ and the distribution of land and sea is not pronounced (Guier and Newton [**24**], Kaula [**40**]) and the nonhydrostatic components in the low degree harmonics of Ψ $(r \simeq R)$ are in no way directly related to those expected from the difference in structure between oceanic and continental regions (Cook [**17**]); hence the horizontal density variations responsible for the regional gravitational field must lie well below the surface, in the upper and lower mantle (and possibly even in the solid inner core).

The Earth's magnetic field, unlike its gravitational field, undergoes complicated changes with time. The most rapid fluctuations at and near the surface occur on time-scales ranging from fractions of a second ("sub-acoustic oscillations") to several days (magnetic storms). Owing to their small amplitude, less than 1% of the total field, these fluctuations were not discovered until comparatively recently in the history of geomagnetic observations. They are due to varying electric currents flowing well above the Earth's surface, in the ionosphere and beyond. When, by taking annual mean values, these rapid fluctuations are eliminated from the magnetic record, the "main geomagnetic field" remains. The main field originates within the Earth and undergoes variations on time-scales of decades upwards. This short time scale, geologically speaking, of the "geomagnetic secular variation" is good evidence that the main field is produced largely in the liquid core and not in the solid parts of the Earth.

The potential Φ in terms of which the main geomagnetic field of the Earth can be expressed above its region of origin may (like Ψ, see Equation (3.1)) be expanded in a spherical harmonic series, each term of which represents the contribution to Φ due to a hypothetical magnetic multipole source placed at the Earth's centre of mass. Thus

$$\Phi = \frac{L}{R^2} \sum_{n=1}^{\infty} \sum_{m=0}^{n} S(a_n, a_n^m, b_n^m) \qquad (3.3)^3$$

(cf. Equations (3.1) and (3.2)), $a_1 L$ being the resolute along the Earth's rotation axis of the (hypothetical) centered magnetic dipole. The coefficients a_1^1, and b_1^1, typically about 0.2 times a_1 in magnitude, specify the amplitude and orientation of the "centred equatorial dipole" and constitute a leading measure of the imperfections in the principle of the magnetic compass. The coefficients of the so-called "nondipole" field, namely a_n^m and b_n^m with

[3] See pages 312–315.

$n \geqq 2$, are typically of order $0.1a_1$ or less. Thanks to the improvement in geomagnetic data brought about by the introduction of a few years ago of artificial-satellite borne magnetometers, reliable coefficients up to $m = n = 10$ have been determined for recent epochs (for references see Hide and Malin [**32**]).

The evidence that the main geomagnetic field originates largely within the liquid core has already been mentioned. The field is completely determined by the distribution of electric current—density $\boldsymbol{j}(r, \phi, \lambda, t)$ where t denotes time—in the Earth, but perfect knowledge of $\Phi(r \geqq R)$ would not suffice to determine $\boldsymbol{j}$ from geomagnetic measurements alone. Moreover, $\Phi\,(r \simeq R)$ provides no more than a lower limit to the magnitude of $\boldsymbol{j}$ since the lines of magnetic force of any toroidal magnetic field present within the Earth cannot penetrate beyond regions where $\boldsymbol{j} = 0$ and therefore remain confined to the deep interior. Certain, though not all, modern theories of the Earth's magnetism suggest strong toroidal fields, of 100 gauss at least (see Braginskiĭ [7], Bullard and Gellman [**11**], Hide [**25**], Lilley [**43**], Lortz [**44**], Malkus [**46**], Parker [**54**], Rikitake [**55**], various articles in Runcorn [**60**]). Reversals in sign of the comparatively weak main dipole field—as revealed by studies of the magnetization of rocks, see Bullard [**9**], Cox et al. [**18**]—need not imply significant fluctuations in the Earth's total magnetic energy (see Lilley [**43**], Hide [**26**]).

We have seen that while the liquid core of the Earth is the only likely location of the electric currents responsible for the main geomagnetic field, potential Φ, the core is the most *unlikely* place to find density variations of sufficient magnitude to produce the observed distortions of the gravitational field, potential Ψ, which must arise largely in the mantle. Therefore, any correlation between $\Psi\ (r \simeq R)$ and $\Phi\,(r \simeq R)$ would reflect processes at the core-mantle interface. Now it is readily shown (Hide and Horai [**31**]) that core-mantle interface topography with $h \simeq 1$ km would produce a measurable though not dominant contribution to $\Psi\ (r \simeq R)$, but only to those global-scale features associated with spherical harmonics of degree not greater than 4. Moreover, it has been argued (Hide [**25**], [**27**]) that, owing to Coriolis forces associated with the Earth's rotation, such topography might interact strongly with magnetohydrodynamic motions throughout the core and thereby measurably distort the Earth's magnetic field. Hence, the discovery of any statistically-significant correlation between $\Psi\ (r \simeq R)$ and $\Phi\,(r \simeq R)$ would constitute supporting evidence for the hypothesis we are investigating and, conversely, a demonstration that no correlation exists would constitute evidence against the hypothesis.

Correlations between different geophysical quantities (e.g. Ψ, Φ, heat flow, seismic data, etc.), have been sought by several workers, often with vague and inconclusive results. Coode and Runcorn [**16**] found no significant correlation between Ψ and $\partial\Phi/\partial t$, thus refuting a claim by Egyed [**19**] to the contrary, based on poorer data. Recently, however, Hide and Malin [**32**] introduced a new element into the study of correlations between Φ and Ψ, with encouraging results. Theoretical models of the magnetohydrodynamic processes by which "bumps" on the core-mantle interface might distort the geomagnetic field, the subject of recent and contemporary research, are as yet too primitive for detailed predictions to be possible (Hide [**25**], [**27**]). Qualitatively, however, distortions in the magnetic field could be displaced relative to bumps through distances at least comparable with the horizontal scale of the bumps—thousands of kilometres. Therefore, Hide and Malin [**32**] studied the degree of correlation of those large-scale nonaxisymmetric features of Ψ (the gravitational potential) which can be described by a spherical harmonic series truncated at terms of degree n^* with corresponding features of Φ (the potential of the geomagnetic field) when Φ is twisted eastward in longitude relative to Ψ through a variable angle $\bar{\lambda}$ (see Figure 1).

They established, first, that when $n^* = 4$ the correlation coefficient k for epoch 1965 attains a maximum value $\hat{k} = 0.84$ when $\bar{\lambda} = \hat{\lambda} = 160°$ (to be compared with $k = 0.1$ when $\bar{\lambda} = 0$) and that this value of $\hat{k}$ is unlikely to occur by chance. They then showed (amongst other things) that under the reasonable assumption that Ψ does not change appreciably on the time-scale of the geomagnetic secular variation, the angle $\hat{\lambda}$ has increased linearly with time t as follows

$$(3.4) \qquad \hat{\lambda} = (126.2 \pm 0.2)^0 \pm (0.273 \pm 0.005)(t - 1835 \pm 10)^0$$

(where t is the epoch (year AD)) since 1835, the date of the earliest reliable spherical harmonic analysis of the geomagnetic field, by Gauss (see Figure 2). This time-dependence of $\hat{\lambda}$ on t is associated with the familiar westward drift of the geomagnetic field (see above), a property first noted by Halley in the seventeenth century.

If the correlation between Φ and Ψ is a manifestation of magnetohydrodynamic interactions between core motions and topographical features of the core-mantle interface (Hide [**25**], [**27**]), then the main features of the drifting part of the nondipole component of the present geomagnetic field might have been produced by an interaction that occurred, or started up, about 500 or 600 years ago when, if we can assume that Equation (3.4) was at least roughly applicable at and since that time, $\hat{\lambda}$ was still positive but very small (see Figure 2). This conclusion is tentative, but it is also consistent both with theoretical ideas concerning the rates of propagation and dispersion of hydromagnetic waves in the core (Hide [**25**]) and with certain observational evidence, notably archaeomagnetic

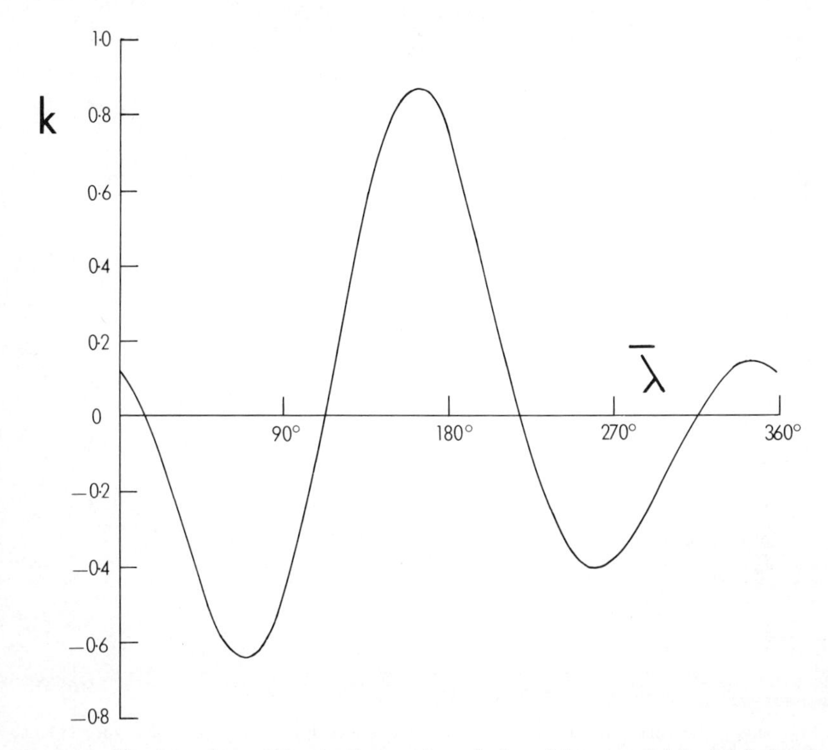

FIGURE 1. The dependence of k—the degree of correlation of the potential Φ of global-scale (i.e. $n^* = 4$) features of the main geomagnetic field for epoch 1965 with corresponding features of the potential Ψ of the gravitational field—on the eastward displacement in longitude of Φ relative to Ψ through an angle $\bar{\lambda}$. ($k = 1$ would correspond to perfect correlation.) At 0.84, the maximum value of k, namely $\hat{k}$, found at $\bar{\lambda} = \hat{\lambda} = 160°$, is so high that the odds against its occurrence by chance exceed 40 : 1, but no statistical significance attaches to the two minima in the k versus $\bar{\lambda}$ curve (see [**32**] and [**71**]).

data, which indicates that 0.23 ± 0.06 degrees of longitude per year has been the westward drift rate since about 1400 AD but that the drift was much slower and possibly even eastward between 1100 AD and 1400 AD (see Aitken and Weaver [**3**], Kawai et al. [**42**], Burlatskaya et al. [**14**]; cf. Bullard et al. [**10**], Nagata [**51**], Yukutake and Tachinaka [**69**]).

4. **Concluding remarks.** It is, of course, entirely possible that the correlation found by Hide and Malin can be explained without invoking bumps on the core-mantle interface. For instance, the effect on core motions of temperature variations over the interface might be sufficiently pronounced to produce measurable distortions of the geomagnetic field (Hide [**27**]) and if these temperature variations also reflect the thermal

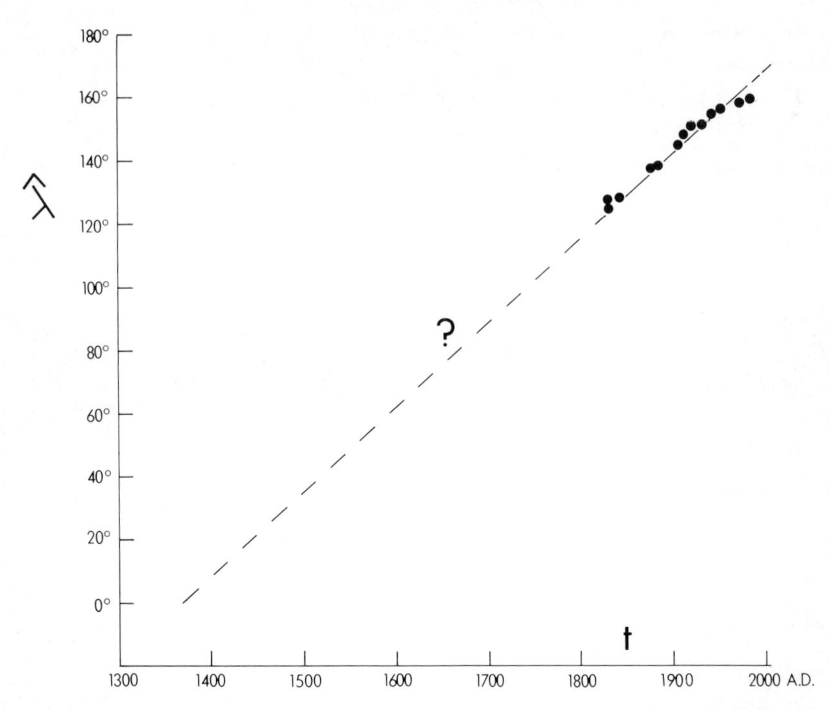

FIGURE 2. The variation of $\hat{\lambda}$ with t when $n^* = 4$ (see Equation (3.4) and Figure 1) since 1835 AD. The variation is due to the fairly linear westward drift of low-degree harmonics of the nondipole field at 0.273 ± 0.005 degrees of longitude per year (after Hide and Malin [**32**]). If the correlation between Φ and Ψ is a manifestation of magnetohydrodynamic interaction between core motions and topographic features of the core-mantle interface, then the main features of the drifting part of the nondipole component of the present geomagnetic field might have been produced by an interaction that occurred, or started up, about 500 or 600 years ago when, if we can extrapolate the observed variation backwards in time, $\hat{\lambda}$ was still positive but close to zero. This conclusion is tentative, but it is consistent with theoretical ideas concerning rates of propagation and dispersion of hydromagnetic waves in the core and with archaeomagnetic evidence (Aitken and Weaver [**3**], Kawai et al. [**42**], Burlatskaya et al. [**14**]) that from 1100 AD to 1400 AD the drift of the geomagnetic field was very slow (and possibly even eastward) in the comparison with its behaviour since 1400 AD (see [**32**] and [**71**]).

and therefore the density structure of the lower mantle then Ψ and Φ would be correlated. (Associated with horizontal temperature variations in the mantle will be electrical conductivity variations whose pattern might be reflected in $\partial\Phi/\partial t$; Hide and Malin [**32**] found a strong correlation between Φ and $\partial\Phi/\partial t$.) Such a situation could arise if thermal convection occurs in the lower mantle, but the depth of penetration and the cause of mantle convection are controversial questions at the present time; some geophysicists consider that mantle convection must be confined to the top few hundred kilometres (see Orowan [**53**] and various articles in Blackett et al. [**6**], Gaskell [**20**], Hurley [**34**], Marsden and Cameron [**47**], Runcorn [**59**], [**60**] Tozer [**66**] also [**70**]). If, however, deep convection *does* occur then it is unlikely that concomitant viscous stresses will leave the core-mantle interface undisturbed, but whether the amplitude of these stresses would be sufficient to give topography with $h \simeq 1$ km (and thus provide the "topographic coupling" discussed in §2) is not clear (see Hide and Horai [**31**]). Other mechanisms might distort the core-mantle interface, such as mechanical, chemical or electrochemical erosion of the mantle by the core, but their discussion lies beyond the scope of the present paper.

It is likely that changes in core-mantle interface topography (and temperature field), however produced, would have occurred over geological time. Theoretical work on the "geomagnetic dynamo" makes it plausible to suppose that the sign of the geomagnetic dipole depends on quite minor features of the pattern of core motions (Bullard and Gellman [**11**], Lilley [**43**]) so that comparatively minor changes in these motions, such as might be produced by changes in the topography (or thermal field) of the core-mantle interface should be reflected in changes in frequency and other properties of the polarity reversals studied by paleomagnetic workers. It has been argued on this basis that correlations might be expected between reversals and geological processes, including polar wandering and continental drift, and the need for paleomagnetic studies designed with these possibilities in mind has been emphasized (Hide [**27**]); it might be significant in this connection (a) that while two dozen reversals have occurred during the past 4×10^6 years, the polarity of the field apparently remained fixed (and opposite to its present sense) throughout the 50×10^6 years occupied by the Permian, which ended 230×10^6 years ago, and (b) that petrological associations apparently occur in certain reversely magnetized rocks (Ade-Hall and Watkins [**1**], Ade-Hall and Wilson [**2**], Blackett [**5**], Bullard [**9**], Cox et al. [**18**], Irving [**36**], Wilson and Watkins [**68**]). Fluctuations in core motions must in general be of two types, namely "forced fluctuations"—associated directly with departures from axial or spherical symmetry and/or time variations in the "boundary conditions" (e.g. topography of the core-mantle interface, changing size of core, etc.)—and "intrinsic fluctuations"—which can occur even under steady and axisymmetric or spherically symmetric boundary conditions (see Hide [**27**])—and the absence of reversals in the Permian indicates that the former are probably more important than the latter. Departures from axial or spherical symmetry in the boundary conditions should be reflected in the amplitude and form of the geomagnetic nondipole field and secular variation, so that paleomagnetic studies of these properties of the field during reversal of the main dipole and of the extent to which they are correlated with the frequency of reversals would also be of direct theoretical importance.

I have tried, hopefully with some success, to make a case for taking seriously the concept of global-scale core-mantle interface topography with a vertical amplitude of a kilometre or so. By pivoting the discussion about well-established quantitative geophysical data of various kinds, the use of arguments that are sensitive to specific theoretical models of poorly-understood dynamical processes in the core and mantle has been avoided. The discussion makes clear that interactions between core motions and the typography of core-mantle interface will deserve detailed consideration in future theoretical studies; indeed it was first suggested several years ago that hydromagnetic waves of a new type might be excited by such interactions ([**25**]; see [**45**], [**63**], [**21**], [**29**], [**65**] for subsequent work).[4] The magnetohydrodynamics of core motions is a comparatively new branch of geophysical fluid dynamics with many challenging mathematical problems for the geophysical scientist.

Appendix: Plane hydromagnetic waves in a stratified rotating incompressible fluid. (Based on reference [**29**].) In [**25**] (see also pages 278 and 327–328) an approximate dispersion relationship for free hydromagnetic oscillations of the fluid core of the Earth was proposed, and it was conjectured that when

$$|N|^2 \ll |2\mathbf{\Omega}|^2 \tag{1}$$

effects due to vertical gradients of density, ρ, can safely be ignored. Here $\mathbf{\Omega}$ is the angular velocity of the Earth's rotation and

$$\mathbf{N} \equiv \frac{\mathbf{g}}{g}\left(\frac{g}{\rho}\frac{\partial \rho}{\partial z}\right)^{1/2} \tag{2}$$

$|N|$ is the Brunt–Väisälä frequency (effects due to compressibility being negligible in the theoretical model) and $\mathbf{g} = (0, 0, g)$ is the acceleration due to gravity and centrifugal effects. The purpose of this appendix is to investigate the validity of the criterion expressed by Equation (1).

The most acceptable procedure would be to find an accurate dispersion relationship for the case $\mathbf{N} \neq 0$, but this has not yet proved feasible. Owing to the (nearly) spherical geometry of the system, even when $\mathbf{N} = 0$ the problem is mathematically intractable (see [**25**], [**21**], [**29**], [**45**], [**63**], [**65**]), except in very special cases. For this reason we shall examine an elementary

[4] See pages 327–328.

but related problem whose solution should indicate, in part at least, how effects due to rotation, density stratification and magnetic fields interact with one another. Thus, we shall consider plane, small amplitude, harmonic waves propagating in an inviscid, perfectly conducting, incompressible, rotating fluid of indefinite extent in all directions when both $\boldsymbol{N}$ (based on the undisturbed density field $\rho = \rho_0(z)$) and

$$V \equiv \boldsymbol{B}_0/(\mu\rho_0)^{1/2}, \tag{3}$$

the Alfvén velocity, are uniform, where $\boldsymbol{B}_0$ is the undisturbed magnetic field vector and μ is the magnetic permeability.

The equations of the problem referred to a frame which rotates with steady angular velocity $\boldsymbol{\Omega}$ relative to an inertial frame are, when all transport processes (viscosity, electrical resistivity, thermal conduction, etc.) are negligible, the following (in rationalized MKS units):

$$\partial\boldsymbol{u}/\partial t + (\boldsymbol{u}\cdot\nabla)\boldsymbol{u} + 2\boldsymbol{\Omega}\times\boldsymbol{u} = -(1/\rho)\nabla p + (\boldsymbol{j}\times\boldsymbol{B})/\rho + \boldsymbol{g}, \tag{4}$$

$$\nabla\cdot\boldsymbol{u} = 0, \tag{5}$$

$$\partial\rho/\partial t + (\boldsymbol{u}\cdot\nabla)\rho = 0, \tag{6}$$

$$\nabla\cdot\boldsymbol{B} = 0, \tag{7}$$

$$\nabla\times\boldsymbol{B} = \mu\boldsymbol{j}, \tag{8}$$

$$\nabla\times\boldsymbol{E} = -\partial\boldsymbol{B}/\partial t, \tag{9}$$

$$\boldsymbol{E} + \boldsymbol{u}\times\boldsymbol{B} = 0. \tag{10}$$

Here $\boldsymbol{u}$ denotes the Eulerian flow velocity, p pressure, $\boldsymbol{j}$ current density, $\boldsymbol{E}$ electric field, and t time. Equations (4)–(6) express conservation of momentum, matter and density of individual fluid elements, respectively; Equations (7)–(9) are the laws of Gauss, Ampère and Faraday, respectively; Equation (10) states that in a perfect conductor of electricity, the electric field acting on a moving element must vanish because otherwise, by Ohm's law, electric currents of infinite strength would be implied.

If we write

$$\begin{aligned} \rho &= \rho_0(z) + \rho_1(x, y, z, t), & \boldsymbol{u} &= \boldsymbol{u}_0 + \boldsymbol{u}_1(x, y, z, t), \\ p &= p_0(z) + p_1(x, y, z, t), & \boldsymbol{B} &= \boldsymbol{B}_0 + \boldsymbol{B}_1(x, y, z, t) \end{aligned} \tag{11}$$

where $\boldsymbol{u}_0 = 0$ and $\nabla p_0 = g\rho_0$ and we assume that $|p_1| \ll |p_0|, |\boldsymbol{B}_1| \ll |\boldsymbol{B}_0|$ $|\rho_1| \ll \rho_0$ and spatial variations in ρ_0 are small in comparison with ρ_0, then to first order of small quantities p_1, $\boldsymbol{B}_1, \rho_1$ and $\boldsymbol{u}_1 = (u_1, v_1, w_1)$ Equations (4)–(7) become

$$\partial u_1/\partial t + 2\boldsymbol{\Omega}\times\boldsymbol{u}_1 = -(1/\bar{\rho}_0)\nabla p_1 + ((\nabla\times\boldsymbol{B}_1)\times\boldsymbol{B}_0)/\mu\rho_0 + g(\rho_1/\bar{\rho}_0) \tag{12}$$

(where $\bar{\rho}_0$ is the mean density of the fluid),

$$\nabla \cdot \boldsymbol{u}_1 = 0, \tag{13}$$

$$\partial\rho_1/\partial t + w_1(d\rho_0/dz) = 0, \tag{14}$$

$$\nabla \cdot \boldsymbol{B}_1 = 0, \tag{15}$$

—Equation (8) having been used to eliminate $\boldsymbol{j}$ from Equation (4)—and the equation that results when $\boldsymbol{E}$ is eliminated between Equations (9) and (10) becomes

$$\partial\boldsymbol{B}_1/\partial t - (\boldsymbol{B}_0 \cdot \nabla)\boldsymbol{u}_1 = 0. \tag{16}$$

Eliminate p_1, $\boldsymbol{B}_1$ and ρ_1 between Equations (9)–(16) and show that

$$[\partial^2/\partial t^2 - (\boldsymbol{V}\cdot\nabla)^2]\nabla \times \boldsymbol{u}_1 - 2(\boldsymbol{\Omega}\cdot\nabla)\partial\boldsymbol{u}_1/\partial t - \boldsymbol{N} \times \nabla(\boldsymbol{N}\cdot\boldsymbol{u}_1) = 0 \tag{17}$$

(cf. Equations (2) and (3)).

Substitute a plane wave solution

$$\boldsymbol{u}_1 \propto \exp\{i(\omega t - \boldsymbol{K}\cdot\boldsymbol{r})\} \tag{18}$$

(where ω is the angular frequency, $\boldsymbol{K} = (k, l, m)$ is the wavenumber vector and $\boldsymbol{r} = (x, y, z)$), in Equations (13) and (17) and thus find, after a little manipulation, the dispersion relationship

$$\omega^2 = (\boldsymbol{V}\cdot\boldsymbol{K})^2 + \frac{1}{2}\left\{\frac{(\boldsymbol{N}\times\boldsymbol{K})^2 + (2\boldsymbol{\Omega}\cdot\boldsymbol{K})^2}{K^2} \pm \left[\left(\frac{(\boldsymbol{N}\times\boldsymbol{K})^2 + (2\boldsymbol{\Omega}\cdot\boldsymbol{K})^2}{K^2}\right)^2 + \frac{4(\boldsymbol{V}\cdot\boldsymbol{K})^2(2\boldsymbol{\Omega}\cdot\boldsymbol{K})^2}{K^2}\right]^{1/2}\right\}. \tag{19}$$

The phase velocity, $\omega K/K^2$, and group velocity, $\partial\omega/\partial\boldsymbol{K}$ follow directly from Equation (19). When $N^2 = \bar{\rho}_0^{1} g\, d\rho_0/dz > 0$, ω is always real; we restrict attention in what follows to this case of stable density stratification. (In the other case, $d\rho_0/dz < 0$, the density stratification is unstable, as evinced by the result that ω may then take imaginary values.)

Particle displacements are transverse with respect to the wave fronts, i.e. $\boldsymbol{u}_1 \cdot \boldsymbol{K} = 0$, the occurrence of nonzero values of $\boldsymbol{u}_1 \cdot \boldsymbol{K}$ being incompatible with the assumption of incompressibility (see Equation (13)). Details of the shapes of particle orbits (linear, circular or elliptical) and of concomitant disturbances of the magnetic field and of the fields of density and vorticity, are readily derived from Equations (12)–(19).

In three limiting cases, Equation (19) reduces to particularly simple forms. Thus, when $\boldsymbol{V} = \boldsymbol{N} = 0$ we have "inertial waves," for which

$$\omega^2 = (2\boldsymbol{\Omega}\cdot\boldsymbol{K})^2/K^2 \tag{20}$$

and in which rotation provides the restoring forces; when $\boldsymbol{V} = \boldsymbol{\Omega} = 0$ we have "internal waves," for which

$$\omega^2 = (\boldsymbol{N} \times \boldsymbol{K})^2/K^2, \tag{21}$$

and in which buoyancy effects provide the restoring forces; and when $\boldsymbol{\Omega} = \boldsymbol{N} = 0$ we have "hydromagnetic (magnetohydrodynamic or Alfvén) waves," for which

$$\omega^2 = (\boldsymbol{V} \cdot \boldsymbol{K})^2 \tag{22}$$

and in which the magnetic field provides the restoring forces. Inertial waves and internal waves are highly dispersive, having group velocities that depend on $\boldsymbol{K}$. Alfvén waves, however, are nondispersive; they propagate with group velocity equal to $\pm \boldsymbol{V}$, which is independent of $\boldsymbol{K}$.

It has occurred to a number of workers (see [**23**]) that in some respects rotating fluids and stratified fluids exhibit analogous hydrodynamic behaviour (cf. Equations (20) and (21), especially when $l = 0$ and $\boldsymbol{\Omega} = (0, 0, \Omega)$, but Equation (19) shows that the analogy does not extend to magnetohydrodynamic behaviour. Thus, when $\boldsymbol{\Omega} = 0$

$$\omega^2 = (\boldsymbol{V} \cdot \boldsymbol{K})^2 + \tfrac{1}{2}\{(\boldsymbol{N} \times \boldsymbol{K})^2/K^2\}[1 \pm 1] \tag{23}$$

(the negative sign corresponding to modes for which particle displacements are horizontal and are therefore unaffected by buoyancy forces), whereas when $N = 0$

$$\omega^2 = (\boldsymbol{V} \cdot \boldsymbol{K})^2 + \frac{1}{2}\left\{\frac{(2\boldsymbol{\Omega} \cdot \boldsymbol{K})^2}{K^2}\right\}\left[1 \pm \left(1 + \frac{4(\boldsymbol{V} \cdot \boldsymbol{K})^2 K^2}{(2\boldsymbol{\Omega} \cdot \boldsymbol{K})^2}\right)^{1/2}\right]. \tag{24}$$

The important difference between the last two equations is the presence in the latter of a term containing the (square of the) ratio of $\boldsymbol{V} \cdot \boldsymbol{K}$ to $(2\boldsymbol{\Omega} \cdot K)/K$, which has no counterpart in the former. (When $\boldsymbol{V} = 0$, $\omega^2 = (\frac{1}{2} \pm \frac{1}{2}) \times \{(\boldsymbol{N} \times \boldsymbol{K})^2 + (2\boldsymbol{\Omega} \cdot \boldsymbol{K})^2\}/K^2$.)

This "ratio term," which arises because Coriolis forces act at right angles to $\boldsymbol{u}_1$ and thus prevent the occurrence of "decoupled" modes, is of great physical importance, especially when $4(\boldsymbol{V} \cdot \boldsymbol{K})^2 \ll (2\boldsymbol{\Omega} \cdot \boldsymbol{K})^2/K^2$ (cf. Equations (20) and (22). The roots of Equation (24) are then

$$\omega^2 \doteqdot (2\boldsymbol{\Omega} \cdot \boldsymbol{K})^2/K^2 \quad \text{and} \quad \omega^2 \doteqdot [(\boldsymbol{V} \cdot \boldsymbol{K})^2 K/(2\boldsymbol{\Omega} \cdot \boldsymbol{K})]^2, \tag{25}$$

the latter being the dispersion relationship for a hybrid type of wave discussed by Lehnert and Chandrasekhar (see [**25**]) which has no direct analogue in a stratified fluid.

The discussion of Equation (19) is less straightforward when none of the terms $(\boldsymbol{V} \cdot \boldsymbol{K})$, $(2\boldsymbol{\Omega} \cdot \boldsymbol{K})/K$ and $(\boldsymbol{N} \times \boldsymbol{K})/K$ vanishes, but when these terms are widely separated in magnitude then Equation (19) can be simplified by neglecting small quantities. There are six such limiting cases, corresponding to

(26a) $|\boldsymbol{V} \cdot \boldsymbol{K}| \gg |2\boldsymbol{\Omega} \cdot \boldsymbol{K}|/K \gg |\boldsymbol{N} \times \boldsymbol{K}|/K,$

(26b) $|\boldsymbol{V} \cdot \boldsymbol{K}| \gg |\boldsymbol{N} \times \boldsymbol{K}|/K \gg |2\boldsymbol{\Omega} \cdot \boldsymbol{K}|/K,$

(26c) $|2\boldsymbol{\Omega} \cdot \boldsymbol{K}|/K \gg |\boldsymbol{V} \cdot \boldsymbol{K}| \gg |\boldsymbol{N} \times \boldsymbol{K}|/K,$

(26d) $|2\boldsymbol{\Omega} \cdot \boldsymbol{K}|/K \gg |\boldsymbol{N} \times \boldsymbol{K}|/\boldsymbol{K} \gg |\boldsymbol{V} \cdot \boldsymbol{K}|,$

(26e) $|\boldsymbol{N} \times \boldsymbol{K}|/K \gg |\boldsymbol{V} \cdot \boldsymbol{K}| \gg |2\boldsymbol{\Omega} \cdot \boldsymbol{K}|/\boldsymbol{K},$

(26f) $|\boldsymbol{N} \times \boldsymbol{K}|/K \gg |2\boldsymbol{\Omega} \cdot \boldsymbol{K}|/K \gg |\boldsymbol{V} \cdot \boldsymbol{K}|$

and the solutions are readily written down. Of particular interest in connection with the Earth's interior are the cases (26c) and (26d). When Ω is so large that both $|\boldsymbol{V} \cdot \boldsymbol{K}|$ and $|\boldsymbol{N} \times \boldsymbol{K}|/K$ are very much less than $|2\boldsymbol{\Omega} \cdot \boldsymbol{K}|/K$ then, to first order of small quantities, Equation (19) reduces to

(27) $$\omega^2 \doteqdot \frac{(2\boldsymbol{\Omega} \cdot \boldsymbol{K})^2}{K^2} \quad \text{and} \quad \omega^2 \doteqdot \left[\frac{(\boldsymbol{V} \cdot \boldsymbol{K})^2 K}{2\boldsymbol{\Omega} \cdot \boldsymbol{K}}\right]^2 + \frac{(\boldsymbol{V} \cdot \boldsymbol{K})^2 (\boldsymbol{N} \times \boldsymbol{K})^2}{(2\boldsymbol{\Omega} \cdot \boldsymbol{K})^2}$$

the dispersion relationships for the "inertial" and "magnetic" modes (see page 327). Evidently the criterion expressed by Equation (1) is probably both necessary and sufficient for inertial modes, at least for oscillations with wavelengths that are much smaller than the radius of the Earth's core, but in the case of the magnetic modes, Equation (1) must be supplemented by the more stringent condition that

(29) $$(\boldsymbol{N} \times \boldsymbol{K})^2/K^2 \ll (\boldsymbol{V} \cdot \boldsymbol{K})^2.$$

(The last point had been overlooked until, prompted by a question from the audience, I amplified the discussion given in the original version of these notes.) When $(\boldsymbol{N} \times \boldsymbol{K})^2/K^2 \gg (\boldsymbol{V} \cdot \mathrm{K})^2$ (but is much less than $(2\boldsymbol{\Omega} \cdot \boldsymbol{K})^2/K^2$), we have

(30) $$\omega^2 \doteqdot \frac{(2\boldsymbol{\Omega} \cdot \boldsymbol{K})^2}{K^2} \quad \text{and} \quad \omega^2 \doteqdot \frac{(\boldsymbol{V} \cdot \boldsymbol{K})^2 (\boldsymbol{N} \times \boldsymbol{K})^2}{(2\boldsymbol{\Omega} \cdot \boldsymbol{K})^2}$$

in place of Equations (25) for the respective dispersion relationships for the inertial and magnetic modes.

REFERENCES, LECTURE V

1. J. M. Ade-Hall and N. D. Watkins, *Absence of correlations between opaque petrology and natural remanence polarity in Canary Island lavas*, Geophys. J. Roy. Astronom. Soc. **19** (1970), 351–360.

2. J. M. Ade-Hall and R. L. Wilson, *Opaque petrology and natural remanence in Mull (Scotland) dykes*, Geophys. J. Roy. Astronom. Soc. **18** (1969), 333–350.

3. M. J. Aitken and G. M. Weaver, *Recent archaeomagnetic results in England*, J. Geomag. Geoelect. **17** (1965), 391–394.

4. G. E. Backus, *Kinematics of geomagnetic secular variation in a perfectly conducting core*, Philos. Trans. Roy. Soc. London Ser. A **263** (1968), 239–266.

5. P. M. S. Blackett, *On distinguishing self-reversal from field reversed rocks*, J. Phys. Soc. Japan, **17** (B1) (1962), 699–710.

6. P. M. S. Blackett, E. C. Bullard and S. K. Runcorn (editors), *A symposium on continental drift*, Philos. Trans. Roy. Soc. London Ser. A **258** (1965), 1–323.

7. S. I. Braginskiĭ, *Kinematic models of the Earth's hydromagnetic dynamo*, Geomagnetism and Aeronomy **4** (1964), 572–583.

8. G. R. R. Buchbinder, *Properties of the core-mantle boundary and observations of* PcP, J. Geophys. Res. **73** (1968), 5901–5923.

9. E. C. Bullard, *Reversals of the Earth's magnetic field*, Philos. Trans. Roy. Soc. London Ser. A **263** (1968), 481–524.

10. E. C. Bullard, C. Freedman, H. Gellman and J. Nixon, *The westward drift of the Earth's magnetic field*, Philos. Trans. Roy. Soc. London Ser. A **243** (1950), 67–92.

11. E. C. Bullard and H. Gellman, *Homogeneous dynamos and terrestrial magnetism*, Philos. Trans. Roy. Soc. London Ser. A **247** (1954), 213–278. MR **17**, 327.

12. K. E. Bullen, *An introduction to the theory of seismology*, Cambridge Univ. Press, New York, 1963. MR **29** #5626.

13. K. E. Bullen, F. Press and S. K. Runcorn (editors), *Phase transformations and the Earth's interior*, Phys. of Earth and Planetary Interiors **3** (1970), 1–518.

14. S. P. Burlatskaya, T. B. Nechayeva and G. N. Petrova, *Characteristics of secular variations of the geomagnetic field as indicated by world archaeomagnetic data*, Izv. Earth Physics **12** (1968), 62–70.

15. C. G. Collier and R. Hide, *Transient motion of a fluid in an irregular container*, (in prep).

16. A. M. Coode and S. K. Runcorn, *Satellite geoid and the structure of the Earth*, Nature **205** (1965), 891.

17. A. H. Cook, *Sources of harmonics of low order in the external gravity field of the Earth*, Nature **198** (1963), 1186.

18. A. Cox, G. B. Dalrymple and R. R. Doell, *Reversals of the Earth's magnetic field*, Sci. Amer. **216** (1967), 44–60.

19. L. Egyed, *The satellite geoid and the structure of the Earth*, Nature **203** (1964), 67–69.

20. T. F. Gaskell (editor), *The Earth's mantle*, Academic Press, New York and London, 1967.

21. P. A. Gilman, *Baroclinic, Alfvén and Rossby waves in geostrophic flow*, J. Atmospheric Sci. **26** (1969), 1003–1009.

22. P. Goldreich and S. J. Peale, *The dynamics of planetary rotations*, Ann. Rev. Astronom. Astrophys. **6** (1968), 287–320.

23. H. P. Greenspan, *The theory of rotating fluids*, Cambridge Univ. Press, New York, 1968.

24. W. H. Guier and R. R. Newton, *The Earth's gravity field as deduced from the Doppler tracking of five satellites*, J. Geophys. Res. **70** (1965), 4613–4626.

25. R. Hide, *Free hydromagnetic oscillations of the Earth's core and the theory of the geomagnetic secular variation*, Philos. Trans. Roy. Soc. London Ser. A **259** (1966), 615–647.

26. ———, *Planetary magnetic fields*, Planetary and Space Sci. **14** (1966), 579–586.

27. ———, *Motions of the Earth's core and mantle and variations of the main geomagnetic field*, Science **157** (1967), 55–56.

28. ———, *Interaction between the Earth's liquid core and solid mantle*, Nature **222** (1969), 1055–1056.

29. ———, *On hydromagnetic waves in a stratified rotating incompressible fluid*, J. Fluid Mech. **39** (1969), 283–287.

30. ———, *Dynamics of the atmospheres of the major planets*, J. Atmospheric Sci. **26** (1969), 841–853.

31. R. Hide and K.-I. Horai, *On the topography of the coremantle interface*, Physics of the Earth and Planetary Interiors **1** (1968), 305–308.

32. R. Hide and S. R. C. Malin, *Novel correlations between global features of the Earth's gravitational and magnetic fields*, Nature **225** (1970), 605–609; Nature Physical Science **230** (1971), 63.

33. W. R. Hindmarsh, F. J. Lowes, P. H. Roberts and S. K. Runcorn (editors), *Magnetism and the cosmos*, Oliver and Boyd, Edinburgh, 1965.

34. P. M. Hurley (editor), *Advances in Earth sciences*, M.I.T. Press, Cambridge, Mass., 1966.

35. E. Irving, *Paleomagnetism*, Wiley, New York, 1964.

36. ———, *Paleomagnetism of some carboniferous rocks from New South Wales and its relation to geological events*, J. Geophys. Res. **71** (1966), 6025–6051.

37. H. Jeffreys, *The Earth*, 4th ed., Cambridge Univ. Press, London, 1959.

38. H. Kanamori and F. Press, *How thick is the lithosphere?*, Nature **266** (1970), 330–331.

39. W. M. Kaula, *An introduction to planetary physics: Terrestrial planets*, Wiley, New York, 1968.

40. ———, *A tectonic classification of the main features of the Earth's gravitational field*, J. Geophys. Res. **74** (1969), 4807–4826.

41. ———, *Earth's gravity field: relation to global tectonics*, Science **169** (1970), 982–985.

42. N. Kawai, K. Hirooka and K. Tokieda, *A vibration of geomagnetic axis around the geographic north pole in historic time*, Earth and Planetary Sci. Letters **3** (1967), 48–50.

43. F. E. M. Lilley, *On kinematic dynamos*, Proc. Roy. Soc. London Ser. A **316** (1970), 153–167.

44. D. Lortz, *Exact solutions of the hydromagnetic dynamo problem*, Plasma Phys. **10** (1968), 967–972.

45. W. V. R. Malkus, *Hydromagnetic planetary waves*, J. Fluid Mech. **28** (1967), 793–802. MR **36** #4869.

46. ———, *Precession of the Earth as the cause of geomagnetism*, Science **160** (1968), 259–264.

47. B. G. Marsden and A. G. W. Cameron (editors), *The Earth-Moon system*, Plenum Press, New York, 1966.

48. H. K. Moffatt, *Turbulent dynamo action at low magnetic Reynolds numbers*, J. Fluid Mech. **41** (1970), 435–452.

49. W. H. Munk and G. J. F. MacDonald, *The rotation of the Earth*, Cambridge Univ. Press, New York, 1960.

50. M. W. McElhinny and M. E. Evans, *An investigation of the geomagnetic field in the Precambrian*, Phys. of Earth and Planetary Interiors **1** (1968), 485–497.

51. T. Nagata (editor), *Proceedings of the Benedum Earth magnetism symposium*, Univ. of Pittsburgh Press, Pittsburgh, Pa., 1962.

52. T. Namikawa and S. Matsushita, *Kinematic dynamo problem*, Geophys. J. Roy. Astronom. Soc. **19** (1970), 395–415.

53. E. Orowan, *The origin of the ocean ridges*, Sci. Amer. **221** (1969), 103–109.

54. E. N. Parker, *The occasional reversal of the geomagnetic field*, Astrophys. J. **158** (1969), 816–827.

55. T. Rikitake, *Electromagnetism and the Earth's interior*, Elsevier, Amsterdam, 1966.

56. P. H. Roberts, *An introduction to magnetohydrodynamics*, American Elsevier, New York; Longman, London, 1967.

57. M. G. Rochester, *Geomagnetic westward drift and irregularities in the Earth's rotation*, Philos. Trans. Roy. Soc. London Ser. A **252** (1960), 531–555.

58. R. B. Roden, *Electromagnetic core-mantle coupling*, Geophys. J. Roy. Astronom. Soc. **7** (1963), 361–374.

59. S. K. Runcorn (editor), *Mantles of the Earth and terrestrial planets*, Wiley, New York, 1967.

60. ———, *The application of modern physics to the Earth and planetary interiors*, Wiley, London, 1969.

61. ———, *The rotation of the planets and their interiors*, Ist. Naz. Alta Mat. Sympos. Math. **3** (1970), 193–202.

62. F. D. Stacey, *Physics of the Earth*, Wiley, London, 1969.

63. K. Stewartson, *Slow oscillations of fluid in a rotating cavity in the presence of a toroidal magnetic field*, Proc. Roy. Soc. London Ser. A **299** (1967), 173–187.

64. G. C. J. Suffolk, *Precession in a disk dynamo model of the Earth's magnetic field*, Nature **226** (1970), 628–629.

65. G. Suffolk and D. W. Allan, *Planetary magnetohydrodynamic waves as a perturbation of dynamo solutions*, in Runcorn 1969, pp. 653–656.

66. D. C. Tozer (editor), *Symposium on non-elastic processes in the mantle*, Geophys. J. Roy. Astronom. Soc. **14** (1967), 1–450.

67. R. L. Wilson, *Permanent aspects of the Earth's non-dipole magnetic field over upper Tertiary times*, Geophys. J. Roy. Astronom. Soc. **19** (1970), 417–437.

68. R. L. Wilson and N. F. Watkins, *Further correlations between the petrology and the natural remanent magnetization of basalts*, Geophys. J. Roy. Astronom. Soc. **12** (1967), 405–415.

69. T. Yukutake and H. Tachinaka, *The westward drift of the geomagnetic secular variation*, Bull. Earthquake Res. Inst. **45** (1968), 1075–1102.

70. P. Goldreich and A. Toomre, *Some remarks on polar wandering*, J. Geophys. Res. **74** (1969), 2555–2567.

71. R. Hide, *On the Earth's core-mantle interface*, Quart. J. Roy. Meteorol. Soc. **96** (1970), 579–590.

72. M. G. Rochester, "Core-mantle interactions: Geophysical and astrophysical consequences," In: *Earthquake displacement fields and the rotation of the Earth*, L. Mansinha et al. (editors), Reidel, Dordrecht, 1970, pp. 136–148.

METEOROLOGICAL OFFICE, BRACKNELL, BERKSHIRE, ENGLAND

Permission has been kindly given for the use of copyrighted figures by the following publishers: Academic Press, Inc. (London) Limited; American Meteorological Society; *Journal of Fluid Mechanics*; *Nature*, MacMillan (Journals) Limited; Oliver and Boyd; Pergamon Press Limited; Royal Meteorological Society. The references in brackets with each figure identify the source.

Tropical Cyclogenesis and the Formation of the Intertropical Convergence Zone

Jule Charney

1. **The problem of tropical cyclogenesis.** The great migratory waves and vortices of middle latitudes, whose surface manifestations are the familiar low- and high-pressure systems (cyclones and anticyclones) of the daily weather map, have now been tolerably well explained. They are due to the instability of the mean, axially-symmetric, circumpolar vortex whose gravity and Coriolis forces balance the pressure forces associated with the radiatively induced pole to equator temperature gradient.[1] In contrast, the origins of the migratory low pressure systems of the tropics (depressions and hurricanes) are far less well understood. An explanation analogous to that for the extratropical cyclones does not seem possible. Tropical temperature gradients are weak, and the associated potential energy supply is not by itself a likely source for the kinetic energy of the tropical disturbance. Indeed, tropical cyclones form most often over the tropical oceans where the surface temperatures are particularly uniform.

However, the oceans are also a source of moisture, and what is most characteristic of tropical depressions in oceanic and coastal regions is the copious rainfall they produce. It is very likely that the release of the latent heat of condensation is the primary driving mechanism. But tropical precipitation occurs for the most part in deep cumulus convection cells whose diameters are of the order of 1 to 10 km, and it is not easy to see how these cells are organized so as to supply energy to the large-scale flow whose scale is of the order of 1000 km.

AMS 1970 *subject classifications*. Primary 86A10, 86A35, 80A20, 76D10, 76E15, 76E20.

[1] See §§7 and 8 in [**3**].

Precipitation also occurs in extratropical cyclones, but here it is forced by the rising of the warm light air, which, in conjunction with sinking of the cold dense air, is required for the release of the potential energy associated with the strong horizontal temperature gradients. Since the hydrostatic pressure decreases with altitude, the rising air expands and cools adiabatically until it becomes saturated and is forced to release its moisture as cloud and precipitation. In the extratropical cyclone latent heat is an additional source of energy, not the primary source.

In this lecture it will be shown that there is another mechanism, independent of horizontal temperature gradients, which is capable of organizing the release of heat of condensation. Surprisingly, it involves boundary friction which one thinks of ordinarily as dissipative rather than generative. Although the mechanism is capable of operating alone, it probably never does. Asymmetries of the undisturbed flow may help or hinder the formation of the tropical depression. Indeed, most depressions form in the so-called Intertropical Convergence Zone (ITCZ), a narrow zone paralleling the equator but lying at some distance from it, in which air from one hemisphere converges against air from the other to produce cloud and precipitation. The ITCZ is characterized by low pressure and cyclonic (counterclockwise in the Northern Hemisphere and clockwise in the Southern Hemisphere) shear vorticity in the surface layers. It will be suggested that the mechanism which leads to the formation of the tropical depression may also be responsible for the formation of the ITCZ itself.

2. **Conditional instability.** It will be necessary to consider motion on two scales, that of the cumulus cloud and that of the cyclone. While the detailed dynamics of the cumulus cloud are complicated, the explanation for its formation is simple. A combination of radiative, sensible and latent heat transfer maintains the tropical atmosphere in a state of hydrostatic equilibrium such that the temperature decreases with height at a rate of about 5.5°C per km in its lower part. Since a parcel of air rising adiabatically cools at a rate of 10°C per km, it finds itself colder and denser than its environment and is therefore acted upon by a restoring buoyancy force (resultant of pressure and gravity). If, however, the parcel is saturated, the moisture condenses, and the heat of condensation prevents it from cooling as rapidly as if it were not saturated. Since a rising, saturated parcel cools at a rate of about 4.5°C per km in the lower atmosphere, it finds itself warmer than its environment and is accelerated upward by the buoyancy force. The tropical atmosphere is therefore gravitationally stable for unsaturated air and gravitationally unstable for saturated air. Such an atmosphere is said to be conditionally unstable. Cumulus clouds are a manifestation of gravitational instability and convective overturning

in an atmosphere whose lower part is kept moist by evaporation from the sea-surface.

Is it possible, then, that the tropical disturbance is simply a large-scale overturning in a conditionally unstable atmosphere? We shall answer this question in the negative by showing that such an atmosphere is far more unstable to cumulus cloud scales than to disturbance scales, and that, were it not for friction and entrainment of dry air, the preferred convection cell would have an infinitely narrow ascending branch.

3. **Competition or cooperation?** To show that the cumulus-scale perturbation is favored over the cyclone-scale in a conditionally unstable atmosphere, we must consider the thermodynamics of unsaturated and saturated air. Let p, ρ, T, θ, α, s, c_p, c_v, R, and Q be the pressure, density, temperature, potential temperature, specific volume, specific entropy, specific heat at constant pressure, specific heat at constant volume, specific gas constant, and specific rate of accession of heat, the last five quantities referring to unit mass of dry air. Variations in these quantities due to an admixture of water vapor or cloud will be small enough to be ignored. Since air is very nearly a perfect gas, $p\alpha = RT$, $R = c_p - c_v$, and the first law of thermodynamics takes one of the forms

$$\begin{aligned} ds &= c_v\, dT/T + p\, d\alpha/T = c_p\, dT/T - R\, dp/p \\ &= c_v\, dp/p - c_p\, d\rho/\rho = c_p\, d\theta/\theta = dQ/T. \end{aligned} \tag{3.1}$$

Since the undisturbed atmosphere is in hydrostatic equilibrium,

$$-\partial p/\partial z - \rho g = 0, \tag{3.2}$$

where g is the acceleration of gravity and z is the vertical coordinate.

Let a parcel of air be lifted adiabatically from the height z to the height $z + dz$, and assume that its pressure is always that of the environment. Then from (3.1) and (3.2),

$$\left(\frac{dT}{dz}\right)_s = \frac{RT}{c_p p}\frac{\partial p}{\partial z} = -\frac{RT\rho g}{c_p p} = -\frac{g}{c_p} \equiv -10^\circ/\text{km}. \tag{3.3}$$

The restoring force per unit mass at the height $z + dz$ is $g\delta\rho/\rho$, where δ represents an increase at $z + dz$ as a consequence of the displacement. Since $\delta p = 0$, and $\delta s = s(z) - s(z + dz) = -(\partial s/\partial z)\, dz$, we get from (3.1)

$$c_p\frac{\delta\rho}{\rho} = \frac{\partial s}{\partial z}dz = c_p\frac{\partial \ln\theta}{\partial z}dz = \left(\frac{c_p}{T}\frac{\partial T}{\partial z} - \frac{R}{p}\frac{\partial p}{\partial z}\right)dz,$$

or, if N^2 is the restoring force per unit mass per unit displacement,

$$N^2 = g(\delta\rho/\rho)/dz = g(\Gamma_a - \Gamma)/T = g(\partial \ln\theta/\partial z), \tag{3.4}$$

where $\Gamma_a = -(dT/dz)_s = g/c_p$ is the adiabatic lapse-rate of temperature, and $\Gamma = -\partial T/\partial z$ is the lapse rate of temperature in the environment. The quantity N has the dimensions of a frequency and, in fact, is the frequency of a gravitational oscillation in a stably stratified compressible fluid whose displacements are primarily vertical.

Consider an air parcel which has reached saturation. If the parcel is lifted, the temperature will drop, moisture will condense, and latent heat will be liberated. We have $dQ = -L\,dq_s$, and

$$ds = c_p\,d(\ln\theta) = -L(dq_s/T), \tag{3.5}$$

where q_s is the specific humidity and L is the latent heat of condensation per unit mass. Since fractional changes in q_s are far greater than those of T or L—the latter may be regarded as constant—Equation (3.5) may also be interpreted as stating that the entropy of the moist air

$$\tilde{s} = c_p \ln\theta + Lq_s/T = c_p \ln\theta_E = \text{constant}. \tag{3.6}$$

The quantity $\theta_E = \theta\exp(Lq_s/c_pT)$ is called the "equivalent potential temperature" in analogy to "potential temperature." It is conserved for changes in state in which the latent heat of condensation is retained by the parcel. Such a process is called "moist-adiabatic." Integration of (3.5) after multiplication by T and use of the hydrostatic relation also gives the result that the enthalpy

$$h = c_pT + gz + Lq_s = \text{constant}. \tag{3.7}$$

Suppose now that a saturated air parcel is lifted from z to $z + dz$. Again the pressure at $z + dz$ does not change, and we have from the gas law that the restoring force per unit mass is

$$-N_m^2\,dz = g(\delta\rho/\rho) = -g(\delta T/T) = g((\Gamma_m - \Gamma)/T)\,dz, \tag{3.8}$$

where $\Gamma_m = -(dT/dz)_{\tilde{s}}$ is the lapse rate of temperature for the moist-adiabatic process. Thus if $\Gamma - \Gamma_m < 0$, or alternatively if $\partial\theta_E/\partial z < 0$, the atmosphere is conditionally unstable.

In reality, rising motion must be compensated by descending motion to preserve mass continuity. Let us schematically denote the area of rising motion at a fixed level by A_+ and the area of sinking motion by A_-. Let a typical upward displacement be l_+ and a downward displacement l_-. From mass continuity $l_+A_+ = l_-A_-$. The work W_- done against gravity in the downward displacement is measured by $W_- = N^2l_-^2A_-$, and the work done by the buoyancy forces in the upward displacement is measured by $W_+ = N_m^2l_+^2A_+$. The ratio of the two is

$$\frac{W_+}{W_-} = \frac{N_m^2}{N^2}\frac{l_+^2}{l_-^2}\frac{A_+}{A_-} = \frac{N_m^2}{N^2}\frac{A_-}{A_+}. \tag{3.9}$$

If energy is to be released we must have

$$A_+/A_- < N_m^2/N^2 = (\Gamma - \Gamma_m)/(\Gamma_a - \Gamma) \sim \tfrac{1}{5}.$$

Thus the area of the rising branch of the circulation must be less than approximately one-fifth that of the descending branch. But it also follows from (3.9) that for a given amount of work done against gravity in the downward displacement, an unlimited amount can be realized through conditional instability in the upward displacement providing the ascending branch is of arbitrarily small cross-section. Of course, friction and entrainment of dry air will limit the width of the ascending branch, but it is known from the theory of Bénard convection that these limitations will not be effective until the width approaches that of the depth of the troposphere, i.e., the width of a cumulus cloud. It follows that conditional instability will by itself give rise to cumulus clouds, not depressions or hurricanes.

Nevertheless, conditional instability, by permitting cumulus convection, must play a role in the formation of the tropical cyclone. As we have noted, the most striking characteristic of the depression and hurricane is its enormous rainfall; the latent heat energy released is two orders of magnitude greater than the amount needed to maintain the kinetic energy against frictional dissipation. This suggests that we should look upon the depression and the cumulus cell not as competing for the same energy, for in this competition the cumulus cell must win; rather we should consider the two as supporting one another, the cumulus cell by supplying the heat energy for driving the depression, and the depression by maintaining the moisture supply for driving the convection. The cumulus and cyclone-scale motions are thus to be regarded as cooperating rather than competing phenomena.

4. **The frictional boundary layer.** The clue to the cumulus-cyclone interaction lies in the character of the frictional boundary layer. It is here that most of the moisture is fed into the bases of the towering cumulonimbus clouds in the tropical depression. What causes the low-level convergence of mass and moisture? The tentative answer is that it is caused by boundary friction.

The theory of surface frictional interaction in large-scale rotating systems is presented in this volume by Howard and by Pedlosky. Here we shall only review certain salient features. Circulations of large space and time scale on the rotating earth are quasi-horizontal and are characterized by approximate geostrophic as well as hydrostatic force balance: in the interior of the atmosphere, frictional forces and relative accelerations are small, and the horizontal pressure force is approximately balanced by the horizontal Coriolis force. Let us consider a flow relative to a plane rotating at right angles to gravity. We introduce a right-handed Cartesian system of coordinates with x and y in the plane and z at right angles and opposite to gravity. Let u, v, and w be the x, y, and z velocity components respectively, and let ω be the speed of rotation of the plane. Then the conditions of geostrophic and hydrostatic balance are expressed by

$$-\rho f v = -\partial p/\partial x, \tag{4.1}$$

$$\rho f u = -\partial p/\partial y, \tag{4.2}$$

$$\rho g = -\partial p/\partial z, \tag{4.3}$$

where $f = 2\omega$. The velocity components u, v, and w must vanish at the horizontal boundary, but since friction does not directly affect the density and therefore the vertical pressure variation, the horizontal pressure force components do not necessarily vanish. Hence one must introduce frictional forces near the boundary to reduce the horizontal velocity to zero. Because of the large horizontal scale, only the vertical shearing stress is important, and the horizontal momentum equations become

$$-\rho f v = -\partial p/\partial x + \rho\nu(\partial^2 u/\partial z^2), \qquad \rho f u = -\partial p/\partial z + \rho\nu(\partial^2 v/\partial z^2), \tag{4.4}$$

where, in applications to the atmosphere where the Reynolds numbers are always large, ν is to be interpreted as a turbulent eddy viscosity whose order of magnitude is far greater than that of the molecular viscosity. For simplicity ν is taken to be constant.

The depth of the frictional boundary layer is seen to be of order $(\nu/f)^{1/2}$, and within this depth the horizontal pressure force components and density do not vary appreciably. The solution of (4.4) for the case $\partial p/\partial x = 0$ is then given by

$$u = u_g(1 - e^{-z/D}\cos z/D), \qquad v = u_g e^{-z/D}\sin z/D, \tag{4.5}$$

where u_g is the geostrophic wind $-(1/\rho f)(\partial p/\partial y)$, and D is the Ekman depth $(2\nu/f)^{1/2}$. The theory of the stationary frictional boundary layer was originally presented for the oceans by Ekman [**1**] who also calculated the transient response to an impulsively applied forcing. If, for example, the current above the boundary plane were changed suddenly, the boundary-layer would adjust in a time of the order ω^{-1}. This is the time T for the vorticity to diffuse upward from the boundary through the distance D, i.e., $(\nu T)^{1/2} = (2\nu/f)^{1/2}$.

Integration of (4.5) gives for the components M_x and M_y of the boundary layer mass transport

$$M_x = \int_0^\infty \rho(u - u_g)\,dz = -\rho D u_g/2, \qquad M_y = \int_0^\infty \rho v\,dz = \rho D u_g/2, \tag{4.6}$$

or, in general,

$$M = (\rho D/2)\, \boldsymbol{k} \times \boldsymbol{V}_g, \tag{4.7}$$

where $\boldsymbol{V}_g = (u_g, v_g)$ and $\boldsymbol{k}$ is a vertical unit vector. Suppose now that the geostrophic wind varies horizontally. As we need not here consider space or time variations of ρ, we may write the equation of continuity as

$$\partial(\rho u)/\partial x + \partial(\rho v)/\partial y + \partial(\rho w)/\partial z = 0, \tag{4.8}$$

and obtain by integration

$$M_x + M_y + \rho w_E = 0, \tag{4.9}$$

where w_E is the vertical velocity at the top of the boundary layer. Substitution from (4.7) then gives

$$w_E = (D/2)\zeta_g, \tag{4.10}$$

where ζ_g is the vorticity component of the geostrophic wind:

$$\zeta_g = \frac{\partial v_g}{\partial x} - \frac{\partial u_g}{\partial y} = \frac{1}{\rho f}\left(\frac{\partial^2 p}{\partial x^2} + \frac{\partial^2 p}{\partial y^2}\right). \tag{4.11}$$

Thus cyclonic vorticity produces horizontal mass convergence in the boundary layer and positive vertical mass flow out of the boundary layer.

Charney and Eliassen [**2**] have applied this result to show how the kinetic energy of a geostrophically balanced circular vortex in a homogeneous fluid is destroyed by frictional interaction with the boundary. The frictional convergence in the boundary layer toward lower pressure must be compensated by divergence above the layer. The radial flow above the boundary layer is therefore directed outward toward high pressure, work is done against the inward directed pressure force, and the kinetic energy is reduced. Let U be the radial and V the tangential velocity above the boundary layer. Then by (4.6) and mass continuity, $V \sim UD/H$, and the percentage rate of reduction of U by outward expansion of rings of air with conservation of absolute angular momentum—or, alternatively, by action of the Coriolis force in relative flow on the radial velocity—is $fV/U — fD/H — fE^{1/2}$, where E is the nondimensional Ekman number ν/fH^2. Thus the characteristic "spin-down" time for the vortex is $1/fE^{1/2}$. Another way of looking at it is to say that the cross-sectional areas of the vortex tubes in absolute motion, which are formed mainly by the vertical vortex lines of the rotating plane, expand at the rate w_E/H per unit area, so that the relative vertical vorticity component is reduced by the amount fw_E/H per unit time. The characteristic time for the destruction of the relative vorticity is again

$$\frac{\zeta_g}{fw_E/H} \sim \frac{\zeta_g}{f\zeta_g D_E/H} \sim \frac{1}{fE^{1/2}}.$$

Thus we see that the adjustment time for the flow outside the boundary layer is longer than the boundary-layer adjustment time by the factor $E^{-1/2}$, which ordinarily is large, about 10 in the atmosphere. It is for this reason that the boundary layer itself can be considered to be in steady equilibrium, i.e., to reflect instantly the time-variable geostrophic flow above the boundary layer.

5. **Conditional instability of the second kind.** Consider now a conditionally unstable moist tropical atmosphere and suppose that a large-scale disturbance of small amplitude is somehow imparted. In regions of low-level cyclonic vorticity there will be boundary layer convergence, upward pumping of the moist surface air into the cumulus cells, and therefore enhancement of convection; in regions of anti-cyclonic vorticity there will be boundary layer suction carrying down dryer air from aloft, and therefore quenching of cumulus convection. Moreover, as heat is liberated in the cyclonic region, the pressure will fall and the geostrophic vorticity will increase. Thus we have the possibility of a self-excited disturbance. It is this process that Charney and Eliassen [**2**] have called "conditional instability of the second kind (CISK)".

The CISK process may be studied analytically for circular or slab symmetry. Other cases present problems because of the nonlinear character of the condensation process: the liberation of heat of condensation is proportional to $\frac{1}{2}(w_E + |w_E|)$. We will here consider a zonally and slab symmetric perturbation of a resting atmosphere on the rotating earth. We define locally Cartesian coordinates with x directed eastward, y northward and z upward. Anticipating that the y-scale will be small, we shall ignore effects from the curvature of the earth and the variation with latitude of the Coriolis parameter, $f = 2\Omega \sin \phi$, where Ω is the angular speed of rotation of the earth and ϕ is the latitude. The perturbation equations of motion above the boundary layer become

$$\partial u'/\partial t' - fv' = 0, \tag{5.1}$$

$$\partial v'/\partial t' + fu' = -(\partial/\partial y)(p'/\bar{p}), \tag{5.2}$$

$$0 = -\partial p'/\partial z' - \rho' g, \tag{5.3}$$

$$\frac{\partial \rho}{\partial t'} + \frac{\partial}{\partial y'}(\bar{\rho} v') + \frac{\partial}{\partial z'}(\bar{\rho} w') = 0, \tag{5.4}$$

$$\frac{1}{\bar{\theta}}\frac{\partial \theta'}{\partial t'} + \frac{N^2}{g} w' = \frac{Q'}{c_p \bar{T}}, \tag{5.5}$$

where primes denote perturbations, bars horizontally averaged mean quantities, and the dimensional independent variables are also denoted by primes so that we may drop the primes for the nondimensional independent variables. In Equation (5.5)

$$\frac{\theta'}{\bar{\theta}} = \frac{T'}{\bar{T}} - \frac{R}{c_p}\frac{p'}{\bar{p}} = \frac{c_v}{c_p}\frac{p'}{\bar{p}} - \frac{\rho'}{\bar{\rho}} \tag{5.6}$$

and

$$N^2 = g\partial(\ln\bar{\theta})/\partial z'.$$

From (5.3) and (5.6) we get

$$g\frac{\theta'}{\bar{\theta}} = \frac{\partial}{\partial z'}\left(\frac{p'}{\bar{\rho}}\right) - \frac{\partial \ln\bar{\theta}}{\partial z'}\frac{p'}{\bar{\rho}} \cong \frac{\partial}{\partial z'}\left(\frac{p'}{\bar{\rho}}\right), \tag{5.7}$$

since $\ln\bar{\theta}$ is a slowly-varying function of z in the atmosphere.

From what has been said, we expect u to be geostrophic and v to be driven by friction. Hence we may scale the independent variables as follows: $t' = f^{-1}E^{-1/2}t$, $y' = By$, $z' = Hz$, where H is the scale height $R\bar{T}_m/g$ and $\bar{T}_m$ is a vertically averaged temperature. The dependent variables are scaled by

$$\left(u', v', w', \frac{p'}{\bar{\rho}}, \frac{\rho'}{\bar{\rho}}\right) = \left(Uu, E^{1/2}Uv, \frac{H}{B}UE^{1/2}w, fUB\phi, \frac{fUB}{gH}R\right). \tag{5.8}$$

The equations of motion then become

$$\partial u/\partial t - v = 0, \tag{5.9}$$

$$E\partial v/\partial t + u = -\partial\phi/\partial y, \tag{5.10}$$

$$\frac{f^2B^2}{gH}\frac{\partial R}{\partial t} + \frac{\partial v}{\partial y} + \frac{1}{\bar{\rho}}\frac{\partial(\bar{\rho}w)}{\partial z} = 0, \tag{5.11}$$

$$\frac{\partial^2\phi}{\partial t\partial z} + \frac{N^2H^2}{f^2B^2}w = \frac{gHU}{f^2BE^{1/2}}\frac{Q'}{c_p\bar{T}}. \tag{5.12}$$

The Ekman number is of the order 10^{-2}, and since N^2H^2/f^2, Rossby's "radius of deformation," is the only important horizontal scale in the problem, we choose B so that N^2H^2/f^2B^2 is of order unity. Hence

$$f^2B^2/gH = (N^2H^2/f^2B^2)^{-1}(\partial\ln\bar{\theta}/\partial z) < O(1). \tag{5.13}$$

Equations (5.10) and (5.11) therefore simplify to

$$u = -\partial\phi/\partial y, \tag{5.10$'$}$$

$$\partial v/\partial y + (1/\bar{\rho})(\partial(\bar{\rho}w)/\partial z) = 0. \tag{5.11$'$}$$

The assumption is now made that all the moisture pumped out of the boundary layer condenses into heat and is distributed in the vertical according to a form function F whose vertical average is unity. Equation (5.12) then becomes

$$\partial^2\phi/\partial t\partial z + \lambda^2(w - F\eta w_E) = 0, \tag{5.12$'$}$$

where $\lambda^2 = N^2H^2/f^2B^2$ is the nondimensional radius of deformation (it represents the horizontal influence scale for a spike function) and

$$\eta = \frac{Lq_s}{c_p\bar{T}}\left(\frac{\partial \ln \bar{\theta}}{\partial z}\right)^{-1}, \tag{5.14}$$

$$w_E = -\frac{1}{2^{1/2}}\frac{\partial u}{\partial y}(y, 0) = \frac{1}{2^{1/2}}\frac{\partial^2\phi}{\partial y^2}(y, 0), \tag{5.15}$$

nondimensionally.

Since the statistical mechanics of cumulus convection is not understood well enough to permit a theoretical derivation of F, nor is F even known empirically with any accuracy, nothing is gained by attempting to integrate the perturbation equations in full generality. We shall instead introduce the approximation of expressing vertical derivatives in finite-difference form and write the equations of motion at two levels, centered at the midpoints by weight of the lower and upper halves of the atmosphere. Thus we divide the interval 0 to $\bar{p}(0)$ into four equal subintervals at the points $\bar{p} = \bar{p}(0)[1 - \frac{1}{2}(n - \frac{1}{2})]$ $(n = 1, 1\frac{1}{2}, 2)$ and write (5.9), (5.10), and (5.11)$'$ at the points $n = 1, 2$ by setting

$$\frac{1}{\bar{\rho}}\frac{\partial(\bar{\rho}w)}{\partial z} = -gH\frac{\partial(\bar{\rho}w)}{\partial\bar{p}} = -gH\frac{[(\bar{\rho}w)_{n+1/2} - (\bar{\rho}w)_{n-1/2}]}{\frac{1}{2}\bar{p}(0)}, \tag{5.16}$$

and (5.12)$'$ at the point $n = 1\frac{1}{2}$ by setting

$$\partial\phi/\partial z = -gH\bar{\rho}(\partial\phi/\partial p) = gH\bar{\rho}_{3/2}(\phi_2 - \phi_1)/\tfrac{1}{2}\bar{p}(0).$$

Utilizing the condition that $w = w_E$ at $\bar{p} = \bar{p}(0)$ and $\bar{\rho}w = 0$ at $\bar{p} = 0$ (no mass flux through the top of the atmosphere), and the quite good approximation $\bar{p} = \bar{\rho}R\bar{T}_m = \bar{\rho}gH$, we get

$$\partial u_1/\partial t - v_1 = 0, \qquad \partial u_2/\partial t - v_2 = 0, \tag{5.17}$$

$$u_1 = -\partial\phi_1/\partial y, \qquad u_2 = -\partial\phi_2/\partial y, \tag{5.18}$$

$$\partial v_1/\partial y + w_{3/2} - w_E = 0, \qquad \partial v_2/\partial y - w_{3/2} = 0, \tag{5.19}$$

$$(\partial/\partial t)(\phi_2 - \phi_1) + \lambda^2(w_{3/2} - F\eta w_E) = 0. \tag{5.20}$$

To the present order of approximation, the lower level may be taken to be the top of the boundary layer. Hence

$$w_E = -\frac{1}{2^{1/2}}\frac{\partial u_1}{\partial y} = \frac{1}{2^{1/2}}\frac{\partial^2 \phi_1}{\partial y^2}. \tag{5.21}$$

Also

$$\begin{aligned} F &= 1, \quad w_E > 0, \\ &= 0, \quad w_E < 0. \end{aligned} \tag{5.22}$$

We seek exponentially growing solutions, with rising motion for $|y| < a$ and sinking motion for $|y| > a$. Setting all dependent variables proportional to $\exp(\frac{1}{2}\sigma t)$, we obtain by elimination:

$$\frac{\partial^2 v_2}{\partial y^2} - \frac{1}{\lambda^2}\left(\frac{1 + 2\sigma}{1 + \sigma - \eta}\right)v_2 = 0 \qquad \left(\frac{\partial v_2}{\partial y} > 0\right), \tag{5.23}$$

$$\frac{\partial^2 v_2}{\partial y^2} - \frac{1}{\lambda^2}\left(\frac{1 + 2\sigma}{1 + \sigma}\right)v_2 = 0 \qquad \left(\frac{\partial v_2}{\partial y} < 0\right). \tag{5.24}$$

Mass continuity at $y = a$ gives the jump condition $[v_2] = 0$. Continuity of pressure, and therefore of temperature, gives

$$(\sigma + 1 - \eta)(\partial v_2/\partial y)(a-) = (\sigma + 1)(\partial v_2/\partial y)(a+). \tag{5.25}$$

The conditions at $y = -a$ follow from the anti-symmetry of v.

Equations (5.17)–(5.25) are solved by

$$\begin{aligned} v_2 &= A(\sin y/\lambda_+)/(\sin a/\lambda_+), \quad y < |a|, \\ &= A\exp(-|y - a|/\lambda_+), \quad y > |a|, \end{aligned} \tag{5.26}$$

$$\lambda_+ = \left(\frac{\eta - 1 - \sigma}{1 + 2\sigma}\right)^{1/2}\lambda, \qquad \lambda_- = \left(\frac{1 + \sigma}{1 + 2\sigma}\right)^{1/2}\lambda,$$

provided that σ satisfies the eigenvalue equation,

$$\frac{1}{\lambda_+}\tan\frac{a}{\lambda_+} = \frac{1}{\lambda_-}. \tag{5.27}$$

It may be shown from (5.27) that σ increases monotonically from zero at $a/\lambda = (\eta - 1)^{1/2}\tan^{-1}(\eta - 1)^{1/2}$ to $\eta - 1$ at $a/\lambda = 0$. Hence the criteria for instability are $\eta > 1$ and $a/\lambda < (\eta - 1)^{1/2}\tan^{-1}(\eta - 1)^{1/2}$.

The specific humidity q_s decreases so rapidly with height that by $\bar{p} = \frac{1}{2}\bar{p}(0)$ it is a small fraction of its surface value. Hence from the definition (3.6) we may write $\Delta \ln \theta_E \cong \Delta \ln \theta - Lq_s/c_p\overline{T}$, and, since $\Delta \ln \theta \cong \partial \ln \theta/\partial z$ (nondimensionally), (5.14) becomes $\eta = 1 - \Delta \ln \theta_E/\Delta \ln \theta \cong 1 + 0.1$ in

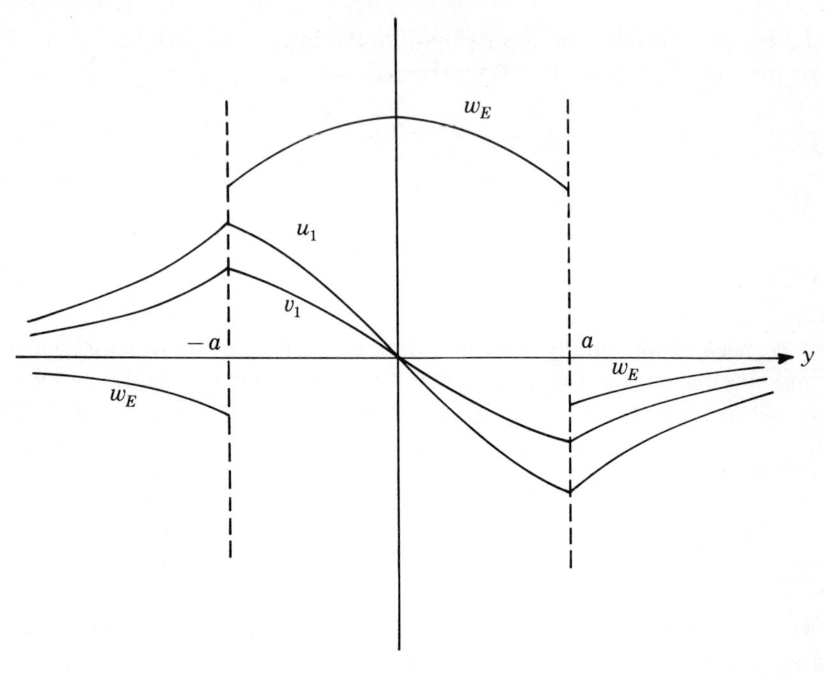

FIGURE 1

the rain areas of the oceanic tropics. Thus $\sigma_{\max} \cong 0.1$, and the cut-off $a/\lambda = (0.1)^{1/2} \tan^{-1} (0.1)^{1/2} \cong 0.1$. Taking $\phi = 15°$, we get $\lambda \sim 3000$ km, and the cut-off $a \sim 300$ km. This is indeed the right order of magnitude for the width of the rain area in a tropical depression or hurricane.

A schematic graph of u_1, v_1 and w_E is shown in Figure 1. We see that the low-level shear vorticity is positive in the region of ascent ($w_E > 0$), and negative in the region of descent.

6. **The Intertropical Convergence Zone.** It was mentioned in the introduction that the ITCZ is a zone of cyclonic shear vorticity which parallels the equator but lies at some distance from it. It was partly for this reason that the stability analysis was given for a zonally symmetric flow, the shear in the flow being expected to determine the symmetry of the disturbances which form in it. The existence of easterly winds in the tropics decreasing to nearly zero at the equator, and therefore of zonally directed cyclonic shear, is a necessary property of the thermally-driven, zonally-symmetric circumpolar vortex. Disturbances in this region should also have zonal symmetry, and the CISK process can therefore be expected to intensify the cyclonic shear vorticity into a narrow zone, i.e., CISK

should produce an ITCZ within the region of weak cyclonic vorticity. According to our analysis, its width should be approximately 300 km, and this is what is observed.

The process that we have described seems plausible, but many questions remain. Why is it that, contrary to popular notion, the zone of maximum tropical rainfall is not at the equator—at least not over the oceans where, unlike over continents, convection requires dynamical pumping of moisture? A possible explanation comes from the nature of the CISK process itself. In dimensional form, $\sigma_{\max} = f(\eta - 1)E^{1/2}$, and if $f = 2\Omega \sin \phi$ is now considered to vary parametrically, we get

$$\sigma_{\max} \cong (2\Omega\nu/H^2)^{1/2}(\eta - 1)(\sin \phi)^{1/2}.$$

Thus $\sigma_{\max}$ vanishes at the equator because the efficiency of the Ekman pumping goes to zero. At the same time q_s, and therefore η, strongly depends on temperature and decreases with increasing latitude until $\eta - 1$ eventually becomes negative and CISK no longer possible. It appears to be for this reason that the ITCZ is close to, but not at, the equator.

A numerical calculation (Charney [**5**]) of the radiatively-driven, zonally-symmetric circulation for a finite amplitude two-level model of the kind used in the present analysis did indeed produce a sharp ITCZ near to, but not at, the equator.

7. **Cloud bands.** Observations reveal that cumulus convection in the rain areas of tropical disturbances is not uniform but tends to be concentrated in bands of the order 20–50 km in width. The ITCZ itself usually consists of one or more such bands paralleling the equator. A remaining problem is to explain the origin of these bands. Here the clue may lie precisely in the fact that the CISK analysis gives no lower limit to the size of the region of convection. Do the geostrophic and Ekman dynamics we have assumed apply on scales as small as the band widths? To answer this question, let us again consider a zonally symmetric perturbation but this time perturb a horizontally shearing flow, $\bar{u} = \bar{u}(y)$, such as we know exists in the ITCZ and is produced by the simple CISK process. The perturbation Equations (5.1)–(5.5) remain the same as before, but (5.1) becomes

$$\partial u'/\partial t' - (f - \partial \bar{u}/\partial y)v' = 0,$$

and we must solve an additional problem: to find the boundary-layer structure for the mean flow when the shear vorticity $-\partial \bar{u}/\partial y$ is no longer small relative to f. In general, the boundary-layer equations are nonlinear and parabolic (second-order in z and first-order in y) and cannot be solved except by numerical methods. However, in the case of uniform horizontal shear, the y-variation can be removed, and the ordinary differential equations in z are similar to those obtained by von Karman in the analysis of the boundary-layer flow over a rotating plate. Solutions have been found and justify qualitatively the approximations in the next paragraph.

We must ask: Does the time scale for the adjustment of the boundary layer remain small compared to the time scale of the modified CISK process, and does the zonal flow remain quasi-geostrophic? A naive Oseen type of approximation which simply replaces f by $Z = f - \partial \bar{u}/\partial y$ in the x-momentum equation, both above and within the frictional boundary layer, can be justified qualitatively. The new space and time scales are then obtained by replacing f by $(fZ)^{1/2}$. The modified Ekman depth becomes $\nu^{1/2}(fZ)^{-1/4}$, the boundary-layer adjustment time $(fZ)^{-1/2}$, and the spin-up time $(fZ)^{-1/2}E^{-1/2}$, where E is now $\nu(fZ)^{-1/2}H^{-2}$. The ratio of the boundary-layer adjustment time to the spin-up time thus becomes even smaller, since $Z > f$. The ratio of the relative acceleration to Coriolis force in the y-momentum equation is $O(E)$ as before, and therefore still smaller, so that the zonal flow remains quasi-geostrophic. Thus the CISK process continues to apply, and it is plausible that the cloud bands themselves are a manifestation of a modified CISK process. The ultimate limitation in size may be due to limitations on geostrophy which have not been considered, to horizontal shear instabilities in the bands, or, as seems more likely, to the fact that the quasi-Ekman dynamics no longer operates in the presence of concentrated cumulus convection.

References

1. V. W. Ekman, *On the influence of the Earth's rotation on ocean currents*, Ark. Mat. Astronom. Fys. **2** (1905), no. 12, 1–12.

2. J. G. Charney and A. Eliassen, *A numerical method for predicting the perturbations of the middle latitude westerlies*, Tellus **1** (1949), no. 2, 38–54. MR **12**, 555.

3. ———, *On the growth of the hurricane depression*, J. Atmospheric Sci. **21** (1965), no. 1, 68–75.

4. J. Pedlosky, *Geophysical fluid dynamics*, Lectures in Appl. Math., vol. 13, Amer. Math. Soc., Providence, R.I., 1971, pp. 1–60.

5. J. G. Charney, *The Intertropical Convergence Zone and the Hadley circulation of the atmosphere*, Proc. WMO/IUGG Symposium on Numerical Weather Prediction in Tokyo, Meteor. Soc. Japan, Tokyo, 1969, pp. III 73–79.

Massachusetts Institute of Technology

Author Index

Roman numbers refer to pages on which a reference is made to an author or a work of an author.

Italic numbers refer to pages on which a complete reference to a work by the author is given.

Boldface numbers indicate the first page of the articles in the book.

Subject Index